SUSTAINABLE GARDENS

ROB CROSS · ROGER SPENCER

CSIRO
PUBLISHING

CSIRO PUBLISHING GARDENING GUIDES

© Royal Botanic Gardens Board 2009

All rights reserved. Except under the conditions described in the *Australian Copyright Act* 1968 and subsequent amendments, no part of this publication
may be reproduced, stored in a retrieval system or transmitted in any form
or by any means, electronic, mechanical, photocopying, recording,
duplicating or otherwise, without the prior permission of the copyright owner. Contact
CSIRO PUBLISHING for all permission requests.

National Library of Australia Cataloguing-in-Publication entry

>Cross, Robert.
>Sustainable gardening/authors, Rob Cross; Roger Spencer.
>Collingwood, Vic. : CSIRO Publishing, 2008.
>
>9780643094222 (pbk.)
>
>CSIRO Publishing gardening guides
>
>Includes index.
>Bibliography.
>
>Sustainable horticulture
>Sustainable agriculture
>Organic gardening
>Gardening
>
>Spencer, Roger.
>
>333.7616

Published by
CSIRO PUBLISHING
150 Oxford Street (PO Box 1139)
Collingwood VIC 3066
Australia

Telephone: +61 3 9662 7666
Local call: 1300 788 000 (Australia only)
Fax: +61 3 9662 7555
Email: publishing.sales@csiro.au
Web site: www.publish.csiro.au

Front cover photos by (clockwise, from top right): Rob Cross, Janusz Molinski, Janusz Molinski, Rob Cross.

Back cover photos by (clockwise, from top right): Andrew Laidlaw, Andrew Laidlaw, Vivien Spencer, Andrew Laidlaw, Janusz Molinski.

All figures and tables supplied by the authors unless otherwise specified.

Set in 10.5/14 Adobe ITC New Baskerville
Edited by Janet Walker
Cover and text design by James Kelly
Typeset by Desktop Concepts Pty Ltd, Melbourne
Printed in China by 1010 Printing International Ltd

The paper this book is printed on is certified by the Forest Stewardship Council (FSC) © 1996 FSC A.C. The FSC promotes environmentally responsible, socially beneficial and economically viable management of the world's forests.

CSIRO PUBLISHING publishes and distributes scientific, technical and health science books, magazines and journals from Australia to a worldwide audience and conducts these activities autonomously from the research activities of the Commonwealth Scientific and Industrial Research Organisation (CSIRO).

The views expressed in this publication are those of the author(s) and do not necessarily represent those of, and should not be attributed to, the publisher or CSIRO.

CONTENTS

Acknowledgements   *iv*
Introduction   *vii*

1	**Introduction to sustainability**	1
2	**The origins of sustainable horticulture**	13
3	**Sustainability accounting – how do we know what is sustainable?**	31
4	**Energy and emissions**	51
5	**Water**	73
6	**Materials**	99
7	**Food**	113
8	**Biodiversity and ecology**	127
9	**Designing low impact gardens**	137
10	**Sustainability in the broader landscape**	215
11	**Constructing landscapes sustainably**	229
12	**Landscape maintenance**	251
13	**Sustainable gardens, landscapes and lives**	291

Appendix   *307*
Endnotes   *323*
Index   *330*

ACKNOWLEDGEMENTS

Sustainability is a multi-disciplinary subject covering many subject areas outside our particular expertise in horticulture. We owe a special debt of gratitude to the many people with specialist knowledge who have generously donated their time and experience in assisting us to piece this story together.

We owe a major debt of gratitude to Paul Tregenza who worked with us as a volunteer for a period of about eight months. His encouragement, ideas and contribution to the project gave us the impetus needed to get the project underway.

Staff at the Royal Botanic Gardens Melbourne helped in many ways; their support and ideas about sustainability have contributed to the development of this book. We make special mention of Executive Director Dr Philip Moors, Professor Jim Ross, Professor David Cantrill and Dr Frank Udovicic who encouraged us throughout our journey; Andrew Laidlaw for his suggestions concerning garden design and construction; Peter Symes for discussions of water management in times of drought; Amy Hahs who provided comment on urban biodiversity; Kiah Martin who helped with soil management issues; the expertise of Renee Wierzbicki for her in-depth knowledge of vegetable and food gardening; John Reid and Val Stajsic for sharing their thoughts on sustainability; our Library staff Helen Cohn and Jill Thurlow who always ensure we have access to the information we need; Teresa Lebel and Niels Klazenga for their help with Photoshop; and staff at ARCUE.

We also wish to thank the following: Caoimhin Ardren, who discussed the green roof at the Thurgoona campus of Charles Sturt University; Dr Cara Beal from the Department of Natural Resources and Water in Queensland for her expertise on urine-separating toilets; Sue Berkeley and Mark Coffey for their insights on the productivity of vegetable gardens, pot recycling and general knowledge of sustainable horticulture; Emily Blackwell from the Soil Association for comments on wood accreditation; Fiona Brockhoff for sharing her beautiful garden and thoughts on sustainable landscape. Also: David Braggs from Ecospecifier Pty Ltd about sustainable product listings; Catherine Clowes assisted with research into chemicals; Geoff Connellan for his in-depth knowledge of irrigation; Steven Cramer and Maria Main, Plantic Technologies Ltd, for information about starch based plastics; Michael Dalton from Sentek for information about Enviroscan soil water monitoring; Ana Deletic, Monash University, for information about permeable paving and Water Sensitive Urban Design; Gabrielle Dickens, Lighting Consultant from Custom Lighting, for her advice about landscape lighting; Matt Elliot, Anne Jolic, Robert Lembo and Tom Scholfield from VicUrban for information about the Aurora housing development and rain gardens; Dr Tim Entwisle, Executive Director, and Paula Found, Development Manager, both of the Royal Botanic Gardens Sydney for insight into their garden's conservation programs; Christine Goodwin and Graeme Hopkins from Fifth Creek Studios for their knowledge of green walls and roofs; Juliet Elizabeth for drawing our attention to the work of artist Chris Jordan; Beth Gott, for her knowledge of the First Australians; Bob Green for a photo of Brisbane; Grant Harper for introducing to us his work on greywater filtering systems; Mick Hassett of 2MH Consulting for his input on permeable paving; David Holmgren whose vision for the future has been an inspiration to both of us; Peter Lumley for his valuable comments on drafts of the book; Adam Maxey of the Alternative Technology Association and Dr Barry Meehan, Associate Professor of Environmental Science at RMIT University, for their knowledge of greywater systems; Dr Peter May for his support and horticultural and sustainability insights; Robin Mellon, Technical Manager, Green Building Council of Australia, who introduced their rating programs to us; Warren Marsden-Sayce for information on pools and nesting boxes; Dr David Murray for comments on genetically modified plants; David Oliver, Managing Director Elmich Australia Pty Ltd, for his experience with green walls and roofs; Kirsten Parris from the School of Botany at the University of Melbourne for discussion and information on Melbourne's historical frost data; Dr Robert Patterson of Lanfax Laboratories for his invaluable contribution to controlling greywater quality through the analysis of detergents; Prof. Tony Priestley, Deputy CEO, CRC for Water Quality & Treatment, CSIRO Land & Water for information on embodied energy in water; Marion Raad-Chenailler from Greentech for information about biodegradeable plant fibre pots; John Rayner for his contribution to plant selection; Edwina Richardson, Research Officer, Australian Institute of Landscape Architects, for very informative exchanges on the Institute's sustainability initiatives; Michael Rogers and Australian Native Landscapes Pty Ltd, for information and photograph of green waste recycling; Rob Rouwette, Research Consultant, Centre for Design, RMIT University, for his knowledge on Life Cycle Assessments; Lesley Rowland, National Climate Centre Bureau of Meteorology, for information on rainfall and evaporation; Ian Shears of the City of Melbourne, for information on sustainable water use in public landscapes; The Water Pro Moorabbin staff for their experience in irrigation and thoughts on greywater systems; Peter Watson, BioNova Australia, for insights into natural swimming pools; Dr Leanne Webb from CSIRO Marine and Atmospheric Research for climate change predictions; Bob Williamson, Founder and Chair, Greenhouse Neutral Foundation, for his experience in recycling plastic plant pots.

We share similar goals with Sustainable Gardening Australia and appreciate their help during the project, thanking especially Bruce Plain, Mary Trigger, Frances Saunders, Paul Gibson-Roy and Paul McMahon who provided information on many things including their chemical assessment method.

Thanks go also to all those who contributed images, individually acknowledged through the text.

We very much appreciate the professionalism and friendly industry of the editing team at CSIRO Publishing, especially Janet Walker, John Manger, Tracey Millen and Briana Elwood, and thanks to Rodger Elliot for introducing us to them.

Final thanks go to our families and friends, for without their support our goals could not have been achieved. Thanks Viv, Martin, Stephen, Anne and Anthony and many others.

Any errors and opinions within the text remain our responsibility.

Roger Spencer and Rob Cross

INTRODUCTION

We must join together to bring forth a sustainable global society founded on respect for nature, universal human rights, economic justice, and a culture of peace. Towards this end, it is imperative that we, the peoples of Earth, declare our responsibility to one another, to the greater community of life, and to future generations.

THE EARTH CHARTER

This book explores the way horticulture can make a real contribution to a more sustainable future.

Despite a growing environmental awareness that started in the 1960s, global environmental problems continue to escalate both in number and severity of impact. The urgency for effective management of our limited resources increases daily.

Our task is to harmonise human activity with the cycles of nature, upon which we all depend, working from the individual through to the global levels of human organisation. To do this we need to clearly establish and quantify the web of connections between human activity, resource depletion and environmental degradation.

In an increasingly urbanised world, parks and gardens are, for most of us, the main point of contact with nature: they have a vital role in helping us understand the principles that will guide our transition to a sustainable society.

The challenge

The Earth Charter is like a global mission statement reminding us that we have a duty of care to manage the planet not only for ourselves but also for those that depend on us, including other organisms and future human generations.

Modern urban life has distanced us, literally and metaphorically, from the natural world that is our life support system. But with our attention increasingly drawn to global environmental problems like climate change (with its floods, fires and droughts), poverty, land degradation and biological extinctions our society is beginning to realise that securing an environmentally sustainable future will involve a transformation as socially significant as the Agricultural and Industrial Revolutions. A major part of this cultural change will be a period of environmental accounting, to identify connections between the environment and human resource consumption. Only this way can we establish effective pathways to a sustainable future.

It would be much better if such a momentous social transition were founded not on necessity, but on a genuine and heartfelt connection with, and concern for, the environment. For almost all of our evolutionary history we lived in direct contact with nature as small hunter-gatherer tribes. Now, globally, more than 50% of the world's population are city dwellers and in Australia this is a staggering 88%. Most of us have lost any deep-seated connection with nature and the land. So it is in urban parks and gardens that we are now closest to the two fundamental cycles of life: the planetary cycle of the seasons, and the biological cycle of birth, growth, maturation, reproduction, death, decay and renewal. This is what connects us to our origins and the biosphere – the thin envelope of land, sea and air around the Earth's surface that supports and contains all living organisms.

All things are connected. This is the first lesson in ecology: a change in one part of a system has effects that echo through the system in ways that are often difficult to predict. Gardening, like all human activities, has clear and measurable connections to global environmental problems so every gardener can contribute to environmental stewardship. But gardening is a very special human activity because parks and gardens connect to and interact directly with the biosphere. Though they are consumers of resources, they also have the capacity for primary production. By providing us with food and interacting with the wider environment they can relieve nature of the environmental demands of agriculture.

Our gardens are a microcosm of nature. The ecological processes that occur in a garden mirror those operating on a global scale. The better we understand the cycles and processes that occur in our parks and gardens, the greater will be our awareness of the significance of parallel events happening on a global scale, and the more effective will be our management strategies.

We make demands on nature through our need for water, energy, food and materials. All these factors are important in gardens. Sustainable horticulture attempts to harmonise garden consumption with the cycles of nature and the other organisms on the planet.

The challenge for our generation is to lay the foundations for a sustainable future, and horticulturists must play their part.

The book

This book is an introduction to sustainability science applied to horticulture. We have written it to provide home gardeners, professional horticulturists, landscapers and those interested in sustainability science with the necessary background to make informed decisions about how to manage cultivated land as part of a general strategy for leading more sustainable lives.

Behaving sustainably reaches into all aspects of our lives so a major challenge of this book has been to place horticulture within the complicated context of ecology, environmental and social management, and practical gardening. Readers will find as much in this book about sustainability as they find about their parks and gardens.

The book introduces sustainability, its relationship to landscapes and gardens, and then proceeds to chapters on sustainable garden design, sustainability in the broader landscape, landscape construction and maintenance.

Chapter 1 is a brief introduction to current environmental issues and the importance of sustainability.

Chapter 2 explores the historical development of horticulture, including its relationship to agriculture and the environment. This establishes the cultural relationship between humans and plants and the environmental situation that we face at the start of the 21st century.

Chapter 3 introduces important tools for measuring sustainability, many of which are still in their infancy.

Chapters 4, 5, 6, 7 and 8 look at the garden consumption of water, materials (including hard landscape, tools and machinery, chemicals such as pesticides and synthetic fertilisers, and garden waste) and energy (plants, power tools, structures, products, chemicals and labour), and how environmental impacts can be relieved through local food production and the encouragement of biodiversity.

Chapters 9, 10, 11 and 12 are more practical in the sense that they present ideas that can be applied to make gardens and landscapes more environmentally friendly through their design, construction and maintenance phases.

Chapter 13 introduces sustainability audits as well as summary guides to sustainable horticultural practice. We also visit a garden that successfully combines beautiful design with sustainability, and discuss likely future trends for sustainable horticulture.

Connecting all chapters is the general theme of human consumption as a major driver of environmental impact because this is the area where we can all act and make a difference.

The book is not about 'new' or 'old' gardening methods, or 'right' and 'wrong' practices but simply an aid to gardening more in harmony with the natural world.

As part of the United Nations Decade of Education for Sustainable Development 2005 to 2014, the Australian Government is developing a National Action Plan for Education for Sustainable Development. The objective of the Plan is to contribute to the achievement of a more sustainable Australia through community education and learning. We hope that this book will make a contribution to the new and rapidly evolving discipline of sustainability science as we have tried to decipher and, where possible, quantify the complex connections between the environment, human activity and horticulture.

Living sustainability – the challenge.

INTRODUCTION TO SUSTAINABILITY

> **KEY POINTS**
>
> Humanity is currently living unsustainably.
>
> Global sustainability action attempts to harmonise economic, social and environmental goals.
>
> Three tools for analysing and managing environmental sustainability:
> - management hierarchy
> - production and consumption
> - sustainability accounting.
>
> Human consumption is the driver of environmental impact through the use of the basic resources: water, energy, food and materials.

Sustainable gardening can be defined very simply as: gardening to maximise environmental benefit and human well-being.

This could involve completely different activities such as: encouraging native biodiversity; avoiding the use of synthetic chemicals; growing plants that will not escape into the natural environment; and buying environmentally friendly products. But there is much more to it, as we will see.

Sustainable horticulture is a small part of a global movement whose focus is sustainable living. Although this book will concentrate on the management of urban landscapes, this will make little sense unless we know how this fits into the big sustainability picture. What exactly is sustainability? Why is it important? And how can we become more sustainable?

Living unsustainably

Environmental problems such as climate change, freshwater depletion and species extinction are now global in scale. Nature is no longer keeping up with human consumption of the world's natural resources

Figure 1.1 Living unsustainably.

(see Figure 1.1). This is sometimes expressed through an economic analogy: instead of living off nature's interest, we are drawing down its capital by turning resources into waste faster than nature can turn waste back into resources. Addressing this issue will require a major human effort through the 21st century.

The Millennium Ecosystem Assessment

The Millennium Ecosystem Assessment (MEA) is the most comprehensive summary of the condition of living systems on our planet and the current biological challenges that we face. It was prepared by over 1360 biological scientists from 95 countries between the years 2001–2005 as an overview of the state of the biosphere. It was scrutinised by governments and published in March 2005.

The MEA was compiled explicitly as a scientifically reliable account of the biological state of the planet, to be used as a guide for decision-makers and those developing environmental policy. Two of the main findings from this report express the stark situation that confronts us:

1. *Over the past 50 years, humans have altered ecosystems more rapidly and extensively than in any comparable period of time in human history in response to rapidly growing demands for food, fresh water, timber, fibre and fuel. This has resulted in a substantial and largely irreversible loss in the diversity of life on Earth.*

2. *The changes that have been made to ecosystems have contributed to substantial net gains in human well-being and economic development, but these have been achieved at growing costs in the form of the degradation of many ecosystem services* [benefits derived from nature, such as a regular supply of rainwater, clean air and nutrient cycling], *increased risks of nonlinear changes* [suddenly triggered large changes, such as shifts in the circulation pattern of ocean currents], *and the exacerbation of poverty for some groups of people. Unless addressed, these problems will substantially diminish the benefits that future generations obtain from ecosystems.*

The MEA assessed the condition of 24 ecosystem services and found that only four have shown improvement over the last 50 years, 15 are in serious decline, and five are in a precarious condition. The Report draws

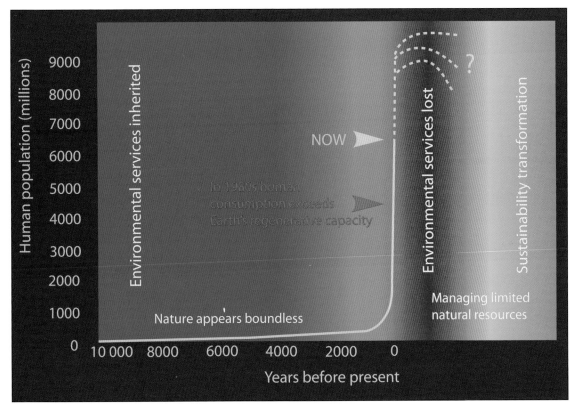

Figure 1.2 Human use of the Earth.

attention to the escalating effects of a global population that is growing in number and increasing in affluence.

Figure 1.2 is a simplified summary of human resource use, including our use of nature. When the human population was relatively small, the Earth would have seemed boundless, and nature productive beyond human need and therefore in limitless supply. The use of the concentrated energy in fossil fuels combined with advanced technology and modern medicine during the Industrial Revolution sparked a human population explosion until, in the 1980s, human harvesting of the Earth exceeded the Earth's capacity to regenerate. This trend is still gathering momentum.

The task for sustainability is to break the link between, on the one hand, human population increase and economic growth and, on the other, environmental degradation and resource depletion. This is an enormous undertaking.

What is sustainability?

In 1972, the Club of Rome (a scientific think-tank) published *Limits to Growth* – a compelling scientific case for the common-sense argument that, on a planet with limited resources, there are limits to growth. The book sold more than 30 million copies worldwide. Now, over 35 years later, we have yet to develop strategies that will secure our future. We are currently approaching a broad range of environmental limits. Natural systems have diminished to the point where we now have little additional arable land for crops. In many parts of the world, readily available freshwater is becoming scarce and this in turn creates difficulties for irrigated food production and therefore food security. As a global community we cannot simply move on to somewhere more bountiful as our ancestors did.

We are the custodians of planet Earth and this requires a protective attitude to nature and the resources that remain.

The starting point for sustainability as a special goal for humanity is generally accepted as the publication in 1987 of the *United Nations Brundtland Commission Report on Sustainable Development*. It states:

Sustainable development is development that meets the needs of the present without compromising the ability of future generations to meet their own needs.

Since the Report's publication, sustainability has come to mean many things, but a broadly held definition involves the promotion of a dynamic balance between three key factors:

1. protection of the natural environment
2. maintenance of economic security
3. respect for social values.

More specifically, sustainability implies that none of these factors can be pursued in isolation; to function properly as a society we must treat them as interdependent and this may require the reconciliation of competing interests.

The challenge to live sustainably involves consideration of two important 'stakeholders', neither of whom have a political voice: the non-human organisms on the planet, and future human generations.

As an idea, sustainability has now been accepted by corporations, government and the general public. More and more people now perceive the Earth as a 'global village', an international community with a shared fate and the need for a cooperative international approach to resolve problems such as hunger, poverty and global environmental issues. But there is hard work ahead. Here in Australia, we are still on average travelling further, living in larger houses with fewer occupants, using more energy, consuming more resources and producing more waste.

The environmental component of sustainability, which is the focus of this book, requires us to live in greater balance with the rest of the living world, leaving enough resources after human harvesting, to maintain healthy and diverse living systems. Ensuring environmental sustainability is one of the eight *Millennium Development Goals for 2015* set out in the United Nations *Millennium Declaration* in September 2000 and endorsed by 189 countries.

Sustainability management

Saving life on Earth is not a spectator sport.

ANON.

Figure 1.3 'Fingers crossed'. (© Andrew Weldon)

The future is not fixed; it is in many ways something we create ourselves.

ANON.

We know that things are not right on Planet Earth but at present we are not sure how best to help. What can we do to lead more sustainable lives? This is not always obvious, as the following newspaper extract shows.

Sometimes you just feel like Kermit the frog. With so many green messages pulling us this way and that, it's hard to know how green your green choices really are. Should I go solar or wind? What about GM-free, chemical-free or free-range? Are they better than CFC-free, triple-A rated or energy-efficient? Even in your quietest, energy-saving moments you can wear yourself out just thinking about minimising your ecological footprint on the planet. Is it enough that I only flush for Number 2s and always turn the tap off during teeth brushing or should I overhaul the plumbing and reticulate the grey water via planet-killing plastic tubing all through my garden? ... I've come to think of my backyard as a mini carbon-sink – the eucalypts, wattles and coastal tea-trees breathing as fast as they can to compensate for the car fumes ... And I always take a canvas bag to the supermarket, which offsets the fact that I've taken the car to carry the shopping home without overbalancing on my bike ... Does the Australian flag on the label mean Australian-made or Australian-packaged and does it really matter? Are these tomatoes gleaming with natural health or is it really cochineal? ...

TRACEE HUTCHISON[1]

All things are connected

This excerpt illustrates how hard it is to make decisions about how to live more sustainably.

Ways of tackling sustainability seem infinite, from international legislation to changes in individual lifestyle, from political action to innovative technology, and much more.

Tracee's questions are hard to answer partly because there is at present a lack of information. But they are also difficult to answer because of the complexity of the web of connections that we deal with unwittingly every day. Direct environmental impacts like land clearing, logging and water pollution are easy to understand, but the connection between our consumption patterns and environmental deterioration is much less clear, and the constant barrage of alarming statistics can produce a feeling of helplessness that stifles any enthusiasm for positive change.[2]

How do I decide what is best? Am I 'sweating the small stuff'?

Answering this question will take up the rest of this book. But in coming to grips with sustainability there are three important management tools that we must introduce, because they underpin everything that follows. These three tools are:

1. the sustainability management hierarchy
2. the cycle of production and consumption
3. sustainability accounting.

Sustainability management tools

The sustainability management hierarchy

As the world becomes more environmentally, socially and economically integrated, so the future becomes more a global responsibility.

ANON.

The world's greatest advances often begin with one person or small groups of people following an idea.

ANON.

Living systems consist of a number of organisational levels (scales, contexts or frames of reference). For example, animals and plants can be studied as complete individuals or as a collection of organs, tissues or cells. Each level has its own particular set of properties and relations and because the levels become progressively more inclusive they are generally referred to as being a hierarchy, the more inclusive levels being 'highest'. These organisational levels are not separate from one another but completely interdependent. To separate them into levels is simply a convenient way of examining progressively smaller parts of an integrated whole.

This biological example provides a good analogy for social, political and economic systems. Management, too, can be carried out at different levels of human organisation, from the global down to the national, regional, local and individual. In managing climate change, for example, it is possible to act at a worldwide level through the UN, the global 'organism', by introducing international legislation. But at the other extreme, and in a much narrower context, individual people, which we could call global 'cells', can help by reducing their personal greenhouse gas emissions. The level(s) of management needed will be determined by the particular problem. Climate change, a global problem, requires a response at all levels. A local oil spill can be dealt with at a lower level.

There are many action levels. Figure 1.4 shows some of the more familiar ones and

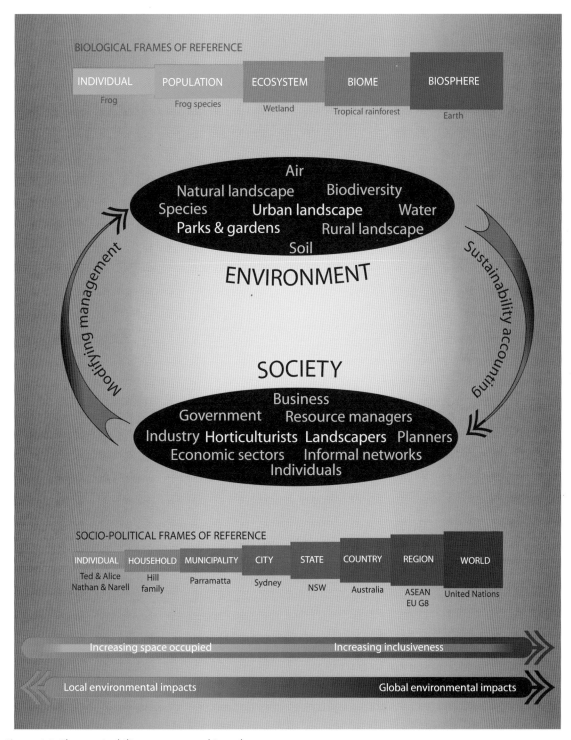

Figure 1.4 The sustainability management hierarchy.

indicates how, in general, as you pass from low to high levels in the hierarchy, the units become more inclusive and occupy more physical space. There are other features of this hierarchy: adjacent levels tend to have the greatest similarity, levels can be sub-divided, and management operations initiated between levels may occur over different time periods.

For our purposes it is the general idea that is important. The sustainability management hierarchy of action levels is a convenient way of thinking about, explaining and managing ecological, political and social complexity. Sustainability is managed at many levels of biological organisation (say home garden to the biosphere) and human organisation (say family home to the United Nations). An efficiently functioning hierarchy is like an organism whose cells, tissues and organs are totally integrated and contributing to the health and well-being of the entire organism. This book will therefore consider how parks, gardens and urban space affect the wider biological environment (from local to global) and how this can be managed by different levels of human organisation (individuals, families, local councils etc.)

Integrating the many levels of the world community is sometimes referred to as Earth system governance. Key players include:

- *international organisations*, to coordinate biosphere stewardship across political and geographic scales, and assist in tracking progress and resolving disputes.
- *national governments*, to monitor ecosystem services through national accounts that enable civil society to set minimal standards, track progress, and hold decision-makers accountable and provide environmental regulation and incentives.
- *civil society* (general public), to distribute environmental information, and hold both public and private sectors accountable while promoting partnerships and innovations on the path to sustainability reform by aligning economic management with environmental sustainability.
- *business*, to realise environmental stewardship as a source of new markets that can take advantage of new technologies and products that reduce environmental impacts, while including sustainability in performance reviews.
- *research communities*, to focus on bridging social, economic and environmental disciplines towards sustainability through the protection of ecosystem services.
- *local communities*, included in decision-making and informed of influences impacting on ecosystem services while ensuring maximum local sustainability.
- *individuals*, to lead sustainable lives and provide the impetus for social change.

Although sustainability must be addressed by all levels of human organisation, it is as individuals that we have the greatest control. We can all make a difference by adopting sustainable lifestyles and encouraging others to do the same.

Production and consumption

Sustainability attempts to minimise the negative effects of human activity on natural systems. A knowledge of the key processes on which all life depends can reveal ways to secure our future.

The biological world is driven by the energy of the Sun. Plants capture the energy of sunlight during photosynthesis and store it in their tissues. This energy is taken up by other organisms as it passes through the food chain. Because plants lie at the bottom of the food chain they are referred to as *primary producers* (they are the planet's life-support system, underpinning everything else in the biological world), and other organisms are *consumers*. Energy passes through the biosphere to be eventually dissipated as heat, but on the way it passes through the organic cycle of production and consumption. Without energy there is no activity. All those processes and materials that maintain this cycle of life we can call the *natural economy* (Figure 1.5), and the large number of benefits we derive from the natural economy are called *ecosystem services*. It is these ecosystem services that must be protected if human development is to be sustainable.

Running within (and ultimately dependent on) the natural economy is the *human economy* with its production and consumption of goods and services. It too depends on primary production – mostly the primary production of crops that provide the food (energy) for ourselves and our livestock which in turn allows us to make use of other resources – materials, water, chemicals and especially the ancient stored primary production of fossil fuels. It is human consumption that threatens nature's environmental services.

To live is to consume. Basic human needs have remained the same throughout time. We can imagine our hunter-gatherer ancestors' lives preoccupied with their basic need for food, shelter, water and materials. As they roamed in small 'consuming' bands their impact on the environment would have been negligible. As human numbers increased, along with scientific and technological expertise, these simple needs have progressively taken over the natural economy and its services. Our requirement for food has turned into industrial agriculture; the need for shelter into the building and construction industry; the advantages of mobility transformed into modern transport systems; the necessity for water into vast dams, pipelines, treatment plants and irrigation systems; and our use of materials into the manufacturing industry. This is the system we must manage more effectively (see Figure 1.6), and the principles of production and consumption are as significant in the garden as they are in the biosphere.

The emphasis of environmental management has changed. In the early days it was mainly reactive, correcting the situation at the point of impact: tidying up oil spills, revegetating degraded land, and so on. This is sometimes

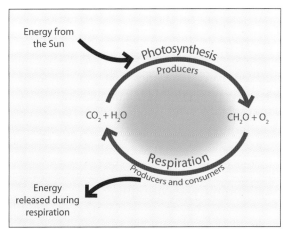

Figure 1.5 Natural economy – the cycle of life.

1 – INTRODUCTION TO SUSTAINABILITY

Figure 1.6 The planet and human needs.

called 'end of pipe' management because it occurs at the end of a long chain of events that gave rise to the problem: it deals with the symptoms not the disease. Sustainability management is now becoming much more proactive, working to prevent or reduce environmental problems at the 'start of the pipe'. This is demand management, and in sustainability terms the demand is human consumption.

Sustainability accounting

To prepare for a secure future we must live sustainably: to find out what is sustainable we must establish effective, standardised environmental accounting at all management levels.

ANON.

All organisms make demands on the environment. In itself, this is not a problem. The important question to ask is whether this impact on the environment is sustainable. To find this out requires measuring and monitoring. Armed with quantitative information it is then possible to establish benchmarks, to set management goals, assess trends, anticipate problems and measure progress. This is similar to the careful way we manage our financial lives so it is known as *environmental* or *sustainability accounting*.

As a discipline, sustainability accounting is very new. Increasing numbers of agencies at various levels of the management hierarchy have begun recording environmental statistics, or improving their methods of data collection and the resolution of their findings. But there is a universal need for more precise accounting. Data must be gathered over a long period to get reliable results, and it helps when the data can be applied across management levels. Above all, it needs to be accessible to policy-makers.

In Australia, data produced by the Australian Bureau of Statistics, the Australian Bureau of Agricultural and Research Economics, and the National Heritage Council now provide much of the benchmarking and many of the indicators that are useful for sustainability management.

Water, energy, food and materials

There is at present a direct proportional link between human consumption and resource depletion and environmental degradation. The more we consume, the greater the environmental impact.

There are three broad options for reducing this increasingly heavy burden on nature:

- reduce population
- reduce consumption
- manage consumption in a more sustainable way.

We will discuss population later. Our focus here is on consumption. The consumption connection must be broken by developing a green economy based on sustainability principles, essentially an economy that uses resources in smarter ways, uses different resources (especially renewable ones), and different technology.

Effective sustainability accounting will be needed to help us through this process and sustainability science is currently exploring ways in which this can be done.

Environmental impact can be assessed by analysing the activities of broad economic sectors like mining, agriculture and industry, or it can be related to the individual goods and services that are produced by these sectors. It can even go right back to the consumer wants and needs that drive the whole process. This is so important that we devote Chapter 3 to the accounting methods, like the Ecological Footprint, that are being used to measure sustainability.

For simplicity, and to allow comparisons across the sustainability management hierarchy, we have reduced all this complexity to a consideration of the environmental consequences of the human use of four interrelated and fundamental resources:

- energy
- water
- materials
- food.

These are useful categories to analyse because they apply to all human activity regardless of place and time. They can also be used to compare the human and biological processes of production and consumption at any organisational level, from individuals to households, neighbourhoods, nations and the world.

Garden and landscape consumption is no different from any other form of human consumption. It is possible, for example, to analyse all the environmental impacts of our use of urban space by looking at how we use these resources. It is clear that, in general, if we use less water, energy and materials then we will be reducing environmental impact. It is also important to note that almost all the activities involving these simple commodities produce waste.

But before we discuss how we can measure and then reduce the environmental impact of our horticultural practices, let's examine the agricultural and horticultural traditions and ideas that have already established a path for the future.

THE ORIGINS OF SUSTAINABLE HORTICULTURE

> **KEY POINTS**
> Sustainable land management practised by nomadic hunter-gatherers.
> The benefits and costs of early settled agrarian communities.
> The environmental history of agriculture.
> The GM debate.
> Emergence of sustainable agriculture and horticulture.

Living sustainably in harmony with nature is not a new idea. There are valuable lessons to be learned from the past. In this chapter we take a historical glance at the relationship between humans, the land and nature. We will also look at the way important questions about the land have been handled: questions about land ownership; how the Earth is to be harvested; how the Earth's resources are to be shared; and the role that the land and its plants and animals have played in different cultural traditions. This will help in understanding the events leading to the current environmental situation and provide a context for the challenges that lie ahead for sustainable horticulture.

Aboriginal gardening

Different attitudes to the land, nature and its management are well illustrated through the contrasting perceptions held by Australia's Indigenous people and the early settlers. Both communities needed to obtain food from the land but they used very different methods.

Settlement of Australia by the British after over 40 000 years of Aboriginal occupation was justified by the precept of *terra nullius* (land owned by no-one) and sovereignty was acquired on the basis of occupation alone. To the settlers' eyes, the land was not being 'used'. Aboriginal people were not cultivating the land, or at least not in a way that the settlers understood. The settlers were accustomed to boundaries of fences and hedges. Influential English garden chronicler J.C. Loudon, in his *Encyclopaedia of Agriculture* (1835), urged the planting of gardens 'as proof of possession'.

We now know that Aboriginals were using fire in a highly effective, skilful, controlled and sustainable system of selective harvesting that

Figure 2.1 Beth Gott, 2007.

has become known as 'fire-stick farming'. The timing, extent and frequency of these fires was determined by the type of vegetation, accumulation of litter and the season. It was used to drive out food animals. It also encouraged lush new growth that would attract these animals back, and would induce some plants to produce fruit. By burning in a mosaic pattern, it was possible to get patches of new growth in areas of older growth and this would allow the ecosystem to regenerate. In south-eastern Australia this kept both dry sclerophyll woodlands and the grassy plains in an open condition favourable for the growth of the major food plants: small herbaceous perennials with fleshy rootstocks like lilies, orchids, Yam Daisy, Murrnong (*Microseris*), and water plants like the Bulrush (*Typha* spp.).

Beth Gott, an eminent Australian ethnobotanist (a person who studies how people of a particular culture and region make of use of indigenous plants), has tabulated Aboriginal agro-horticultural land management as 'natural cultivation' of the land (see Table 2.1).[1] It was unlike those kinds of cultivation elsewhere in the world that marked the beginnings of agriculture, and totally different from the tilled, enclosed fields, domesticated animals and pastures that the settlers remembered in England. Yet it was clearly more than a simple process of food 'gathering'.

Not all food was obtained while 'on-the-move'. Stone fish traps were built, fruit and meat were dried, and pelicans were confined in pens.

The Aboriginal approach demonstrates the manipulation of plants within the whole environment by making use of them where they grow naturally rather than growing them in special enclosures or plantations. Thousands of native species were used as food plants, mostly tubers in southern Australia, seed in arid regions, and fruits in the tropics.

Table 2.1 Comparison of Aboriginal (Koori) environmental management with European horticulture and agriculture[1]

European agriculture/horticulture	Koori environmental management/gathering
Preparation of soil, cultivation	Digging, loosening soil, incorporating litter and ash
Fertilising	Burning at specific times, producing ash
Thinning of perennials	Clumps separated, tubers, etc. removed
Sowing and planting	Some tubers left or replanted; burning timed after seeding
Care of seedlings	Open structure of vegetation, allowing penetration of light, maintained by regular burning
Spread of cultivars	Tubers and seeds carried to camps, traded between tribes

There is no evidence that Aboriginals introduced new plant species. The first confirmed naturalised alien plant in Australia was *Tamarindus indicus*, Tamarind, introduced to the Northern Territory by Macassans from the South Celebes in about 1700.[2]

But even this form of land management left its mark on the land. Over time, constant firing was to alter the species composition of plant communities and even the characteristics of the plants themselves. It has also been suggested that the fire-modified habitat and direct harvesting of animals resulted in the demise of Australia's large mammals.

Settled communities and plant domestication

Domestication of plants and animals is generally assumed to have taken place about 10 000 years ago after the glacial retreat of the last Ice Age. Known to anthropologists as the Neolithic Revolution, this was the transition in human lifestyle from nomadic hunter-gatherer following food sources, to permanently settled communities producing and storing food by farming. Occasionally this took the form of shifting agriculture; cultivation of a small area until nutrients were depleted and pests became prohibitive, then moving on to allow the soil time to replenish. Food plants were grown near dwellings in areas that were fenced off and irrigated, and the soil was tilled. This appears to have occurred independently in up to a dozen centres across the world between 6000 and 10 000 years ago. Most of these cultures were based on grains: wheat in Europe, rice in Asia, maize in the Americas, and sorghum in Africa.

The best known of these early agrarian communities was in Mesopotamia. This civilisation thrived in a lush and productive land area known as the Fertile Crescent, the human settlements built on the rich sedimentary soils of the Tigris and Euphrates River Delta in what is present-day Iraq. Though still cultivated, much of this land is now impoverished. This transition from a nomadic lifestyle had a profound social, economic and cultural effect. Permanent settlement and food storage allowed large numbers of people to gather together. It released workers from the land to concentrate on other matters. There was the development of sophisticated government, division of labour, and more complex and permanent technologies. Being restricted to a single location meant that communities were more vulnerable to attack, hence the development of armies. A secure food supply allowed time for the flowering of art, science and commerce. With the development of written symbolic languages, the historical record was established for future generations: it encompassed all aspects of human existence from the abstractions of mathematics, religion and philosophy to historical accounts and day-to-day commercial transactions.

The Neolithic Revolution marked the advent of modern culture and the path from temporary dwellings to villages and then, about 6000 years ago, to cities, coined money and international expansion. This change was once viewed as a form of social evolution or progress, a triumphant transition from savagery or barbarism to civilisation. There is no doubt that for many it resulted in a more diverse, intellectually challenging and comfortable physical environment. But from an environmental perspective, the great human transitions – the Neolithic Revolution of 12 000–6000 BP (Before Present), the Industrial Revolution (1650–1850 AD), and the technological, electronic and communications revolution (1940s on) – cannot be viewed as a simple steady advance in human well-being. Civilised lifestyles have generated costs as well as benefits.

One of the most serious of these is the spread of human pathogens. Bacteria, viruses and other infectious microbes spread relatively easily in densely populated communities. Many are transferred to humans from animal hosts, jumping from closely related species; others from close contact with domesticated animals. Modern medicine tries to stay a step ahead of these pathogens. Temperate diseases thought to have a domestic animal origin include measles, mumps, smallpox, influenza A and tuberculosis. However, many tropical diseases, such as AIDS, came from wild non-human primates, such as chimpanzees. Though not as abundant as domestic animals, these primates are our closest cousins and therefore pose the weakest species barrier. In both tropical and temperate zones, virtually all other diseases came from mammals and occasionally from birds. For example, rodents, though genetically removed from us, have spread diseases like the Black Plague, which wiped out a third of Europe's population in the Middle Ages.

Adapting to an entirely different physiology is not an easy thing for a pathogen to do. The process is facilitated now by modern behaviour including blood transfusions, international travel and intravenous drug use.

This illustrates in stark human terms the environmental problem of moving organisms around the world. Feral animals and invasive plants (the result of human transference) constitute one of the top three threats (along with climate change and land clearance) to global biodiversity and are therefore a major part of any sustainability program.

Settled communities have sometimes placed such high demands on the resources of the surrounding countryside that this has brought about their own destruction. Examples include the deforestation by Mayan and Easter Islander civilisations, and the degradation of the southern Mesopotamian Fertile Crescent soil through the salinisation resulting from prolonged irrigation.

Pleasure gardens

We can assume that, as part of the emerging artistic expression resulting from the increased leisure time in these early settlements, it was possible to enjoy plants for their beauty as well as their practical value as sources of fibre, medicine, materials and food. There are records of gardens constructed over 7000 years ago in Sumeria, one of the Mesopotamian centres. The Hanging Gardens of Babylon, one of the Wonders of the Ancient World, were in the same region. These are among our first records of ornamental pleasure gardens which nurtured the 'soul' rather than the body.

Of course, pleasure gardens were not confined to the Mediterranean region. Records of Asian gardens are rare, but Chinese writings suggest gardening is an ancient art in a tradition extending back well before 2500 BP.

What we now call ornamental horticulture or 'gardening' probably arose out of crop cultivation. Although many gardens to this day contain sections devoted to food, pleasure gardens and food gardens have followed their own paths of development. Food gardening on a grand scale became agriculture and production horticulture.

Development of agriculture

Once the possibilities of agriculture were recognised, there was a progressive clearing of natural vegetation. This provided the land for the crops and pasture needed to support the livestock for the expanding populations. Much of the land was used for cereal crops and much of this crop was needed to feed the domesticated animals. Crop yields were improved by selection of higher yielding varieties, a slow process taking place over thousands of years.

Many of today's environmental challenges relate to agriculture and its practices, and horticulture has much to learn from agriculture's past. The most dramatic change in land management after the Neolithic Revolution was the industrial agriculture that emerged with the rapid advances in science and technology in the 19th and 20th centuries.

20th century

The science of genetics, which originated about 100 years ago, initiated a quantum change in the ability to manipulate plant performance. After World War 2, several high-yielding crop varieties were produced which, it was hoped, would increase the food supply and slow down the rapid rate of land clearance. It was also assumed that high yields would alleviate world food shortages and increase the income of poor farmers, giving the non-industrialised world time to tackle birth rates and address some of the social problems contributing to poverty and hunger. This hopeful scenario became known as the Green Revolution. The performance of genetically produced high-yielding crops was dramatically improved by the use of petrochemical fertilisers, controlled irrigation and pesticides. New high-yield dwarf varieties and varieties insensitive to day-length were developed. This allowed for planting across latitudes and for more than one crop each year in suitable climates. Other varieties were exceptionally high-yielding when plied with water and fertiliser.

Large, highly mechanised farms were more profitable than small family farms. They were, for example, able to buy fertilisers and pesticides at cheaper rates through bulk buying, and could rapidly adopt new technologies. These 'superfarms' are now part and parcel of Western agriculture. Since World War 2, the number of farms has decreased by two-thirds and the average farm size has doubled. The benefits of mass production were not spread equally through the farming community. Disenchantment followed as more people lost their livelihoods because of the increased efficiency gained by using machinery, and family based farming communities were swamped by giant agro-industry.

In the 1960s, environmentalists focused their attention on two environmental issues: world food shortage due to unrestricted population growth, and pollution as a side-effect of technology. Gradually it became evident that the alleviation of hunger was not just a matter of food production. The poor of the world did not have access to the technology or money to help produce or buy food, and it became apparent that the solution to hunger and poverty lay as much in socio-economics, politics and the logistics of distribution, as in supply.

There were other problems. It had been known for some time that monocultures of genetically identical plants grown on a large scale could lead to genetic erosion; the loss of a natural genetic variation. As a consequence, crops became extremely susceptible to pests and diseases, drought and temperature extremes, and they lacked the genetic resilience of their wild ancestors.

To cope with pests, a chemical industry strengthened by the technological advances made during World War 2 manufactured a wider range of synthetic pesticides in increasing quantities. Pests evolved resistance to the new chemicals and so further chemicals were needed. So began a chemical cat-and-mouse game that continues to this day. Pesticides and other chemicals alter the balance of micro-organisms in the soil, thereby affecting its fertility, and other soil problems like salination became more prevalent.

The Green Revolution alleviated many problems, but feeding the human population came at an environmental and social cost. The list of unfortunate side-effects is a long one: the clearing of vast tracts of land, along with its plants and animals, often resulted in topsoil depletion, erosion and conversion to desert, overgrazing, salination, sodification, waterlogging, high levels of fossil fuel use, reliance on inorganic fertilisers and synthetic organic pesticides, reductions in genetic diversity, water resource depletion through excessive and inappropriate use of irrigation water, pollution of waterbodies by run-off and groundwater contamination by fertilisers and toxic chemicals. (See Figure 2.2.)

A second Green Revolution

With the advent of global agribusiness and sophisticated biotechnology, especially genetic engineering (known as GM, genetically modified), food production has taken another quantum leap. For some time now there has been a global cost–benefit debate on GM organisms. The new technology has the potential to offer many environmental and economic benefits. For example, genetically engineered cultivars of crops like canola and wheat can be grown with the use of less herbicide and water and with built-in resistance to pests and salinity. Implanting genes coding for Omega 3 fatty acids could give extra health benefits; fruit and nut trees may produce earlier; rice cultivars may be enhanced with additional minerals and vitamins that can alleviate dietary deficiencies and even prevent blindness; anti-allergenic Rye-grass is being investigated.

The prevailing view of the scientific establishment, including the International Council for Science, Australian Academy of Science, and the Office of the Gene Technology Regulator, is that the technology carries no substantiated evidence of ill effects on human health – it requires fewer pesticides and uses greener farming practices.

Although this book is not the place for an extended treatment of this controversial subject, biotechnology is likely to assume a greater role in both agriculture and horticulture. Many of the developments have environmental consequences, and readers will benefit from some background information on the issues currently being debated.

Plant genetic engineering

Genetic engineering is a relatively new extension of genetics whose techniques, developed in the 1970s, became a commercial reality in the 1990s. It may be defined as: *the use of various laboratory techniques to produce molecules of DNA containing new genes or novel combinations of genes, usually for insertion into a host cell for cloning.*

GM plants are different from those produced by conventional breeding methods because of the method of gene insertion. There is generally a reliance on bacterial and viral DNA sequences that accompany the transferred genes. The first plant gene used in an animal was in January 2002 when a spinach gene was inserted in pigs to reduce saturated fats. The US is by far the greatest producer of GM crops, especially soybeans, cotton and maize. Imported processed foods like GM soybeans, corn products, sugar beet, canola oil and cottonseed oil are consumed in Australia, but at present little else.

There seems no reason why we should not benefit from biotechnology. However, past experience with technological advances suggests we use maximum care in striving for a balance between short-term economic and scientific enterprise and the long-term safety of both humans and the environment. Less controversial technology may emerge sometime in the future.

In Australia, human health and environmental issues fall under the purview of the Federal Office of the Gene Technology Regulator, while the states have responsibility for marketing, production and trade. Food Standards Australia and New Zealand is responsible for food safety and labelling in Australia.

Because it is involved with global food production, biotechnology carries with it many broad political, social, ethical and health issues apart from the actual gene technology. This is one reason for the protracted global debate.

The fundamental and contentious issue in the debate centres around what constitutes an acceptable level of risk in the pursuit of potentially beneficial technologies. Is the cure worse than the disease? In making assessments about the merits of particular technologies it appears sensible to treat each case separately. The *Cartagena Protocol on Biosafety*, arising out of the *United Nations Convention on Biological Diversity*, was established in 2000 as a global regulatory system for ensuring the safe transfer, handling and use of genetically modified organisms (GMOs), especially in relation to their transport between countries.

The following list of concerns is adapted from the popular press and critical literature. It is not intended as an argument against GM but as an indication to the reader of the tenor of the debate, and how wide-ranging the issues of biotechnology have become. We have divided concerns into three groups: business and bio-ethics, health, and the environment.

1. Business and bio-ethics

There is a perception that, in the search for profits, consumers and the environment are often the losers and that much greater emphasis must be placed on the exploration of ecologically sustainable and socially equitable practices for sharing seeds, crops and food.

One historical example of the way a section of the global community lost out was 'genetic piracy': when genetic material donated or taken from a developing nation was used by a developed country to engineer new varieties that generated large profits without there being any compensation offered to the country of origin.

It is claimed that GM crops and the chemicals needed for their cultivation are owned and controlled by a few large multinational corporations whose primary motivation is making a profit. These transnational corporations wield enormous power over the world's food supply. Using GM entails complex global legislation on intellectual property – patents, copyright and plant variety protection – all of which lay a potential foundation for future corporate control of food and farming that may undermine attempts to maintain biodiversity, ensure food security and meet the needs of developing countries. Any public concerns raised about their activities can be countered by expensive lawyers and lobbyists that only rich corporations can afford.

Biotechnology now has the potential to change life in far-reaching ways with consequences affecting everyone, but how, when and where these technologies are applied is not decided at the ballot box, and antagonists claim that the majority of public opinion is still opposed to the introduction of GM.

Certainly the speed of change is an important factor. Biotechnology as a profitable enterprise, it may be claimed, is surging ahead of public information and discussion. Has there been time for a full public exploration and consideration of all the consequences? As GM seeds are patented, farmers may be 'locked in' to using the products of a single company: they may also be encouraged to use herbicides and fertilisers when these may not be ecologically desirable or the best solutions to particular problems. There is a tendency to see a chemical solution to every biological problem when biological solutions such as Integrated Pest Management may be possible. This may be seen as the fine-tuning of chemical-industrial agriculture rather than a search for ecologically sustainable solutions. Use of herbicides *does* lead to increases in herbicide residues in grain products, and degradable pre-emergent herbicides are clearly preferable.

Is our freedom of choice threatened when mandatory labelling of foods containing ingredients produced from genetically modified plants is not fully effective when, for example, meat from animals that have been fed on GM plants is not so labelled?

Concerns are sometimes raised about gene ownership when intellectual property has the potential to act against the common interest.

Companies see ownership of plant genes as a legitimate way to increase profits. However, gene ownership can have the effect of threatening the livelihoods of farmers in many countries by denying them access to specific GM crops. Private intellectual property in some cases seems to lack social justice. GM foods are often presented as a solution to Third World poverty and poor health. Detractors point out that this is not a simple issue and should be debated in the context of environmental repair, equitable land tenure, and the encouragement of ecologically sustainable systems of agriculture and horticulture. For example, shortages of vitamin A in the diet are just as effectively addressed by growing pumpkins as growing GM rice.

A great deal of money is at stake in the gamble as to whether consumers will accept gene technology. Countries adopting GM foods may be leading the world in one sense, but they could also be left out of international trade should GM be rejected by large parts of the world.

2. Health

The debate over the health safety of GM foods has a long way to go. In the rush to get new products on the market, it is claimed that there have been few studies of long-term effects on health. There are no absolute guarantees as to the effects of inserted genes, and major biotechnology companies are reluctant either to undertake proper animal feeding studies, or to publish the results if they do.

If chemical residues are considered a problem then at present the only sure way to avoid ingesting them is to eat food from plants grown in the complete absence of pesticides and in uncontaminated soil. Concerned consumers support organic production systems and grow at least some of their own food. Certainly, if genetic engineering poses even the minimum risk of whatever kind, it would appear prudent to restrict its application to enterprises that have obvious benefits. Do we really need fish that glow in the dark, or blue roses?

3. Environment

Until now, most commercial genetic engineering in plants has been directed towards herbicide resistance, so herbicide use must be included as part of the GM debate. This has been promoted as a means of improving weed management and reducing herbicide use, but it may be asserted that both these claims are contradicted by what has actually happened.[3] GM crops do not always produce better yields than conventional crops. In recent times concerns have been raised about the potential of glyphosate (Roundup^{TM}) to enhance *Fusarium* wilt and other fungal crop diseases.[4]

There is also concern over the transfer of herbicide-resistant genes from crop-plants to weedy relatives to produce herbicide-resistant 'super-weeds'. Herbicide resistance is certainly a problem. It is now apparent that repeated application of the same herbicide in the same location is providing selection pressure in favour of herbicide resistant genotypes (15–30 applications of glyphosate is sufficient in the case of *Lolium rigidum*).[5]

In Australia, 23 weed species have acquired herbicide resistance – and this is in advance of herbicide-resistant crop plants. Canola is a food plant that is also a weed, and containment boundaries are questionable when seed can be transferred long distances and pollen blown for many kilometres. As has happened with antibiotics, it may be claimed that this could be the start of a treadmill of staying ahead of superpests that have developed immunities to the latest chemicals.

There is also the problem of quarantining conventional crops from GM crops. There has been disagreement over the legal liability in the event of genetic contamination of conventional crops with GM crops. If a farmer's crop becomes contaminated with GM crops then technically s/he can be held liable for growing these plants illegally – a situation which seems unjust.

Mainstream sustainable agriculture

Mainstream agriculture has not been indifferent to the many social and environmental concerns raised about its philosophy and practice. There is now a well-established international movement for sustainable agriculture which began in the 1980s and which has as its goals environmental health, economic profitability, equity, and an awareness and preparedness to act on social issues. With kindred philosophies known as 'farming with nature', 'eco-agriculture' and 'whole farm planning', this approach is more systems-based (taking greater account of the interactions between the various elements of farming, especially interactions with surrounding ecosystems) than conventional agriculture, partly as a reaction to the trend toward specialisation in industrial agriculture.

Among other things, sustainable agriculture addresses:

- Environmental considerations such as limited natural resource use, selection of crops that are appropriate to the site and to conditions.
- Diversification of crops (including livestock) and cultural practices to enhance the biological and economic stability of the farm.
- No over-production.
- Care for land and water, improvement of soil quality.
- Protection of biodiversity.
- Caution in introducing new technology.
- Putting a monetary value on natural resources.
- Community participation.
- Social equity (asking whether new technology threatens livelihoods or treats people unfairly).
- A global perspective on the impacts of the activity.
- Inter-generational equity – providing for the future.

Other approaches

Disenchantment with conventional agriculture and horticulture has resulted in the exploration of various alternative philosophical and practical approaches that try to integrate with nature.

In Australia, an early example of small-scale farming was the market gardens and small mixed farms of early settlement, notably those

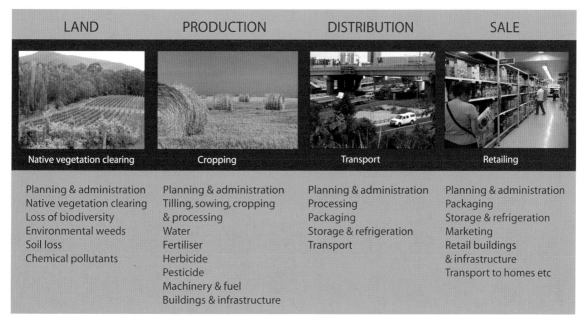

Figure 2.2 Environmental demands of agriculture.

managed by Chinese coming to Australia during the Gold Rush. On 1–4 hectare plots in the suburbs, they used labour-intensive techniques with an average of one person per half hectare supplemented by additional seasonal labour. Assorted waste was used for nutrient, including animal manure, abattoir waste and human faeces, and vegetables were often sold door-to-door from purpose-built barrows. From the 1920s, market gardening saw an influx of Italian and Slavic family businesses that were eventually replaced by suburban residential development.[6]

Several alternative systems have explicitly outlined philosophies concerning relationships between the land, commerce, farmers, modern agriculture and agribusiness, and local communities. We include a brief account of some of these even though sometimes their scientific credentials may be questioned. They all explored sustainability long before the word 'sustainability' gained general currency in the 1980s.

1. Biointensive gardening

This is a method of food and crop production originally based on ancient practices, and promulgated by English horticulturist Alan Chadwick (1909–1980), a charismatic teacher and strong influence in California where he prompted the foundation of the University of California Santa Cruz Center for Agroecology. As practised now, it emphasises the use in food production of: double-dug raised beds; composting; intensive planting; companion planting; carbon farming (reducing the transfer of organic matter and nutrients from one place to another by trying to develop

Figure 2.3 Alan Chadwick (1909–1980). (Source: University of California Santa Cruz Photo Collection)

Figure 2.4 Rudolph Steiner (1861–1925).

'closed systems'); calorie farming (growing high calorie plants with high yield per area therefore requiring minimal space – such as potatoes, sweet potatoes, garlic, leeks, burdock and parsnips); avoiding GM cultivars and using conventional selection methods for seed stock; and incorporating these ideas into a whole-system approach. Intensive farming like this greatly reduces water use, human input and mechanical energy consumption.

2. Biodynamic gardening

This originated in the school of thought termed anthroposophy, developed by philosopher Rudolph Steiner. His ideas were formulated in the 1920s and are difficult to distil, but emphasise a holistic and spiritual approach which treats the farm with its crops, livestock, recycling of nutrients and soil maintenance as all part of an integrated whole, a 'farm organism'. Thinking in this way leads to management practices that address the environmental, social and financial aspects of the farm. Lunar and astrological cycles play a key role in the timing of biodynamic practices (hence the alternative name, 'astrological agriculture'), and emphasis is placed on the use of special biodynamic preparations that enhance soil quality and stimulate plant life. The preparations consist of mineral, plant or animal manure extracts, usually fermented (often in hollow cow horns) and applied in small proportions to compost, manures, the soil, or directly onto plants, after dilution and stirring procedures called 'dynamisations'.

3. Fukuoka farming

This is perhaps the purest expression of these approaches and is based on the teachings of Masanobu Fukuoka who began

Figure 2.5 Masanobu Fukuoka (1913–) and Bill Mollison (1928–). Permaculture Convergence, Washington, 1984. (Photo: Dave Blume from www.permaculture.com)

his professional life as an agricultural scientist. It uses 'natural farming' ('natural gardening', 'natural agriculture') as both a path to personal enlightenment and an inspirational guide on how to grow food and fibre in an ecologically beneficial and sustainable way. Fukuoka notes that all farming is human-dominated since it depends on human knowledge for plant selection, soil manipulation, watering, and weed and pest control. He calls current farming practice 'scientific' farming, which has a high input of labour, energy and resources, and which breaks up farming into discrete components (pest control, nutrient requirements, etc.), each to be studied or managed separately with little consideration of the complex interrelationships that exist between them. Fukuoka advocates 'do-nothing' farming which is based on four major principles – no cultivation, no fertiliser, no weeding, no pesticides. In practice, all of these recommendations have some provisos, but they add up to minimal interference with nature while gaining a livelihood from the land.

4. Organic gardening

This international movement was established in the 1940s when the *Organic Farming and Gardening Society* was formed (the *Australian Organic Farming and Gardening Society* was formed in 1945). Organic gardening is a mainstream variant of the above philosophies. The primary objective of organic gardening has been healthy, uncontaminated food and this has generally been associated with a concern about the effects of chemical residues. The process of organic gardening also treats plants as part of the whole ecosystem and attempts to work in harmony with natural systems. Plants that grow well on the site are deliberately selected in an attempt to minimise and replenish the resources the garden consumes. This might entail the use of biological controls for pests, diseases and weeds. More recently, starting in the 1970s, there has been a concern with self-sufficiency and the collection of old 'heirloom' garden varieties to retain the gene pool of the past as a way to safeguard our genetic heritage in the face of a narrow range of mass-produced varieties introduced through modern breeding techniques.

There has been a groundswell of interest worldwide in organic foods. In Australia, this has resulted in the formation of the National Association of Sustainable Agriculture Australia (NASAA) which certifies organic growers. Certified organic produce carries the labels of authorisation by NASAA and Biological Farmers of Australia.

5. Organic agriculture

Organic agriculture (defined as a holistic production management system that avoids the use of synthetic fertilisers, pesticides and GM organisms, minimises pollution of air, soil and water, and optimises the health and productivity of interdependent communities of plants, animals and people) is practised in 120 countries with 31 million hectares of certified croplands and pastures, 62 million hectares certified wild land (for organic collection of nuts, bamboo shoots, wild berries, etc.) and a global market worth US$40 billion.

A recent UN Food and Agriculture Organization (FAO) report supports organic agriculture related to food security as part of the UN goal of sustainable food security for all.

Because of the low chemical input, it is claimed that energy consumption of organic systems can be up to 70% less than in non-organic systems in European countries and up to 32% less in the USA. Also, carbon sequestration efficiency of organic systems in temperate climates is almost double (575–700 kg C/ha/yr) compared with conventionally cultivated soils.[7]

6. Permaculture

Australians Bill Mollison and David Holmgren coined the term permaculture (*perma*-nent agri-*culture*) to signify the ecological approach to agriculture that they have advocated since the 1970s. This is now a highly influential and popular approach that has gained an international reputation. Their aim has been to promote the development of self-sustaining agricultural ecosystems that generate products useful to humans. It was established as a reaction to Western agriculture. It is also seen as an alternative for communities of the Third World where agricultural practices have degraded land, sometimes producing severe erosion and desertification. It is also promoted for small rural land holdings, and for currently unproductive city land in areas associated with urban buildings, or in under-utilised areas such as transport routes. There is an emphasis on perennial, rather than annual, plants and crops.

The principle behind permaculture is the establishment of stable, long-lived artificial ecosystems containing a diversity of useful plants and animals. The species selection for the ecosystem, and their spatial placement within it, lead to complex interactions so that not only food and fibres are produced, but each component contributes to the well-being of the others, resulting in a balanced self-sustaining system requiring only low levels of maintenance. Through careful choice of species, for example, microclimates can be created and used for growing species that may not otherwise flourish. Trees can protect more delicate plants from strong prevailing winds or harsh summer sun. The system requires little soil disturbance, which in turn encourages a healthy soil structure. Permaculture also helps to avoid pest

Figure 2.6 David Holmgren in 2003.

epidemics through the series of checks and balances that evolve in a multi-species system.

Yields from individual plants may not be as high in permaculture systems, but the overall yield from an area of land can be increased. The mixed species plantings of permaculture use more of the available energy and nutrients compared with the same area of land planted with only a single species, the diversity of species harnessing a wider range of resources. A good combination of species will differ in both the size and form of the root systems and above-ground parts. Healthy growth is sustained by exploiting different components of the resources available, or the same components from different locations within the soil profile or aerial space above.

Permaculture, although still practised by a minority, is now well established in Australia and overseas in both urban and rural communities. David Holmgren has more recently defined permaculture as: *'Consciously designed landscapes which mimic the patterns and relationships found in nature, while yielding an abundance of food, fibre and energy for the provision of local needs.'*[8] He sees permaculture maturing from a sustainable approach to *agriculture* to a more broadly encompassing permanent or sustainable *culture* where the relationships between people and buildings and the way they are organised are also included in the system.

Forest and woodland farming is often considered part of the permaculture movement. It promotes multi-layered woodlands where all the plants are chosen to have a broad range of practical uses (food, medicine, fibres, oils, fauna-attraction, etc). The methods of the movement are attributed to Robert Hart in the 1960s who published his thoughts in the 1980s. Robert Hart died in 2000.

Towards sustainable horticulture

What does this historical legacy teach us about sustainable living?

Our human ancestors migrated out of Africa about 200 000 years ago. Making their way across ancient land bridges, they crossed the Arabian Peninsula moving around the coast of present-day India and down the Malayan archipelago to arrive in Australia 40 000 to 50 000 years ago. This was well before human arrival and occupation of Europe and about 25 000 to 30 000 years before the occupation of North America. These nomadic hunter-gatherers lived sustainably on life's basic resources simply by moving on. When communities settled in one place, the surrounding countryside would only have been harvested for a brief period before becoming depleted, survival then depending on community industry, the farming of domesticated strains of animals and plants, and by trading goods with distant communities. In principle, this environmentally precarious system has remained the same until today, but with a recent human population explosion and massive industrialisation that has lead to a highly sophisticated globalisation of trade.

Over the last 100 years or so horticulture has adapted for its own use many of the techniques, technology, chemicals and fertilisers that were developed for mainstream agriculture. It has also taken on board many of the ideas coming from non-conventional agriculture and horticulture; ideas that tend to fuse gardening and farming in a form of small-scale mixed farming grounded in 'natural' principles that see the garden-farm as an integrated ecological system that can use the land as a sustainable source of diverse products. One aspect of this 'holistic' or 'systems' approach has been the use of fewer chemicals and fertilisers (a major objective of the organic gardening and organic agriculture movements) and the productive use of degraded or unproductive public land. Among a whole suite of other ideas we can list: companion planting, calorie farming (growing high-energy food in a small space), use of heirloom plant cultivars, and various creative and mysterious composting techniques. Many of these movements strive for self-sufficiency (sometimes with intensive cultivation, at other times low maintenance), emphasising the idea of the garden-farm as a closed system (meaning that maximum sustainable use is made of on-site materials, water and the Sun's energy, avoiding as much as possible the introduction of materials from elsewhere). Healthy biological systems are encouraged and sometimes native fauna is supported by providing habitat, shelter and food which conserves local gene pools and helps to develop a sense of place.

Recently drought has encouraged creative ways of overcoming water shortage. Garden green waste management has become mainstream. The dangers of plants escaping from horticulture into crops and the natural environment has become more evident. We can add to this list, following the push for sustainable development, the idea of monitoring the use of all garden resources as being an important part of restraining consumer demand.

Sustainable gardening

After the birth of mainstream sustainable agriculture in the 1980s, sustainable horticultural practices (as crop production rather than ornamental horticulture) gathered momentum. In 2002, the *First International Symposium on Sustainability in Horticulture* was held at the International Horticultural Congress at Toronto.

Here in Australia, the organisation Sustainable Gardening Australia (SGA) began its activities in 2002. It is dedicated specifically to the promotion of sustainable horticultural practices in ornamental horticulture. Many prestigious organisations support its aims, including the collaborative Sustainable Landscapes project in South Australia, Australian Institute of Landscape Architects, and botanic gardens throughout

Australia. Towards 100 retail nurseries have already been accredited.

There is no question that sustainable gardening is attracting increasing interest and support, and the educational gardening programs, information sheets and website developed by SGA continue to encourage its growth. The key areas where environmental impact can be lowered have been identified by SGA as: water, environmental weeds, indigenous plants, chemicals, waste, green purchasing and sustainable design.

Managing our Ecological Footprint requires balancing development, agriculture and biodiversity. (Photo: Skyworks)

SUSTAINABILITY ACCOUNTING – HOW DO WE KNOW WHAT IS SUSTAINABLE?

> **KEY POINTS**
>
> Sustainability addresses 'start of pipe' drivers of environmental change.
>
> Global population to increase from 6 billion now to 9–10 billion by 2050.
>
> Sustainability accounting indicators include:
> - Ecological Footprint, Life Cycle Assessment, Input-Output Analysis, governance indicators.
>
> Our goal is a standard of living that is environmentally sustainable.

The most obvious way of tackling environmental deterioration is to deal directly with threats to nature and biodiversity by careful management of the land, water and atmosphere.

However, it is the 'start of pipe' drivers of environmental change that are now receiving more attention, especially consumer demand and environmentally unfriendly technology.

Between 1950 and 2000 the global economy grew by a factor of 10, energy use increased 13–14 times, use of freshwater increased nine times, the area of land under irrigation increased five times, and the extinction rate of organisms reached alarming levels.[1]

We have certainly thrived economically, but as indicators of economic growth have moved in a steady long-term upward trend, so the indicators of the planet's biological health have pointed steadily down.

We have already introduced the idea of sustainability accounting as a useful tool to use in managing human impact on the planet. Because sustainability reaches into all aspects

Figure 3.1 Path to sustainability? (Source: Christina Reitano, Melbourne University Postgraduate Environment Network)

of life, the potential methods of measuring and monitoring seem infinite. So, for example, calculations can be made at different levels of human organisation, calculating the impacts of individuals, families, organisations, economic sectors, countries, or the whole of humanity; for a particular product, service, ecosystem or resource; for short-term or long-term; or custom-made for particular user groups such as educators, academics, planners, decision-makers, the general public and so on.

In this chapter we look at some of the sustainability monitoring tools, like the Ecological Footprint (EF), that are becoming popular. These tools are just as relevant to the sustainable management of urban space as they are to any other human activity.

We return to the general question we asked earlier: How can we measure our environmental impact?

In 1974, Ehrlich and Holdren derived a formula:

$$\text{ENVIRONMENTAL IMPACT} = P \times A \times T$$

where: P = number of people (population);
A = resource use per person (affluence);
T = impact per resource unit (technology).

This is a powerful equation pointing to population control, reduced consumption and the use of environmentally friendly technology as ways of addressing environmental impact. But ecology is never simple and the conclusions we derive from the equation may need qualification. Is it affluence or poverty that produces environmental degradation? Doesn't wealth provide the resources needed to tackle environmental problems? Surely the way societies treat their natural resources and the way they organise themselves socially and economically is important too, but can this be translated meaningfully into this sort of equation? And then there is always the complicating factor that the negative impact of resource use will vary according to circumstances: water is not a precious commodity where rainfall is high and population is low.

Regardless of these reservations, it is clear that every additional human being brings an increase in resource use and more demand on land and nature. Population numbers, rates of consumption and degree of technological sophistication are clearly important drivers of environmental impact and key factors for sustainability, but population is the place to start.

Population

On present projections, Australia's population in 2050 will be 27 million and the world population

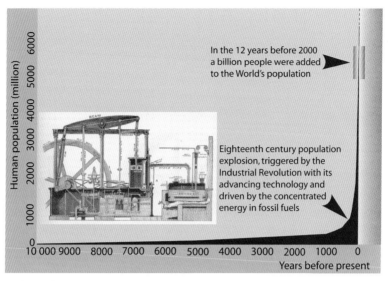

Figure 3.2 World population from 10 000 years ago to present.

9 billion. Yet in 2004 the Earth is believed already to be 25% over the limit of its regenerative and absorptive capacity ... [and yet we must become sustainable] ... while lifting an existing two billion out of poverty and coping with an additional two billion people in the next half century.

JENNY GOLDIE[2]

Global population

Estimates of the human population size at the time when the ice sheets of the last ice age receded about 10 000 years ago suggest about 5 million people. By 5000 BP, the time of the first Egyptian dynasty, world population had increased to approximately 100 million, then jumped to nearly 250 million about 2000 years ago. In 1650 AD, the total was about 400–500 million people just before a major population explosion in Europe and the New World associated with the Industrial Revolution. By 1950, the figure had escalated to 2.5 billion, increasing dramatically to about 6 billion in the year 2000 (see Figure 3.2 and Table 3.1). World population more than doubled between 1950 and 2000 in the lifetime of a single generation. The population in mid-2007 was 6.592 billion.

The fossil fuel energy-based population explosion of the last 150 years has seen human numbers increase from about 1 billion in 1850 to about 6 billion in 2000. The uniquely human capacity to accumulate written knowledge and to skilfully manipulate the environment has allowed *Homo sapiens* to overcome environmental barriers that have confined other species to specific habitats. In biological terms, humans have assumed plague proportions as, over the last 125 000 years the species has spread from Africa across the planet.

In the industrialised world the average population growth has slowed, but consumption levels are extremely high. This places a heavy demand on the environment.

Table 3.1 Human population doubling times

Population doubling times		
Year AD	Population (billions)	Doubling time (yrs)
0–1650	0.25–0.5	1650
1650–1850	0.5–1	200
1850–1930	1–2	80
1930–1975	2–4	45

Source: Goudie A (2006)[3]

In the non-industrialised parts of Africa, the Middle East and Asia, consumption per person is low but population numbers are still surging ahead – and that too places a high demand on the environment. Emerging economies like those of China and India aspire to the living standards of the Western world as does the non-industrialised world in general (Figure 3.3).

A conservative population estimate for the year 2050 is about 9 billion, assuming an annual growth rate of 0.33% and also assuming efforts at family planning continue. Most demographers predict that by 2100 the world population will have levelled off at about 10 billion, which is slightly less than twice the current population.

More than half the annual population increase occurs in just six countries (see Table 3.2).

As population increases, there is a shrinking proportional supply of food and water per person and arable land becomes more scarce. In many countries this is associated with aquifer depletion and sanitation problems, compounded by political and economic instability and a rising incidence of AIDS.

Globally, a general change in trend has recently emerged. In about 1965–70 the global population growth rate peaked at about 2.1% p.a. and by 2002 this had fallen to 1.2% (Figure 3.4). The average woman is having fewer children than ever before but, even with smaller families, there are near-record births. This is because the number of woman of child-bearing age is increasing and global life expectancy at birth is continuing to rise. The average human life expectancy today is 67 years, 10 years greater than it was in 1970.

Hope for population control is now targeted at the improvement of women's health and their economic, educational and political status relative to men, as there is a clear correlation between high relative female status and low fertility. In addition to voluntary family planning this approach is

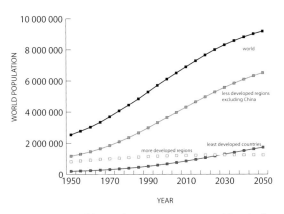

Figure 3.3 World population projections and level of development 1950–2050.[4]

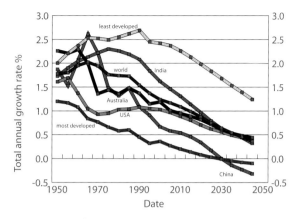

Fig. 3.4 Rate of population change with projection 1950–2050.[4]

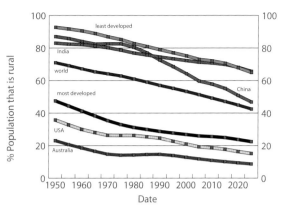

Fig. 3.5 Proportion of population that is rural in a selection of regions.[4]

seen as a major step towards environmental sustainability as well as social equity.

Urbanisation

With increasing population has come increasing urbanisation. In 2007 it was announced that world-wide, for the first time in human history, more people were now living in cities than in the country. This trend towards urban living was established several centuries ago and will continue into the future (Figure 3.5).

Large urban populations create equally large ecological footprints although, through economy of scale, city dwellers require fewer resources per person than people in the country.

The heavy environmental demand created by the high consumption lifestyles of affluent societies is not necessarily felt locally because environmental impact is now embedded in global trade. However, it is clear that affluent nations must learn to live more sustainably while assisting resource-poor nations towards

Table 3.2 Countries in order of population size in 2005, and as projected to 2050[5]

Country	Population in 2005 (millions)	Country	Estimated population in 2050 (millions)
China*	1311	India	1628
India*	1122	China	1437
USA*	299	USA	420
Indonesia*	225	Nigeria	299
Brazil*	187	Pakistan	295
Pakistan*	166	Indonesia	285
Bangladesh	147	Brazil	260
Russia	142	Bangladesh	231
Nigeria	135	Dem. Rep. Congo	183
Japan	128	Ethiopia	145

* = countries contributing the greatest population increase.

self-sufficiency through birth control and sustainable technologies.

In 1798, the English curate Thomas Malthus published his *Essay on the principle of population* in which he claimed that human population increases until it is reined in by (birth) control, famine, war or disease. This prediction has been borne out in some regions and at some times in history but, in general, humanity has flourished by harnessing cheap energy, adopting new technologies, using mass production (especially of food) and by controlling disease using modern medicine. Of course Malthus could yet be proved correct.

Australian population

Australia by world standards is, perhaps surprisingly given its vast area, a highly urbanised society. In 2005, more than four out of every five people (88% and increasing) were urbanites living in seaboard cities.

At Federation in 1901 Australia's population was 3.7 million. On 2 June 2008 22:01:44 (Canberra time), the resident population of Australia was estimated as **21 316 564**.[6] Over this period the highest fertility rate was 13.7/1000 during the baby boom of the 1950s and lowest at 6/1000 in 2002–03. In 2005, it was 6.6. Based on assumptions about future levels of fertility, mortality, internal migration and net overseas migration, by 2051 the ABS estimates a population of 28 million with a high proportion of older people due to a decrease in the birth rate and an increase in life expectancy.

With such a large urban population, Australia has been pursuing a more

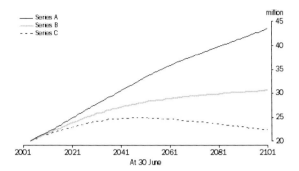

Figure 3.6 Projected Australian population to 2100.[7]

sustainable approach to urban planning and design. In line with other OECD nations, Australia has attempted to move toward more compact cities with increased population density in built-up areas, the intention being to reduce the unnecessary suburbanisation of valuable agricultural land, reduce fossil fuel usage and therefore greenhouse gas emissions, reduce reliance on cars, making more efficient use of infrastructure and improving equity.

Sustainability accounting toolbox

In the last couple of decades, there has arisen a crowded toolbox of quantitative methods to assess sustainability, including various benchmarks, indicators, indexes and audits as well as a host of reporting, modelling and monitoring procedures. Many of these have only just been developed.

Here is a sample:

1. Those that are strictly environmentally based, including: traditional environmental impact assessment and assessments of biodiversity such as the

Living Planet Index and Ecological Footprint (EF).
2. Those that extend beyond economic indicators (like the Gross National Product) into broader, more 'humane' indicators including: the Genuine Progress Indicator, Human Wellbeing Index, Ecosystem Wellbeing Index, Human Development Index.
3. Tools that measure materials, resource and energy flows, including: Materials Flow Analysis, Substance Flow Analysis, Input Output Analysis (I-O Analysis), Product Energy Analysis, Process Energy Analysis.
4. 'Cradle-to-grave' product impact analysis: Life Cycle Analysis (LCA), Life Cycle Costing (LCC).

Some of these methods will doubtless become standard tools while others will lose popularity, and new ones will emerge.

Most of the creators of these tools complain of the need for more reliable and complete baseline data, i.e. the need for better sustainability accounting. Those indicators with the greatest appeal combine simplicity, a good integration of nature and society, and the ability to be used at all levels of the action hierarchy over both long and short terms. Some indicators effectively combine several different methods.

We shall touch on the many paths to sustainability accounting by looking at just four sustainability tools:

1. The Ecological Footprint (EF) – which measures the amount of land needed to sustain a particular activity or resource use.
2. Life Cycle Assessment (LCA) – which measures the environmental impact of a good or service throughout its history.
3. Input-Output Analysis – which relates environmental impact to expenditure, consumption patterns and lifestyle.
4. The Environmental Sustainability Index – which is specific to the country level of the action hierarchy, and uses a combination of sustainability performance measures to assess environmental governance.

Ecological footprint

Perhaps the most popular sustainability tool is the Ecological Footprint (EF, or Footprint).

The Footprint concept was developed in the mid-1990s by academics William Rees and Mathis Wachenagel. It is a management and communication tool that measures humanity's demand on the biosphere in terms of the area of biologically productive land and sea required to provide the resources we use and to absorb our waste. The unit used to express this is the global hectare (gha).

The way of calculating the Ecological Footprint is constantly being refined. It uses both product lifecycle data (see LCA) as well as government statistics on consumption patterns (see I-O Analysis) and it can be used for many different situations. Footprinting gives a clear indication of trends, and summarises a complex situation in a simple graphic way, but for figures to be comparable it is important that there is a standardised method of Footprint calculation and this is now being done.[8]

We will look at the Ecological Footprint at four levels of the sustainability management hierarchy: global, national, household and individual.

All living things have an Ecological Footprint; it is the size, sustainability and particular circumstances of the situation being investigated that must guide management.

Global Ecological Footprint and biocapacity

The global Ecological Footprint varies with population size, average consumption per person and resource efficiency. The Earth's biocapacity or biological carrying capacity is the amount of biologically productive area that is available to meet humanity's needs; it varies with the amount of biologically productive area and its average productivity. The *Living Planet Report*, produced by the World Wide Fund for Nature (WWF), was first produced in 1998 with data starting in 1961, and it summarises the state of the world's ecosystems using Footprint analysis. The WWF also produces a *Living Planet Index* which is a measure of changes in global biodiversity.

In 2003 the global Ecological Footprint was 14 billion gha, equivalent to about 2.2 gha/person. This demand on nature was compared with the Earth's biocapacity, based on its biologically productive area and the result was 1.8 gha/person which meant that humanity's Ecological Footprint exceeded global biocapacity by 0.4 global ha/person (2.2–1.8 gha) or 22%. This global overshoot began in the 1980s and has been growing ever since. Overshoot means that we are using natural resources faster than they are being replaced. At present, this is 1.22 times the productive capacity of the Earth. If we proceed as we are, by 2050 our demand on nature will have risen to twice the biosphere's productive capacity and this will inevitably lead to an exhaustion of natural assets to the point where they are unable to regenerate.

Figure 3.7 presents an optimistic scenario showing the effect of a widespread greening of the global economy.

Different regions of the world have different biocapacities. Global trade is a way of distributing resources from those regions that are resource-rich to those that are not. Over time the world will become more clearly differentiated into ecological debtors (countries that depend on net imports of ecological services to maintain their economies) and ecological creditors (countries with ecological reserves), and as ecological assets become more scarce they will increase in economic significance (Figure 3.8). Australia is a strong ecological creditor. The relationship between population, biocapacity and the Footprint of each of the world's major regions is shown in Figure 3.9 and Table 3.3.

The Australian Ecological Footprint

Of the 147 countries analysed in the *Living Planet Report* of 2006, Australians have

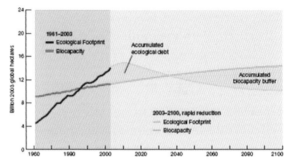

Figure 3.7 The global Ecological Footprint and overcoming overshoot, an optimistic scenario.[9]

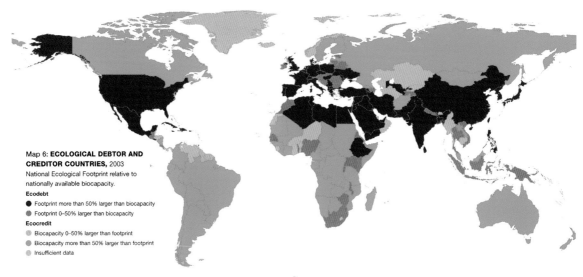

Figure 3.8 Global ecological debtor and creditor countries.[9]

the sixth largest individual Ecological Footprint at 6.6 gha/person, three times the global average of 2.2 gha. This high figure can be attributed to the fact that we live in large houses with few occupants, use a large number of goods and services, travel long distances, and depend heavily on fossil fuels (52% of the footprint) for our energy needs.

Household

The household is a convenient economic unit because heating, cooking, purchase of food and other goods and services are done collectively. With the minimum of effort, household consumption patterns can be calculated. The passage of food, energy, water, goods and waste through the household unit is very similar to the biological activity of an organism and for this reason these flows are sometimes referred to as household metabolism or household ecology.

House size, number of occupants and per capita expenditure all affect consumption. Each year the ABS produces figures for household expenditure on goods and services, and these can be used to obtain a general overview of the potential environmental impact according to spending patterns. In general, the volume of goods and services we consume is closely related to the amount of money we spend. The higher the number of house occupants, the lower the Footprint per person due to shared resources (Table 3.4).

Households are also amenable to analysis of household expenditure using Input-Output Analysis. (See I-O Analysis section below).

Individual

A breakdown of the Ecological Footprint components for an Australian compared with the country's biocapacity are given in Table 3.5.

Table 3.3 Ecological Footprint for a selection of countries in 2003, ranked in order of size of per capita Ecological Footprint

Country	World ranking	Total EF (M 2003 gha)	Per capita EF (gha/person)	Biocapacity (gha/person)	Ecological reserve/deficit (gha/person)
United Arab Emirates	1	–	11.9	0.8	–1
USA	2	2819	9.6	4.7	–4.8
Finland	3	39.52	7.6	12.0	4.4
Canada	4	240	7.6	14.5	6.9
Kuwait	5	18.25	7.3	0.3	–7.0
Australia	6	130	6.6	12.4	5.9
UK	14	333	5.6	1.6	–4.0
World	–	14 073	2.2	1.8	–0.4
China	69	2152	1.6	0.8	–0.9
India	124	802	0.8	0.8	–0.9
Pakistan	140	92	0.6	0.3	0.2
Afghanistan	147	2.4	0.1	0.3	0.2

Source: WWF *Living Planet Report* 2006.[9]

It is soon apparent from sustainability indicators that at present, community effort is concentrated on direct resource use (e.g. the energy and water that appears on household bills), but this ignores the resource impact of our general consumption behaviour.

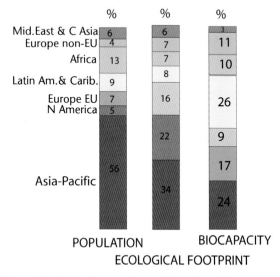

Figure 3.9 Regional proportions of global population, Ecological Footprint and biocapacity.[9]

Figure 3.10 shows the relative proportions of direct and indirect resource use for our water, energy and eco-footprints, demonstrating how small the direct component is, by far the greatest proportion consisting of the embodied resources in the goods and services that make up our general consumption. If every household switched to renewable energy and stopped driving the family car, total emissions would still only decline by about 18%. Figure 3.11 indicates in detail how the average Australian

Table 3.4 Ecological Footprint related to number of occupants per house (gha)

Number of house occupants	EF gha/person
Couple	7.13
Couple + 1 dependent	5.69
Couple + 2	4.49
Couple + 3+	3.61
1 parent with >1 child	3.38
1 person	7.75
Group household (2.2)	7.42

Source: Lenzen (2004).[10]

Table 3.5 Breakdown of the Australian individual Ecological Footprint (gha)[9]

Australia – 2003	
Population	19.7 M
Total Ecological Footprint	6.6
Cropland	1.17
Grazing land	0.87
Forest: timber, pulp, paper	0.53
Forest: fuelwood	0.03
Fishing ground	0.28
CO_2 from fossil fuels	3.41
Nuclear	0.00
Built-up land	0.28
Water withdrawals per person ('000m^3/yr)	1224
Total biocapacity	12.4
Cropland	4.26
Grazing land	1.83
Forest	3.34
Fishing ground	2.73
Ecological reserve	5.9
Footprint change/person 1975–2003	–7%
Biocapacity change/person 1975–2003	–28%

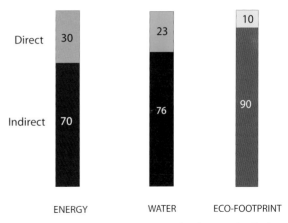

Figure 3.10 Percentage of direct and indirect individual resource use in Australia.[12]

individual's consumption breaks up into its components. These figures are derived from the *Australian Conservation Foundation*'s *Consumption Atlas*, an interactive online tool developed in partnership with the *Centre for Integrated Sustainability Analysis* at the University of Sydney.[11] Based on household expenditure and I-O Analysis, this remarkable tool is able to assess individual resource use as a proportion of household consumption. The Atlas provides a consumption profile for a selected area (state, or region by postcode) together with a map. This means that we now have a nation-wide consumption assessment tool that allows comparisons between the different states and regions.

These figures indicate overwhelmingly that food production is the single greatest environmental impact of human activity in Australia. In countries with lower consumption levels, the proportion of resources dedicated to food production would be much greater.

We cannot stop eating, but we can certainly drastically change the way food is produced, processed, packaged and distributed. At the level of production, there can be a greater emphasis on sustainable agricultural practices. At the level of consumption we can make a huge difference by careful food purchasing and a moderate low meat, low dairy diet. Australia is extremely demanding in land area for livestock, using about three times that of most other OECD countries.

Life Cycle Assessment

Life Cycle Assessment (LCA) is a cradle-to-grave analysis of total resource use needed for a particular product or service. It can measure total environmental, social and

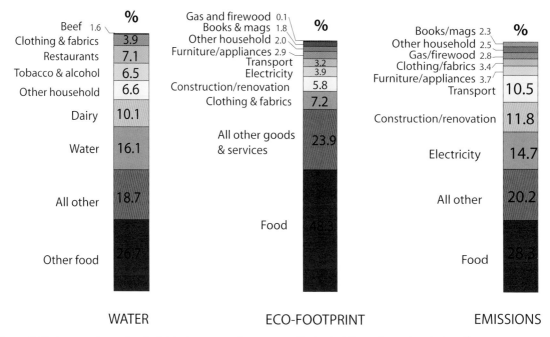

Figure 3.11 Average Australian individual consumption – water, Ecological Footprint and emissions.[13]

economic impact by assessing the quantities of energy and raw materials used, together with solid, liquid and gaseous wastes produced at every stage of a product's life cycle, from its material extraction and acquisition, to its manufacture, use, transport and disposal. In this way, sustainability concerns can be addressed as part of the assessment. This is useful because these impacts are often treated as economic externalities (meaning that the costs or benefits are accrued by a third party that is not producing or supplying the good or service – so the 'true cost' is not included in the financial cost).

Although LCA was introduced in the 1970s, it is only since the early 1990s that international standards have been developed in an effort to harmonise the technique and establish its widely accepted application. There are four international standards specifically related to LCA:

- ISO 14040: Principles and framework
- ISO 14041: Goal and scope definition and inventory analysis
- ISO 14042: Life cycle impact assessment
- ISO 14043: Interpretation.

The Australian Life Cycle Inventory is a public database initiative, led by CSIRO, that will allow users from government and industry to assess and compare products across a number of industries ranging from building to packaging materials, and to choose those likely to give the best performance relative to their environmental impact. This will assist companies and research groups in assessing the life cycle impacts of products and services and reduce the tedious task of gathering data.

Input-Output Analysis

Input-Output Analysis (I-O Analysis) records money flows between economic sectors. These monetary flows are published at regular intervals by government. For Australia this is the ABS *System of National Accounts*. I-O Analysis is a valuable standardised tool using reliable, empirical and comparable data that can be applied around the world.

How can expenditure be related to environmental effects? The resources needed per unit money spent on a product or service are known as resource intensities (see Info Box 3.1). When resource intensities are known it is then possible to relate resource use directly to expenditure and therefore calculate the resource impacts of economic sectors, families (using family expenditure), individuals, and so on. Resource use can then be related to environmental impact.

This form of analysis is especially useful for *Triple Bottom Line* assessment of the economic, social and environmental impacts of consumption patterns.

A recently published report found that for each dollar spent in the Australian economy, the resource intensity was: 1 kg CO_2-e emissions, 7.7 MJ of primary energy, 41 litres of managed water and 3.2 m^2 of land disturbance.[14] The report drew attention to several major factors needing careful management: the effects of an ageing population; the declining availability of oil;

> **INFO BOX 3.1: RESOURCE INTENSITY**
>
> Resource Intensity (RI) is a measure of the resources needed to provide a product or service expressed as the unit of money (usually the dollar) paid for the good or service. Dollar values multiplied by the RI give the total implied resource use, so high resource intensities indicate a high resource cost for an item. The RI can relate to resource use in general or to a particular resource such as water or energy. Resource intensities can be estimated for all goods and services including entire economies. Resource use in general is strongly related to standard of living. Energy and water use is strongly related to climatic conditions and standard of living.

the importance of reducing industrial waste, and the importance of simplifying the production chain. Not surprisingly, primary production has extremely high relative intensities of water, emissions and land disturbance, and the prices paid largely reflect only the costs of production rather than the full resource and environment costs – although the report cautions against drawing simple conclusions from the results. Future reports are likely to take into account environmental exports/imports.

I-O is a relatively imprecise procedure but can, once resource intensity figures are known, give rapid approximations of resource use without the time-consuming detail of LCAs.

Resource intensity is a very simple and effective indicator. We can, for example, express

Australia's water or energy use as quantity of water or energy used per $1 GDP spent. This is a way of measuring progress in decoupling resource use and economic growth. The connection between resource use and environmental impact is more complicated but can be factored into calculations. Australia's first publication on sustainable consumption indicated Australia's level of consumption as being among the highest in the world.

Environmental Sustainability Index and Environmental Performance Index

The Environmental Sustainability Index (ESI) and Environmental Performance Index (EPI) have been developed as a collaboration between the Yale Center for Environmental Law, the Center for International Earth Science Information Network, the World Economic Forum, and the Joint Research Centre of the European Commission. The two indexes are targeted at the national level of the action hierarchy and allow a country-by-country comparison of overall progress in environmental stewardship. The ESI combines 76 data sets into 21 indicators of environmental sustainability expressed through sustainability categories including: environmental systems; environmental stresses; societal capacity to respond to environmental challenges; and global stewardship. Overall it provides an objective basis for policy-making that is aligned to the Millennium Development Goals, and a summary of the findings is produced for policy-makers. (Figure 3.12.)

Seven country 'peer groups' are identified so that leaders and laggards can be seen on an issue-by-issue basis (Figure 3.13). Higher ESI

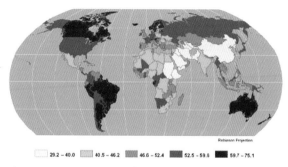

Figure 3.12 Environmental Sustainability Index by country.[15]

scores indicate better environmental stewardship. Australia's ESI is summarised in Figure 3.14. In the 2005 report Australia (ESI 61.0) ranked 13th out of 146 countries, with Finland (ESI 75.1), Norway (ESI 73.4) and Uruguay (71.8) topping the list, and Turkmenistan (33.1), Taiwan (32.7) and North Korea (29.2) at the other extreme. The United States was 45th (ESI 52.9). The EPI puts more emphasis on direct environmental sustainability indicators than the ESI (which is more concerned with issues of governance). Figure 3.15 shows the EPI by country and Figure 3.16 shows the results for Australia.

How does an index like this compare with the Ecological Footprint? Large EFs tend to

Figure 3.13 ESI characteristic-based country groupings.[15]

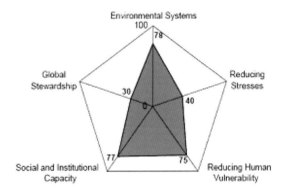
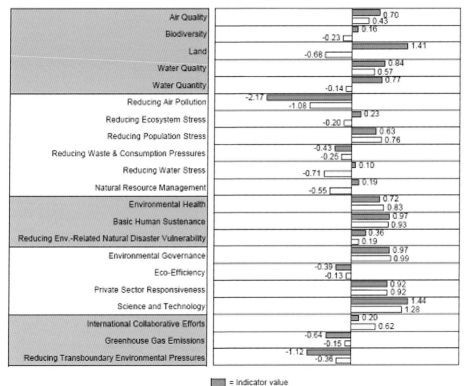

Figure 3.14 Australia's Environmental Sustainability Index.[15]

correspond to high ESI scores, an apparent contradiction as both are regarded as measures of sustainability, and high EFs indicate a high level of consumption while higher ESIs indicate a greater capacity for sustainability. However, the ESI includes consideration of the ability of a country to address over-consumption. Although high levels of consumption are not sustainable, it is also true that in some countries low levels of consumption are also not sustainable. Wealthy nations with large EFs are better

Pilot 2006 Environmental Performance Index

Overall EPI Scores (0–100)

Rank	Country	EPI Score	Rank	Country	EPI Score	Rank	Country	EPI Score
1	New Zealand	88.0	47	Unit. Arab Em.	73.2	93	Kenya	56.4
2	Sweden	87.8	48	Suriname	72.9	94	China	56.2
3	Finland	87.0	49	Turkey	72.8	95	Azerbaijan	55.7
4	Czech Rep.	86.0	50	Bulgaria	72.0	96	Papua N. G.	55.5
5	Unit. Kingdom	85.6	51	Ukraine	71.2	97	Syria	55.3
6	Austria	85.2	52	Honduras	70.8	98	Zambia	54.4
7	Denmark	84.2	53	Iran	70.0	99	Viet Nam	54.3
8	Canada	84.0	54	Dom. Rep.	69.5	100	Cameroon	54.1
9	Malaysia	83.3	55	Philippines	69.4	101	Swaziland	53.9
10	Ireland	83.3	56	Nicaragua	69.2	102	Laos	52.9
11	Portugal	82.9	57	Albania	68.9	103	Togo	52.8
12	France	82.5	58	Guatemala	68.9	104	Turkmenistan	52.3
13	Iceland	82.1	59	Saudi Arabia	68.3	105	Uzbekistan	52.3
14	Japan	81.9	60	Oman	67.9	106	Gambia	52.3
15	Costa Rica	81.6	61	Thailand	66.8	107	Senegal	52.1
16	Switzerland	81.4	62	Paraguay	66.4	108	Burundi	51.6
17	Colombia	80.4	63	Algeria	66.2	109	Liberia	51.0
18	Norway	80.2	64	Jordan	66.0	110	Cambodia	49.7
19	Greece	80.2	65	Peru	65.4	111	Sierra Leone	49.5
20	Australia	80.1	66	Mexico	64.8	112	Congo	49.4
21	Italy	79.8	67	Sri Lanka	64.6	113	Guinea	49.2
22	Germany	79.4	68	Morocco	64.1	114	Haiti	48.9
23	Spain	79.2	69	Armenia	63.8	115	Mongolia	48.8
24	Taiwan	79.1	70	Kazakhstan	63.5	116	Madagascar	48.5
25	Slovakia	79.1	71	Bolivia	63.4	117	Tajikistan	48.2
26	Chile	78.9	72	Ghana	63.1	118	India	47.7
27	Netherlands	78.7	73	El Salvador	63.0	119	D. R. Congo	46.3
28	United States	78.5	74	Zimbabwe	63.0	120	Guin.-Bissau	46.1
29	Cyprus	78.4	75	Moldova	62.9	121	Mozambique	45.7
30	Argentina	77.7	76	South Africa	62.0	122	Yemen	45.2
31	Slovenia	77.5	77	Georgia	61.4	123	Nigeria	44.5
32	Russia	77.5	78	Uganda	60.8	124	Sudan	44.0
33	Hungary	77.0	79	Indonesia	60.7	125	Bangladesh	43.5
34	Brazil	77.0	80	Kyrgyzstan	60.5	126	Burkina Faso	43.2
35	Trin. & Tob.	76.9	81	Nepal	60.2	127	Pakistan	41.1
36	Lebanon	76.7	82	Tunisia	60.0	128	Angola	39.3
37	Panama	76.5	83	Tanzania	59.0	129	Ethiopia	36.7
38	Poland	76.2	84	Benin	58.4	130	Mali	33.9
39	Belgium	75.9	85	Egypt	57.9	131	Mauritania	32.0
40	Ecuador	75.5	86	Côte d'Ivoire	57.5	132	Chad	30.5
41	Cuba	75.3	87	Cen. Afr. Rep.	57.3	133	Niger	25.7
42	South Korea	75.2	88	Myanmar	57.0			
43	Jamaica	74.7	89	Rwanda	57.0			
44	Venezuela	74.1	90	Romania	56.9			
45	Israel	73.7	91	Malawi	56.5			
46	Gabon	73.2	92	Namibia	56.5			

* This column contains sparklines for each of the 6 EPI policy categories showing the relative strengths & weaknesses for each country.

Health Biodiv. Energy Water Air Nat. Res.

www.yale.edu/epi

Figure 3.15 Global Environmental Performance Index by country.[15]

Australia

EAST ASIA AND THE PACIFIC
GDP/capita 2004 est. (PPP) $30,700

Income Decile 1 (1=high, 10=low)

Pilot 2006 EPI	
Rank:	20
Score:	80.1
Income Group Avg.	81.6
Geographic Group Avg.	66.2

Policy Categories

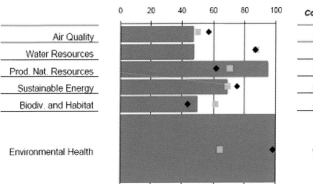

	Country	Income Group	Geographic Group
Air Quality	46.9	57.0	50.1
Water Resources	47.3	87.0	87.6
Prod. Nat. Resources	94.4	62.0	70.4
Sustainable Energy	68.4	75.5	69.3
Biodiv. and Habitat	49.5	43.7	61.6
Environmental Health	99.0	98.2	64.6

Indicator Data

		Value	Target	Standardized Proximity to Target (100=target met)
MORTALITY	Child Mortality (deaths/1000 population 1-4)	0.3	0	98.8
INDOOR	Indoor Air Pollution (%)	0	0	100.0
WATSUP	Drinking Water (%)	100.0	100	100.0
ACSAT	Adequate Sanitation (%)	100.0	100	100.0
PM10	Urban Particulates (µg/m^3)	18.6	10	93.9
OZONE	Regional Ozone (ppb)	60.6	15	0.0
NLOAD	Nitrogen Loading (mg/L)	1,159.3	1	78.0
OVRSUB	Water Consumption (%)	45.7	0	16.6
PWI	Wilderness Protection (%)	12.6	90	14.0
PACOV	Ecoregion Protection (scale 0-1, 1=10% each biome protected)	0.70	1	71.5
HARVEST	Timber Harvest Rate (%)	0.4	3	100.0
AGSUB	Agricultural Subsidies (%)	- 0.8	0	100.0
OVRFSH	Overfishing (scale 1-7)	2	1	83.3
ENEFF	Energy Efficiency (Terajoules / million GDP PPP)	8,961	1,650	69.4
RENPC	Renewable Energy (%)	3.7	100	3.7
CO2GDP	CO$_2$ per GDP (Tonnes / GDP PPP)	209	0	81.7

Figure 3.16 Australia's Pilot Environmental Performance Index.[15]

equipped to deal with environmental problems (pollution, fragile ecosystems, sea level rise, etc.) and can address, though not overcome, their high natural resource consumption levels.

In September 2000, 189 nations adopted the UN Millennium Declaration and its Millennium Development Goals (MDG), designed to alleviate poverty and promote sustainable development. The EPI measures how effective countries are in striving for these goals.

The task ahead

The Ecological Footprint tells us that we in Australia need to reduce our average individual Footprints to one-third of their current level if we are to meet our share of one planet consumption. But what sort of life would that mean? All humanity expects an 'acceptable' quality of life but how does this match up to environmental realities?

The *Living Planet Report* 2006 uses two sustainability indicators to provide a measured assessment of the global sustainability task for the countries of the world. It does this by comparing what would be needed to live sustainably with what constitutes an acceptable lifestyle (Figure 3.17). The United Nations Development Program's Human Development Index (HDI) is plotted on one axis of the graph. This is a measure of what could be called an acceptable living standard – the normalised measure of life expectancy, literacy, education, standard of living and GDP per capita for a country. It is used to measure well-being (quality of life), and also to assess whether a country is developed, developing or underdeveloped. This is plotted against the Ecological Footprint (as a measure of demand on the world's biocapacity). The Asia-Pacific and Africa are currently using less than world average per person biocapacity but they have a low HDI, while Australia, the EU and North America have high HDIs but also a per-capita demand well above the Earth's biocapacity. The ideal is to have a high HDI of between 0.8 and 1.0

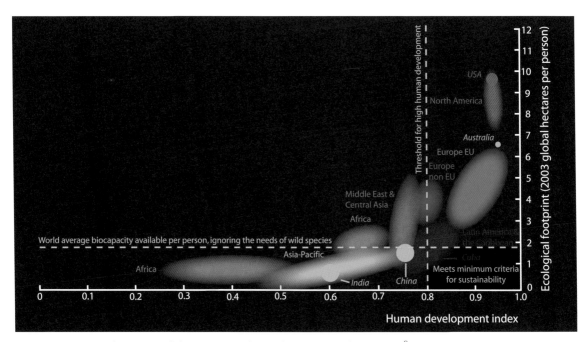

Figure 3.17 Meeting key sustainability criteria. Colour indicates regional groupings.[9]

on a scale of 0 to 1.0, and an average individual Ecological Footprint of less than 1.8 global hectares, which is the world average biocapacity available per person. Only one country falls into this ideal range of sustainable values, and that is Cuba.

We are inclined to think of environmental degradation in terms of the actions of individuals or small groups; perhaps a chemical company releasing toxic pollutants into a river or a timber company logging native forest. In doing this we forget that although improvements in such situations are always possible, these activities are being carried out on our behalf. Very few people wilfully destroy nature. Most environmental degradation is simply the result of our unchallenged need for housing (building materials, infrastructure, heating, electrical appliances), transport (mostly cars and air travel), food (mostly meat and dairy but also plant crops), and water. In the following chapters we will look at each of these four consumption categories in more detail to see how they can be managed more sustainably, especially in gardens and urban landscapes.

Leaves – the solar panels that power humanity and the biosphere.

ENERGY AND EMISSIONS

> **KEY POINTS**
>
> The Sun's renewable energy powers plants and the biosphere.
>
> Climate, fossil fuels, global vegetation and land management all link into the global carbon cycle.
>
> Stabilising climate will require high income countries to reduce emissions by 60–90% over 2006 levels by 2050 with major reductions in place by 2015.
>
> Energy and its environmental impacts are embodied in products and world trade.
>
> To achieve sustainable individual levels of energy use will require dramatic reduction over current levels.
>
> Horticulture must plan for increased urbanisation and climate change.
>
> Parks and gardens are low energy users but can significantly reduce both direct and indirect energy use by local food production.

Chapter 1 showed how sustainable horticulture can become part of a global sustainability effort working towards a secure environmental future. We then looked at some of the many kinds of sustainability accounting that are being used to assist this process. It was also suggested that one simple but effective method of assessing sustainability is to measure the way our use of the resources energy, water, food and materials impacts on biodiversity and ecology.

This and the next four chapters explore each of these resource consumption categories at different levels of human organisation (global, Australian, household, individual) to see how their use in urban space and gardens fits into the big sustainability picture.

In considering environmental impacts we need to distinguish between those that are *direct* and those that are *indirect*.

Direct impacts occur on-site. So in a garden something that affects water flows to the garden beds or encourages native wildlife is having a direct impact. Indirect impacts occur at a distance from the site. So, for example, when we use mains water it has the effect, no matter how small, of diverting water from natural watercourses into reservoirs. This parallels the direct and indirect effects we have on the biosphere through our general consumption of resources, except that in a garden we are also interacting directly with nature itself.

Indirect impacts are said to be 'embodied' or 'embedded' in goods, services and processes. The embodied water of a kilo of beef is the water needed to produce that beef, and the same applies to its embodied energy. When we buy a product we are, in effect, also buying (and, ideally, accepting responsibility for) all the resources, environmental and social impacts that were 'embodied' in that product, except that we rarely have any idea what these are.

Global energy and emissions

Energy is the capacity for 'doing work', for 'making something happen': it is an idea that explains change, and all activity is an expression of energy flow.

To avoid the more extreme effects of climate change, the world must undergo a transition to a secure supply of affordable environmentally benign energy. However, increasingly scarce and emission-producing fossil fuels are likely to be the dominant source of energy until at least 2030, and the increasing demand for gas and oil suggests an energy supply that is vulnerable to price and supply shocks.[1]

Today 2.5 billion people still use biomass (wood, dung, charcoal, agricultural waste) for cooking and heating.[2] Globally, the challenge is to reduce greenhouse gas emissions while providing for the needs of these people as well as the additional 2–3 billion people anticipated from population increase between now (2008) and about 2050.

Global emissions in 2004 totalled 26.6 billion tonnes, increasing at the rate of 1.7% a year. Over the period 1990–2005, use of all energy sources increased in all sectors with natural gas and coal increasing their shares of the total at the expense of oil, nuclear and hydro. Oil remains the leading energy source for most of the world and especially North America but it is losing share in Europe, South and Central America and the Middle East (Figure 4.1).

In 2005, coal was the fastest-growing fuel with global consumption rising by 5%, twice the 10-year average. China accounted for 80% of this global growth. Asia accounts for nearly three-quarters of global growth and China alone accounts for more than a half.

The global Energy Footprint is the fastest growing component of the overall Ecological Footprint.

Climate change will increase the stresses caused by pollution, growing populations and economies, the most vulnerable being arid and semi-arid areas, some low-lying coasts, deltas and small islands.

Stabilising the world's climate will require high income countries to reduce their emissions by 60–90% over 2006 levels by 2050. This should stabilise atmospheric carbon dioxide levels at 450–650 parts per million (ppm) from current levels of about 380 ppm. Above this level, temperatures would probably rise by more than 2°C to

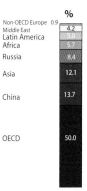

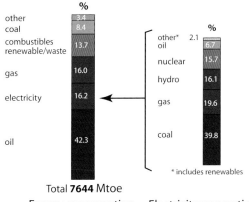

Figure 4.1 Global energy consumption by economic grouping/region and fuel type. Mtoe = megatonnes of oil equivalent.[3]

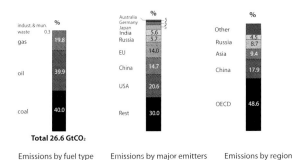

Figure 4.2 Global emissions by fuel type, major emitters and economic grouping 2004.[3] $GtCO_2$ = million tonnes of carbon dioxide.

produce 'catastrophic' climate change.[4] The current acceleration in emissions production must be arrested by 2015 as delays will have a disproportionately large effect later on. The industrialising world (which will, in future, account for the majority of increasing emissions) must be encouraged to leapfrog current fossil-fuel-based energy sources and adopt low- or no-emission technology on the path to sustainability. The way forward will include:

- Improved energy efficiency.
- Increased support for and use of renewable energies.
- Use of advanced technologies that supply energy that is clean and safe.

Renewable and non-renewable energy

We generally think of energy in terms of sources that are suitable for human use – the non-renewable fuels like coal, gas, oil and nuclear power, and renewable fuels like solar, wind, bioenergy and hydro. With the exception of nuclear energy, all these forms of energy can be traced back to the Sun, which is the power generator for life on Earth. It is the energy from the Sun that drives the Earth's climate by influencing winds, the patterns of rainfall and evaporation, and the heating of the oceans to produce climate-affecting water currents.

Primary production

Radiation from the Sun is also captured by plants during photosynthesis and stored in carbon compounds that are formed from the combination of water taken from the soil, and carbon dioxide extracted from the atmosphere. This fundamental life process captures, in plant cells, energy that was formed by nuclear fusion in the Sun. Leaving the surface of the Sun, this radiant energy travels through space at the speed of light for about 8.5 minutes before being caught by the plants on Earth. It is this energy that passes through the food chain from plants to animals and that powers our own bodies when we eat. Energy does not cycle like water and other elements and compounds, but passes through the biosphere to eventually leave in the form of heat.

Oxygen is the life-supporting by-product of photosynthesis – and this is the only significant source on Earth of life-critical oxygen. Geochemical evidence suggests the first rise in Earth's oxygen levels occurred at least 2.3 billion years ago as a result of the activity of photosynthetic bacteria.

Fossil fuels and climate change

More remarkable still is the ancient energy of the Sun that has remained locked up in plants that became fossilised many millions of years ago. This highly concentrated energy, stored in coal, oil and natural gas, now drives human industry. It is released, together with carbon dioxide, when fossil fuels are used, returning the long-stored carbon to a very different atmosphere and world from the one in which it was collected.

Fossil fuels are intimately bound up in the history of the biosphere as well as our own history and way of life. The transition to a post-industrial society and a modern standard of living in the West can be attributed to the use of vast quantities of cheap fossil fuels. We now depend on these in almost every aspect of our daily lives, from our alarm clocks in the morning, to our travel to work, the lighting, heating and cooling of buildings, and the production of the food we eat. But it is this carbon dioxide that also contributes to the enhanced greenhouse effect that is driving climate change. Finding ways to beat our fossil carbon addiction is one of the greatest challenges for sustainability science (see Figure 4.3).

The global carbon cycle

Any management of atmospheric carbon dioxide must start with the knowledge of how carbon cycles between land, plants, the atmosphere and the oceans.

The movement of carbon between the major sinks is known as the carbon cycle and it is closely linked to the flow of energy through living organisms (see Figure 4.4).

Vegetation as a carbon emission sink

Carbon cycles between the atmosphere and vegetation when plants take up CO_2 as they germinate and grow, returning it to the atmosphere when they die and decompose.

Carbon and energy are the major ingredients of a mix that includes plants, carbon dioxide, land management, fossil fuels and climate change. From our knowledge of the carbon cycle, we now know that about 32% (two-thirds of this from the tropics) of the net global annual increase in atmospheric CO_2 currently comes from agriculture and changes in land use. The rest from human use of fossil fuels.

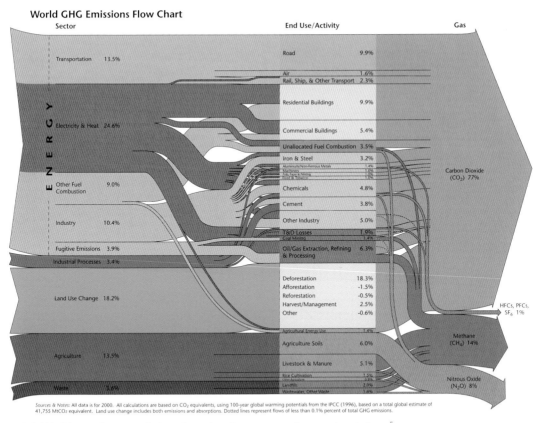

Figure 4.3 World greenhouse emissions flow chart by sector, end use and emissions.[5]

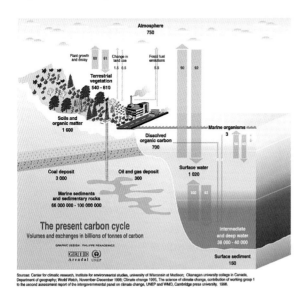

Figure 4.4 The carbon cycle.[6]

Can we store in plants sufficient quantities of CO_2 to influence the levels of CO_2 in the atmosphere and therefore mitigate climate change? What contribution can horticulture make to carbon sequestration?

Global vegetation context

The United Nations Food and Agriculture Organization (FAO) has estimated that about 90% of the carbon stored in land vegetation is locked up in trees and that, over the period 2005–2050, effective use of tree planting could absorb about 10–20% of human emissions.[7]

Forests

The total global forest area in 2005 was just under 4 billion ha, which is about 30% of the land surface area (Figures 4.5 and 4.6).

The biological significance of forests can never be overstated. They are the lungs of the Earth, carry much of its precious cargo of organisms, modify the climate, are moderators of albedo (light reflectance), protect the soil, and much more. Local changes in forest distribution can have global consequences so it is crucial for the planet that the forests of the world are carefully managed.

Because effective use of tree planting could absorb about 10–20% of man-made emissions, we clearly need to monitor the condition of the world's forests very closely as part of any coordinated emissions mitigation strategy. Growing new trees and protecting existing forests can help reduce atmospheric CO_2 but the area of land needed to grow sufficient trees to make a major impact on climate change is extremely large so carbon sequestration in trees can be only one of many approaches to climate change.

Criteria and indicators for sustainable forest management were established by the FAO in 2003. In Australia sustainable forest management has been a priority since 1992.

Trees

When I plant a tree in my garden, for the time it is living it is holding carbon that was previously in the atmosphere. If I plant a lawn where before there was a concrete path, even though the individual grass plants might be

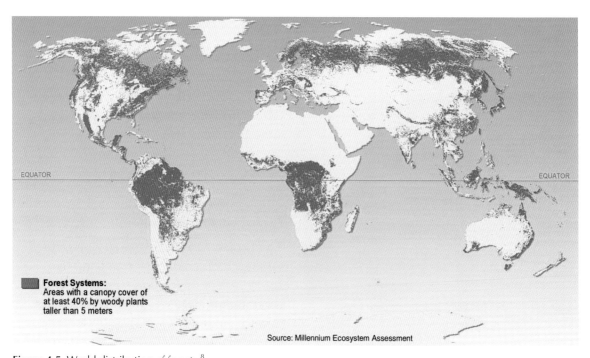

Figure 4.5 World distribution of forests.[8]

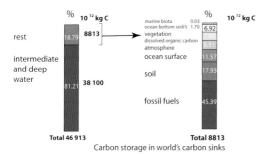

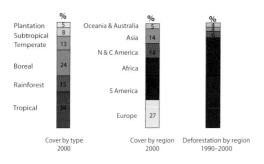

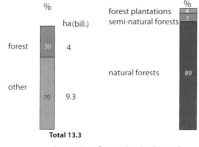

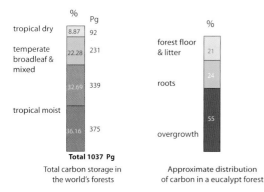

Figure 4.6 The World's forest cover by region and type, deforestation and carbon storage. PJ = 10^{12} joules.[9]

short-lived, I have created a new plant community and have increased the net carbon uptake of my garden. Any new 'permanent' (lasting into the foreseeable future) planting acts as a long-term carbon sink. On the other hand, clearing or burning of vegetation will return CO_2 to the atmosphere and CO_2 is also released from the soil of cleared land.

The carbon held in vegetation at any given time is directly related to its total biomass. It is stored in all parts of a plant and comprises, on average, about 45% of its dry weight. The heavier a plant, the more carbon it has taken from the atmosphere, hence the significance of trees for CO_2 sequestration.

How much carbon is stored in urban tree planting?

Carbon in trees

Forests are only a carbon sink when they are actively growing and 'putting on weight' which, for eucalypts, is roughly the first 50 years. A mature forest has a steady state of carbon exchange because CO_2 uptake is matched by the CO_2 released from death and decay.

The amount of carbon stored in trees will depend on their age, species, growing conditions and management. About 24% of forest biomass is in roots, about 21% in forest floor litter and 55% in overgrowth.

The fresh weight of an 'average' living tree (above and below ground) is about 100 t distributed approximately as follows: fine roots 5%, transport roots 20%, trunk 60%, branches and twigs 15%, leaves 5%.[1] The dry weight of this tree is about 20% of the fresh weight, or about 20 tonnes (of which about

50%, or 10 t, is carbon). It has been estimated that there are about 100 000 trees in public and private open space in inner Melbourne so, using the above figures about 10 t of carbon is sequestered per tree and therefore inner Melbourne trees sequester about 1 million t of carbon.[2]

The most accurate way to calculate the amount of carbon stored in a tree is to fell it, weigh it, and calculate the carbon content (about 45% of its dry weight). More practical methods of approximation include using aerial photography and calculations based on tree circumference at breast height (cbh). The Tree Carbon Calculator of the Australian Cooperative Research Centre for Greenhouse Accounting is available on the web.[10] This is used to estimate above-ground biomass of the tree; the biomass of the roots is then estimated using a root:shoot ratio. These two values are then summed to give the total tree biomass which is then converted to carbon assuming that 50% of the tree biomass is carbon. Sample figures for softwoods and hardwoods are given in Table 4.1.

Actively growing forests sequester 3.5–35 tonnes of CO_2/ha for the first 30 yrs of growth.[4]

Energy in trees

It is likely that in future much more use will be made of biomass energy for power generation. The energy content of fossil fuels is constant for a given weight of fuel. Wood is harvestable biomass energy that is not recovered but lost slowly as heat of decomposition or, possibly, quickly in a fire. However, because timber varies in both density and moisture content, it is not possible to be precise about its energy content. The calorific value of completely dry wood is 19–21 MJ/kg. A freshly cut tree often has a moisture content above 60% and this green wood contains only about 4.65 MJ/kg. After being cut to length and stacked for a year or two, the average moisture content drops to about 20% and this gives an increase in energy content per unit mass of about 14.07 MJ/kg. A 100 tonne tree (fresh weight) would therefore contain 4.65 × 1000 MJ or 4650 MJ/t or 465 000 MJ of energy for the whole 100 tonne tree. The denser the wood the greater amount of energy it contains per unit volume, and density varies from about 750 kg/m^3 in a hard wood like spotted gum (*Eucalyptus maculata*) to about 430 kg/m^3 for a softwood like radiata pine (*Pinus radiata*).[11] When establishing 'carbon plantations' it is important to be aware of the total life cycle of wood production including the energy involved in establishing the plantation, management, felling, transport, processing

Table 4.1 Sample softwood and hardwood tree carbon contents

Tree (circ. cbh cm)	Carbon content (kg)	**CO_2-e
mature pine (200)	1126	4128
medium pine (120)	332	1217
young pine (50)	41	150
mature eucalypt (200)	1940	7113
medium eucalypt (120)	517	1896
young eucalypt (50)	54	197

** To estimate how much a given mass of greenhouse gas contributes to global warming, the gas is compared to a baseline of one unit by weight of carbon dioxide (CO_2) expressed as a 'carbon dioxide equivalent' (CO_2-e). For example, methane (CH_4) has a global warming potential 21 times that of CO_2. Figures are often quoted for carbon (C) alone. To convert C to CO_2 or CO_2-e, multiply by 44/12.

Table 4.2 Land management practices that sequester carbon or reduce carbon emissions

Land type	Expansion of stock	Conservation of stock	Off-site sequestration or emission reduction
Forest	Reforestation: • Modified management, e.g. fertilisation, improved stocking, species mix, extended rotations	• Modified harvesting practices • Preventing deforestation • Change to sustainable forest management • Fire suppression and management	• Wood fuel substitution • Expanded wood products • Extended wood product life • Substitute wood products for concrete/steel • Recycling wood and paper products
Crop	• Afforestation • Agroforestry • Improved cropping systems • Improved nutrient and water management • Conservation tillage • Crop residue management • Restoration of eroded soils • Conversion to grass or other permanent vegetation	• Soil erosion and fertility management • Water management • Maintenance of perennial crops • Residue management	• Substitute biofuels for fossil fuels • Fertiliser substitution or reduction • Other bioproducts substitution
Grazing	• Afforestation • Change in species mix, including woody species • Restoration • Fertilisation • Irrigation	• Improved grazing systems	• Livestock dietary changes • Herd management

Source: Myeong, Nowak & Duggin (2006).[12]

and conversion to final product to make sure that there is worthwhile net energy gain.

There are currently few studies of carbon sequestration in urban landscapes. In the USA it is estimated that urban trees (average of 28% tree cover in urban areas) currently store about 700 million tonnes of carbon with a gross carbon sequestration rate of 22.8 million tC/yr. Carbon storage within cities ranges from 1.2 million tC in New York to 19 300 tC in Jersey City. The national average urban forest carbon storage density is 25.1 tC/ha, compared with 53.5 tC/ha in forest stands.[13] It has been calculated that planting 10 million urban trees annually over the period 1993 to 2003 would sequester (and offset) 363 million tonnes of carbon over a period of 50 years – less than 1% of the estimated carbon emissions in the USA over the same period.[14] Use of satellite image time series can be used to save time and money in urban forest carbon storage mapping.

Carbon and land management

Many factors are at play in the carbon budget of vegetated areas including fire, the carbon in soil, the effect of soil cultivation, and general land management regimes.

Table 4.2 summarises land management regimes that influence CO_2 exchange.

Climate variation in Australia plays a major role in determining the annual variability of net carbon exchange between the land surface and atmosphere and it is driven largely by the El Niño/Southern Oscillation (see Info Box 4.2). Scientists studying carbon sequestration have observed the magnitude of

> **INFO BOX 4.1: USEFUL SUMMARY CARBON STATISTICS**
>
> - The quantity of carbon stored in trees is about the same as the amount of carbon in the atmosphere.
> - The organic carbon in the soil is about twice the amount of carbon in the atmosphere (and also twice the amount stored in plants). About two-thirds of the world's organic non-fossilised carbon is sequestered in forests.[15]
> - 20–30% of human CO_2 emissions are the result of changes in land management (mostly affecting forests, especially those in the tropics).

> **INFO BOX 4.2: EL NIÑO-SOUTHERN OSCILLATION (ENSO), LA NIÑA, AND THE NORTH ATLANTIC OSCILLATION (NAO)**
>
> El Niño (Spanish 'the boy-child') is an expression used by Peruvian fishermen for the appearance, around Christmas, of a warm ocean current off the South American coast. ENSO is an important year-to-year influence on climate variation and is an interaction between sea surface temperatures and atmospheric pressure that occurs across the tropical Pacific region – influencing frequency and intensity of floods, droughts and the location of tropical cyclones. The extensive warming of the central and eastern Pacific leads to a major shift in weather patterns across the Pacific. In Australia it is associated with an increased probability of drier conditions.
>
> La Niña (Spanish 'the girl-child'), the opposite of the better known El Niño, refers to the extensive cooling of the central and eastern Pacific Ocean. In Australia it is associated with an increased probability of wetter conditions.
>
> NAO is a similar phenomenon of the Northern Hemisphere with fluctuating pressure between the low-pressure Icelandic region and the high pressure Azores region.

variation in Australian continental carbon pools over 20 years and found that, depending on climate variability, the landscape of Australia can in some years release up to 75 million tonnes of carbon into the atmosphere in the form of CO_2, and in others can absorb up to 100 million tonnes.

Australia's energy and emissions

Australia is well endowed with most energy sources including solar (Figure 4.7) and wind (Figure 4.8), coal, natural gas, biomass, wave, geothermal energy and uranium and moderate supplies of oil. Fossil fuels supply 95% of Australia's energy needs with coal currently used for 84% of the electricity generation. Renewables hydro, biogas, wind and solar at present make up about 5% of total primary energy consumption (Figure 4.9).

Measurements at Cape Grim, Tasmania show CO_2 levels growing at above-average levels for four consecutive years and twice the rate of the 1980s. Since 1979, all but four years have been hotter than the historical average, with the hottest year being 2005, and average temperatures increasing by about 0.9°C since 1910. This has been combined with multi-year droughts in the south-east. As a rough guide to the potential 'feel' of climate change an average annual temperature increase of 1°C is

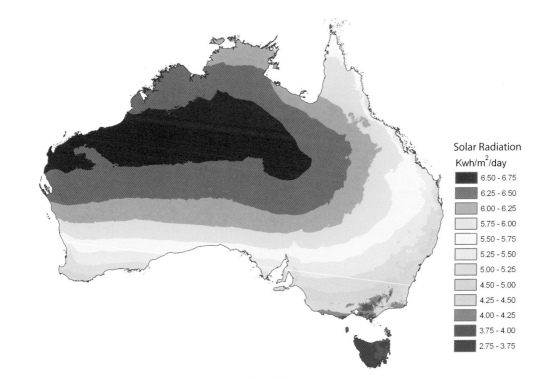

Figure 4.7 Mean annual solar radiation levels kWh/m^2/day.[17]

equivalent to a move in location of about 100 km northwards.

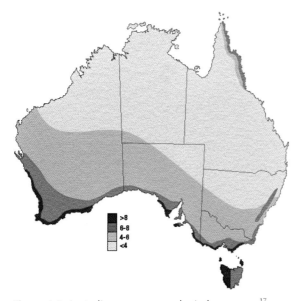

Figure 4.8 Australia – mean annual wind resources.[17]

Australia contributes about 1.5% of the total global greenhouse emissions and is one of the few OECD countries that is a net energy exporter. Since 1986 Australia has been the world's largest exporter of coal. *Australia's National Greenhouse Accounts* provide useful data on emissions and the latest estimates of Australia's greenhouse gas emissions are published by the Australian Greenhouse Office as the *National Greenhouse Gas Inventory*.

Figure 4.10 shows Australian greenhouse emissions by sector.

The largest energy consumers by sector are electricity-generation at 30.8%, transport at 24.3% and manufacturing (especially aluminium, iron and steel) at 22.6%. The greatest relative growth rates over time follow the same order. Road transport accounts for

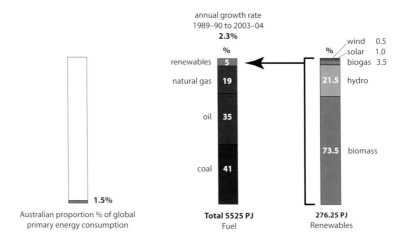
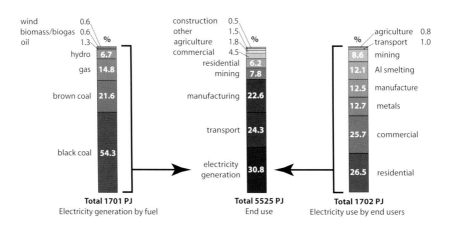
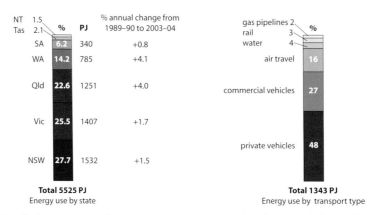

Figure 4.9 Total Australian fuel consumption by type, economic sector and end user, and change between 1990–2004. PJ = 10^{12} joules.[17]

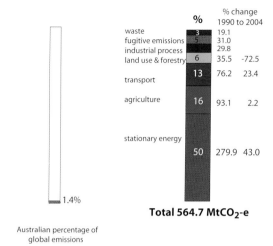

Figure 4.10 Australian greenhouse gas emission by sector. Mt = megatonnes = million tonnes.[18]

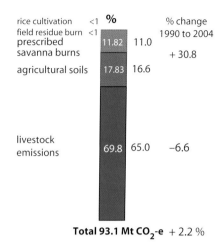

Figure 4.11 Australian greenhouse gas emissions – agriculture.[18]

about 75% of transport energy, two-thirds of which is from passenger cars. Private passenger vehicle travel represents three-quarters of total road travel, and although total petrol consumption continues to increase, there are signs of household consumption stabilising as a result of higher petrol prices and more fuel-efficient cars.

Primary energy consumption is projected to grow by an average 1.5% a year over the period 2004–05 to 2029–30, with stronger growth in the medium term to 2010–11. Since 1990 there has been an overall increase in emissions of 2.3% (strongly influenced by land change emissions, see Figures 4.11 and 4.12). Over this period stationary energy (non-transport energy) has increased by 43% (a result of increasing population, increasing household incomes, increasing resource export) and transport by 23.4% (increasing household income and number of vehicles).

Until the 1990s energy consumption was closely linked to growth in GDP but this energy intensity has subsequently diminished due to, on the one hand, technological improvements and changes in fuel composition and, on the other hand, to the more rapid growth of the less energy intensive service sector.

At present Australia manufactures about 5% of the world's photovoltaic panels,[19] and is

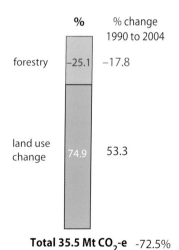

Figure 4.12 Australian land use and land use change. Mt CO_2-e = megatonnes CO_2-e.[18]

4 – ENERGY AND EMISSIONS   63

well situated to explore solar energy with research groups at the Australian National University and the University of New South Wales. As solar technology improves, it will become cheaper while the introduction of carbon credits will make fossil fuel energy more expensive to the point where solar electricity will become economically competitive. North Adelaide is Australia's first of at least four 'solar cities' and will have 17 000 solar panels and 7000 smart meters installed in homes and businesses.

The building sector is responsible for 23% of Australia's total greenhouse gas emissions, mainly driven by its use of electricity, and energy use in buildings is rapidly growing.

By 2050 the building sector as a whole could reduce its share of emissions by 30–35% while accommodating growth in the overall number of buildings. This can be achieved by using today's technology to significantly reduce the energy needed by residential and commercial buildings to perform the same services. In the UK, according to a new *Code for Sustainable Homes*, by 2016, all newly built UK homes must have no net carbon emissions ('Level 6') and the implications for construction techniques are profound. Today, most UK homes are built to about Level 1, or possibly 2. To get to Level 6 will require huge changes in how houses are built, heated and ventilated. And expensive renewable energy technologies will be part of the new era of housing construction. Australia is likely to follow a similar path.

Household

Domestic energy use is generally thought of, and measured, as *direct* energy (sometimes called delivered energy) that is used in the home for heating, cooling and electrical appliances (i.e. usage that appears on the home energy bills). To this can be added the fuel used for domestic travel and transport. However, it is *indirect* (embodied) energy that makes up by far the greater part of our total energy use. Indirect energy is embodied in all our activities and the services we use as well as in the products we buy.

The breakdown of direct emissions for the average Australian household are shown in Figure 4.13.

Sometimes travel and/or waste emissions are excluded from calculations of direct energy use. We have divided household *direct* energy efficiencies into the usual consumption categories.

Total energy use and emission production by households is determined by the income levels of the occupants followed by the physical size of the building, its location and design, the availability of natural resources and infrastructure, the efficiency of the

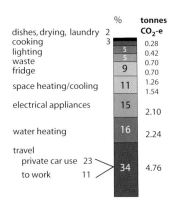

Total 14 tonnes CO_2-e
Household direct energy use emissions

Figure 4.13 Average Australian household direct emissions breakdown.[20]

electricity generation supply that you source, and indirect energy intensities. The household budget categories that use the most energy are: food, heating, electrical appliances, transport and recreation. Low-income households spend about 50% of their budget on heating and electrical appliances and about 20% on fuel, transport and recreation. The same proportions for high income families are, respectively: 25–35% and 30–40%.[21] If we assume the average Australian household has 2.6 people, and total emissions per person is 28 tonnes then the total annual household emissions would be 73 tonnes with *direct* emission production accounting for about 20% of the total emissions. This equates to an average household *direct* energy use of about 14 tonnes CO_2-e p.a.

The indirect energy use which makes up the other 80% of energy use by households will be considered at the individual level. Statistics on household resource use and expenditure are available from the Australian Bureau of Statistics. Household direct energy use is found on energy bills.

Alternatively you can use/borrow an energy meter to check each appliance over a short period and extrapolate to longer periods: this would give you a good idea of the relative efficiencies of your appliances and indicate where improvements can be made. Alternatively you could arrange to have an energy audit.

Individual

Over the period 1950–1973, global energy consumption per person steadily increased, mostly in regions of greatest economic activity: Asia, especially India and China.

The average Australian generates about 28 tonnes of (*direct* and *indirect*) CO_2 emissions a year. Of this about 5.4 tonnes (20%) is direct energy use in the home and for personal travel, and about 22.3 tonnes (80%) the *indirect* energy embodied in the goods and services used. To this total can be added the energy used for boat and plane travel. (See Figure 4.14.)

It has been estimated that to keep emissions at safe levels would require a global average individual emission output of no more than 3 to 4 tonnes/person/yr.[22]

This is about one-eighth of Australia's current per capita levels which are among the highest in the world.

Air travel is rapidly increasing in line with global affluence, population and trade and can be a major contributor to individual emissions as it generates about 2.7 times the CO_2 emissions of the fuel itself due to the additional effects of contrails (the white condensation trails), ice clouds and nitrogen oxides. A return trip to London from Melbourne or Sydney produces about 10 tonnes of emissions which is more than a third of the average Australian individual's yearly emissions and about three times the 'safe level' of annual emissions just mentioned. A return trip between Melbourne and Sydney produces about 0.5 tonnes.

Indirect energy use can be calculated either as the individual's proportion of the total national energy use (subtracting direct energy use) *or* calculated in relation to personal expenditure and the energy

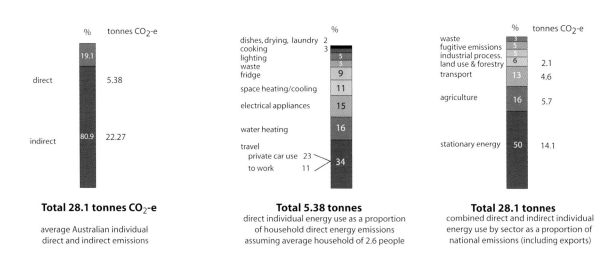

Figure 4.14 Australian individual greenhouse gas emissions. CO_2-e = carbon dioxide equivalent.[23]

intensities of the goods and services we buy which is much more informative.

If, as an individual, you are not adding any carbon to the atmosphere as a result of your direct energy use then your activities are said to be carbon-neutral (climate-neutral). Living a carbon-neutral *direct* energy lifestyle is a great start: it may even be possible to create a negative carbon budget. However, direct energy use is only 20% of our total energy use so much remains to be done.

Following the principles of I-O Analysis your overall (direct and indirect) energy use can be related to money spent (perhaps as a share of the household budget) but you would need to know the energy intensities of the major goods and services you use. Bear in mind that the efficiency of the electricity generation supply that you use is important and that the household budget categories that use the most energy are: heating, electrical appliances, food, transport and recreation.

To live an energy sustainable lifestyle by reducing carbon consumption from 28 tonnes to 3–4 tonnes a year sounds a tall order. However, by using 100% green power, solar water heating, photovoltaic panels if possible, and eating small quantities of meat, substantial inroads can be made. Greater care with general consumption patterns and especially energy intensive air travel will also make a huge difference. And remember, the emissions problem goes away as soon as you source renewable energies.

Energy in the garden

The energy flows in a modern garden are quite complicated (see Table 4.3 and Figure 4.15).

There are both direct and indirect energy *inflows*. Direct energy (energy used on-site) enters the garden as the renewable energy from the Sun which can be captured by the plants. There is also the non-renewable energy of the fossil fuels used to power machinery

Table 4.3 Energy flows in a modern garden

Energy inflows	
Direct	Renewable: Sun (natural solar) energy used in heating, evaporation, transpiration, photosynthesis
	Non-renewable: Fuels, electricity and oil (derived off-site but used on-site) Human labour (if not fuelled by on-site food)
Indirect (embodied)	Non-renewable: In mains water use (water treatment, infrastructure) Hard landscape (fences, paths, sheds, etc.) Garden products (pots, stakes, etc.) Machinery and tools Chemicals Inorganic fertilisers Organic composts, manures, mulches, food scraps (off-site origin) Administration (planning, transporting materials, etc.) Embodied energy in purchased plants
Energy outflows	
	Lighting Garden produce (stored as food energy) Hard waste Green waste (taken off-site) Heat of decomposition, human labour, heat of power use and reflected energy of the Sun Chemical run-off and leaching

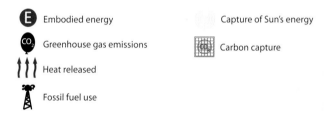

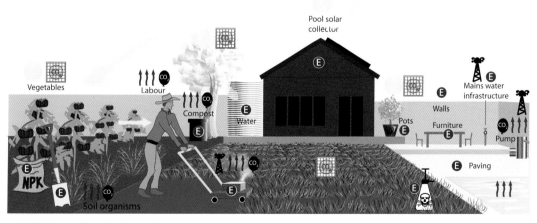

Figure 4.15 Energy flows in the garden.

and tools like mowers and chainsaws. To this we can add the energy of our own labour, derived from the food we eat. Indirect energy (originating from the wider economy) enters from many sources including the energy embodied in tools, chemicals, materials and garden products as well as in the mains water.

Energy *outflows* include the energy in the plants that is eventually dispersed as heat when the plants die and decay. Energy also leaves the garden in organic and inorganic materials such as garden produce (fruit, nuts, vegetables, herbs, cut flowers) and waste (green waste and hard rubbish), chemical run-off and leaching. And then there is the energy that leaves the garden mostly as heat, when we use garden power tools, pool pumps, etc.

Most garden energy use derives from the indirect embodied energy of fossil fuels needed for the manufacture of hard landscape, machinery and tools. This is followed by the energy from fossil fuels required to power them, followed in turn by the energy of our own labour.

All these energy sources combined are very small compared with the energy that is being used in the house, or even the solar energy that falls on the garden over the course of a week.

The average Australian total for direct household energy use per week is 3066 MJ. Our calculations show that the weekly energy use of the average Australian garden is less than 1% of this total (see Appendix A1). Because the amount is so small, minor differences in landscape features can have relatively large energy consequences. For example, a swimming pool pump can use as much energy as a refrigerator and can make up a large proportion of the outside energy use. The amount of energy stored in food plants is also small in relation to total garden energy turnover, but we can take advantage of it nevertheless.

Energy flows can be broken down as follows:

Energy inflows (direct)

Renewable energy

The leaves of plants are like tiny photovoltaic solar panels soaking up the Sun's renewable energy. However, they do not convert sunlight into electricity but instead store the energy as organic compounds. In this way they are more like batteries. For Australia's coastal cities, with the exception of Hobart, average solar radiation is about 4.5 kWh/m^2/day (16.2 MJ/m^2/day). (See Figure 4.9.)

Any energy use involving fossil fuels will produce greenhouse emissions and therefore contribute to climate change, so the Sun's renewable energy falling on a garden is a sustainability opportunity that cannot be missed. It can be used in several important ways:

- Carbon sequestration – removing CO_2 from the atmosphere into permanent vegetation, especially trees, where it can be harvested and stored in wood products.
- Harnessing plants to 'manufacture' our food on-site.
- Organic fuel – green waste for compost, nutrients, and possibly as a power source (in preference to fossil fuels).

We have seen that a 100 tonne tree (fresh weight, including roots and leaves) contains roughly 465 000 MJ of energy.

Experimental studies have shown that the efficiency of conversion of solar energy to chemical energy by plants varies from about 0.1% in poor growing conditions on cloudy days, to about 3% for intensive cropping in good light, although in laboratory conditions, efficiency levels of 25% have been attained.

This compares with the efficiency of photovoltaic panels which is, at present, about 14%. When plants concentrate their harvested energy into certain structures (tubers, leaves, fruits) we can take advantage of this energy as food. Biointensive calorie farming makes the most of a small amount of space by maximising energy productivity. This is done by growing energy-rich produce like potatoes, pumpkins, sweet potato, olives and kiwifruit (see Table A2 in the Appendix) although the advantages of high energy content must be balanced against the nutritional and other benefits of some low calorie foods.

Non-renewable direct energy

Apart from the inflow of the renewable energy of the Sun there is the inflow of energy from human labour and the fuels used in power tools and machinery as well as the embodied energy we introduce in the various materials we use.

Human labour

Human labour does have an energy cost because we are powered by the energy of the food we eat. Like all other animals, we obtain the energy needed to power our bodies either directly from plants (as cereals, fruit, nuts and vegetables) or indirectly from plants through the consumption of animal meat (which ultimately derives its energy from plant matter either directly or through an animal food chain). In a 'whole-of-life' sense we could also include the embodied energy we contain that relates to the garden; such as the food, fuel and other energy we used when we went to the nursery to buy plants and potting mix.

A short diversion into the food-energy requirements of our own bodies when we are resting and going about our daily lives will help us get a feel for the energy flows that occur in other organisms, within ecosystems, and throughout the human food chain. Like a running engine our bodies are constantly using fuel. At rest they tick over at basal metabolic rate.

The approximate human daily energy requirement is about 12 500 kJ, equivalent to an energy generating capacity (at basal metabolic rate) of 80 watts, about the same as an incandescent light (about 20 watts of this energy is being used by the brain). A 100-watt light bulb therefore works 1.25 times harder than our body (100/80 watts).

We could create a human energy unit called a human-equivalent (H-e), and say that the 100-watt light bulb is running at 1.25 H-e.

Every food contains calories and almost every liquid we consume contains calories with the exception of water, coffee (without milk), tea (without milk) and diet soft drinks.

> **INFO BOX 4.3**
>
> A watt (W) (kilowatt, kW = 1000 W, megawatt, MW=10^6 W) is a rate of energy use – 1 joule per second. It is the energy needed to lift a weight of 100 grams a distance of 1 metre in 1 second; or the power from a current of 1 ampere flowing through 1 volt.
>
> The energy used by a 60-watt light bulb running for 1 day would lift a tonne (10^3 kg) a distance of ≈ 530 m; or be equivalent to the energy in a 1000-kg mass moving at ≈370 km/h; or raise the temperature of 1000 kg of water by ≈1.2°C; or is equivalent to the pure chemical energy in ≈0.15 litres of petrol.

Power machinery and tools

In general, with increase in technology comes increased task efficiency, but also increased energy use. This is because the manufacture and use of machinery, the production of synthetic chemicals and fertilisers, and the extraction and processing of materials are all energy-intensive processes and involve the use of fossil fuels. Digging a garden with a rotary hoe uses more energy than digging it with a spade.

Of course, a machine is vastly quicker than a single labourer, more convenient and available at a lower cost because fuel is cheap and labour costs high.

The example in Table 4.4 shows that efficiencies can be made by careful management of both the kinds of tools and machinery we use as well as the methods we employ, human labour being relatively energy efficient.

A list of fossil-fuel-using power tools and equipment would include the following:

- Buggy
- Tractor
- Cherry picker
- Stump grinder
- Mower
- Chainsaw
- Bug catchers
- Garden lighting
- Lighting
- Brushcutter
- Leaf blower
- Hedging shears
- Rotary hoe
- Heaters
- BBQ
- Chippers
- Loaders, etc.

Many aspects of garden construction require power tools and transportation of materials, so by finding ways of reducing their inputs, energy efficiencies can be made.

Indirect (embodied) energy

The last kind of energy we bring into the garden is the energy that is embodied in the full range of garden materials. Apart from the energy used in fuels, this embodied energy makes up most garden energy use. Of particular note here are materials like wood, metals, glass and cement, but also chemicals including synthetic fertilisers, and other garden items including pots, spades and mowers, fences, paving and swimming pools. Embodied energy in products is discussed in more detail,

Table 4.4 Energy used by machines and humans. A comparison of the time and energy used moving 20 m^3 of soil 200 metres

Inputs	Method of moving soil	
	1 earthmover	8 labourers
Time taken (minutes)	2	700
Fossil fuel (diesel) use (L)	3.25	0
Energy use (kJ)	127 078	30 127

Source: Adapted from Thomson & Sorvig (2002).[24]

including estimates of the embodied energy content of materials, chemicals and fertilisers, in the chapter on garden design and Table A3 in the Appendix.

Embodied energy can come from unlikely sources. For example, the provision of mains water (infrastructure, pumping, etc.) carries an embodied energy load. In Melbourne, the energy cost of water delivery to the gravity-fed home tap (excluding energy embedded in infrastructure) is 0.14 KWh/m^3. Pumped water, like that in Brisbane, is about 0.4 KWh/m^3. Seawater desalination requires about 3.5 KWh/m^3, brackish water desalination 1–1.5 KWh/m^3. The embodied energy of water from a proposed Wonthaggi desalination plant in Victoria will be closer to 5 KWh/m^3 when the pumping energy is included. The energy for pumping water from Perth to Kalgoorlie (~560 km) is about 12 KWh/m^3. Melbourne uses about 0.6 KWh/m^3 for wastewater disposal.[25]

Energy outflows

Having covered the rather complicated energy inflows, outflows can be dealt with much more quickly. Energy is removed from the garden in organic matter as green waste and garden produce. Organic matter breaking down on-site releases heat of decomposition. We can regard embodied energy as leaving the garden when materials and products are removed (exchanged, sent to tip, given away, waste entering the sewage system, etc.), and of course energy is lost as heat when power tools and appliances are used, and as we dig. Much of the Sun's energy is either reflected away directly, or absorbed and given off as heat.

We can compare the modern garden with a pre-industrial one. Imagine living in a house made out of local wood and natural products. Using traditional skills, wood and parts of plants from the garden and nearby countryside are used to make fences, garden structures, simple garden tools, furniture, baskets and other household items. This is a subsistence lifestyle using most of the garden to grow herbs, vegetables, fruit and nuts without using any modern machinery. Seasonal food excess is preserved by bottling, pickling, drying, jam and sauce-making. Food variety is obtained by exchanging plant seeds and excess garden produce with neighbours and friends, and some people specialise in cultivating particular crops.

Children share in the growing and cooking of the food plants so they learn all about crop seasons and cultivation techniques and are (hopefully!) keen to eat the food that they have grown and cooked themselves.

Each year some vegetable seed is saved to grow the next year, and for meat and manure chickens, goats and ducks are bred and fed on garden produce. The backyard poultry is great for recycling organic waste, cultivating the soil, producing manure and eggs. As much organic matter as possible is composted to return nutrients to the garden and strategies for pests and diseases are developed that do not use synthetic pesticides and fertilisers. Apart from rainfall, all the water used in the garden has been collected from the roofs.

We mention all this not to necessarily recommend it as a way of living but to show the difference in energy flows between the pre-industrial and modern garden. If we list the energy inputs and outputs of this pre-industrial garden in the form of an energy budget we might well find that it is close to being energy self-sufficient. The plants are the providers, the primary producers collecting energy from the Sun by what might be called *solar horticulture*. This energy is eventually lost as heat when the plants die and the tissues decay, but before this happens we can harness it as food for ourselves, our neighbours and our livestock. This is similar to pre-industrial solar agriculture, which used animal labour and natural materials. In contrast, the modern garden reflects our energy-intensive lifestyle with mass-produced plants bought from nurseries (not swapped or grown from seed), modern energy-intensive materials, tools, chemicals, and relatively recent add-ons such as lighting and swimming pools.

However, compared with most of our other activities, gardens are still extremely low-tech. As a proportion of overall human energy use, the energy demands of gardens are minimal because the use of concentrated energy sources like fossil fuels is so low. Parkland and recreation areas use more, because of their greater use of power tools, chemicals and intensive mowing, although even here it is expended over a wider area, unlike the localised energy use of a household.

The estimates of energy use in the garden given in Tables A1, A2 and A3 of the Appendix can guide energy management by suggesting energy reduction strategies.

By growing more vegetation we are sequestering more of the planet's greenhouse emissions and connecting with the global carbon cycle, although the contribution made by parks and gardens on a global scale would be small compared to serious reforestation projects. Other positive energy links to the garden hinted at in this chapter include the potential cultivation of high energy foods, biofuels and biomass energy. Then there are the energy savings of passive solar heating and cooling that can be built into the garden design.

Sustainable energy use

Suggestions for improving garden energy efficiency:

- Grow high energy foods.
- Minimise use of high embodied energy materials.
- Minimise use of power tools.
- Use renewable energy sources where possible.
- Consider possible use of excess organics for biofuels and biomass energy.
- Consider permaculture.
- Use garden design for passive solar heating and cooling of the house.

WATER

> **KEY POINTS**
>
> Climate, forests and global biogeochemical cycles are all linked to the global water cycle.
>
> Current levels of human water use and diversion threaten food security and ecosystem services.
>
> Water and environmental impacts are embodied in products and world trade.
>
> We must plan for increased urbanisation and climate change.
>
> Water efficiencies come from managing lilac, blue and green water. In urban space this means water-sensitive design that includes:
> - Improving rainwater harvesting and storage linked into buildings.
> - More recycling of greywater and stormwater.
> - Improved management of water flows including water storage in the soil.

Planet Earth has been called the 'Blue Planet' because, when seen from space, this is the colour reflected by the large volume of water that covers 71% of its surface (see Figure 5.1).

Only a planet of the appropriate mass, chemical composition and distance from the Sun could have the range of conditions necessary for life to exist and flourish. Water is one of the vital chemicals needed for life to exist, and nowhere else in the universe has life been found, yet.

The water cycle

Apart from its importance as a component of living organisms, water also has a major influence on the climate, carbon cycle and other biogeochemical cycles.

The hydrosphere is the Earth's total supply of water and the cycling of water between the atmosphere, oceans, waterbodies and biosphere is called the global water cycle or hydrological cycle.

Salty ocean water makes up 97.5% of all the Earth's water. Of the remaining 2.5% freshwater component, the majority, about 40–80%, is locked up in polar ice (most in Antarctica) and underground water, mostly aquifers. The volume of water in aquifers is unknown but estimated at about half that in the polar caps. Of the water in soil, about half is estimated to be in the root zone. Only about

Figure 5.1 Earth, the 'Blue Planet'. (Source: NASA)

0.27% of all fresh water is found in lakes, rivers and wetlands and this is the main source of water used by humans. The annual volumes of the water flows between the various water sinks is summarised in Figure 5.2 which includes estimates of the residence time for water in each sink. Greatest rainfall occurs in the tropics but high temperatures here mean that evapotranspiration is also high.

Global water use

From the 1950s on, global water withdrawal for human consumption has escalated as a result of scientific and technological developments, notably the increase in irrigated land, growth in industrial and power sectors, and intensive dam construction. Irrigation using dams and extraction from river systems leads to fragmentation of the catchment areas and their associated ecosystems. This has altered the water cycle of rivers and lakes, affected their water quality and potential as a human resource, and altered the global water budget. Wetlands, inland water bodies and fossil water aquifers are drying up and renewable aquifers are being drawn down faster than they replenish. Chemical contamination of groundwater, lakes, rivers and the oceans is threatening the quality of the water supply in many parts of the world.

Six major threats to the biology of global river systems can be listed: dams and infrastructure, excessive water extraction, climate change, invasive species, overfishing and pollution. Freshwater ecosystems are among the most species rich on Earth so species extinctions soon follow water depletion.

About 70% of global water use is diverted from the environment to agriculture where, in much of the world, only about 10% reaches the crops (Figure 5.3). More efficient use of water includes: reducing evaporation; more efficient watering systems; developing agriculture away from sites with little water; and reviving old reservoirs and local water-collecting areas rather than depending on large centralised water supplies.

Water shortages, especially those related to food security, can lead to economic and political tensions. In many regions there is competition between urban and rural water use and potential for conflict when systems are shared by two or more nations especially where 'downstream' countries must use the water left by 'upstream' countries. For example, the Nile is shared by Ethiopia, Egypt and Sudan; the Ganges by Pakistan, India and Bangladesh; Tigris-Euphrates by Turkey, Syria

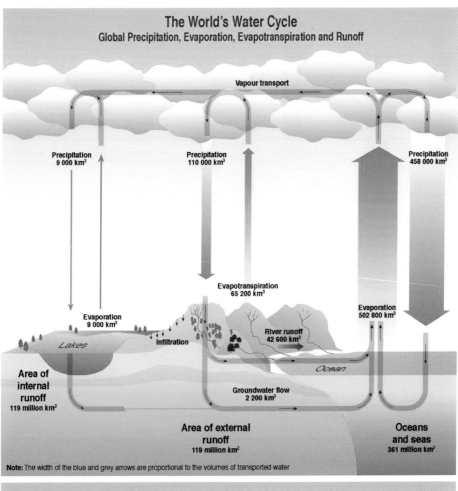
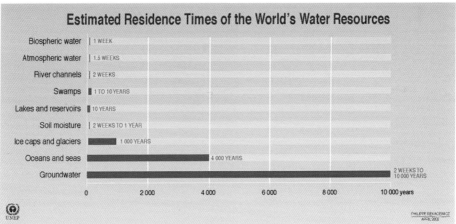

Figure 5.2 Global water cycle.[1]

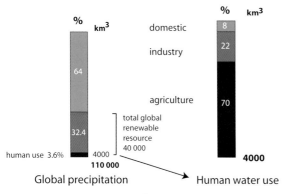

Figure 5.3 Global water use.[2]

and Iraq; Mekong by China, Vietnam, Laos and Cambodia. Figure 5.4 indicates those areas with the greatest unsustainable water consumption. The map shows freshwater withdrawals as a percentage of a country's annual renewable water resources in 2001. If withdrawals exceed a threshold (generally reckoned at about 20–40%), ecosystems will be put under stress. Many countries already exceed this threshold with some withdrawing more than 100% of their annual renewable resources. This is only possible by withdrawing fossil water from underground aquifers, a resource that can only be used once. Large rivers such as the Nile, Huang He (Yellow), Colorado, Amu Darya and Syr Darya are often so depleted by withdrawals for irrigation that in dry periods, they fail to reach the sea.

The effect of human activity on the water cycle can be explained in terms of the 'colour' of the water. Humans have altered the global balance between harvestable surface and groundwater (blue water), non-harvestable soil moisture (green water) and

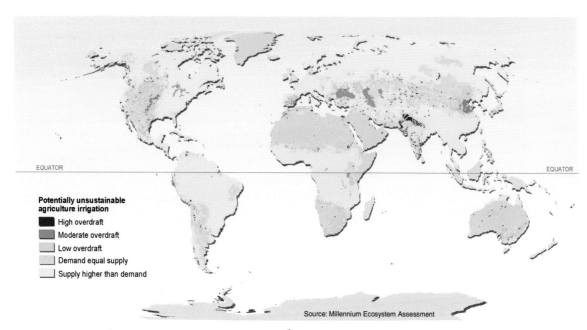

Figure 5.4 Potentially unsustainable agricultural irrigation.[3]

Estimated on a 50 km resolution:
High overdraft = withdrawal of more than 1 km³/yr (water table drawdown > 1.6 m/yr – assuming normal dewatering)
Moderate overdraft = withdrawal of 0.1–1 km³/yr (water table drawdown of < 1.6 m/yr)
Low overdraft = withdrawal of 0–0.1 km³/yr

water produced by evapotranspiration (white water). Most of the world's crops are grown using natural rainfall. Only 15% are produced using blue water for irrigation. Global water conservation is about harvesting more blue water from rainfall, recycling as much of this as possible (lilac water), and using green water more efficiently. This principle is equally applicable to urban landscapes. Until recently water management and monitoring has concentrated on blue irrigation water. But effective management of green water has major implications for global food production. Green water productivity can be improved by increasing rain infiltration and decreasing run-off. We shall consider four issues relating to global water sustainability: deforestation, urbanisation, climate change and world trade.

Deforestation
Changes in one part of the global water cycle can affect situations large distances away.

Deforestation especially has been implicated in quite large-scale climate modification. Forests are climate regulators. Dense tropical rainforest in particular absorbs incoming radiation, keeping the air cool. Trees can absorb and store large volumes of moisture, up to two-thirds of the available water in some regions. This stored water is more than just recycled rainwater; it is the way ecosystems' dissipate the large quantities of heat collected and stored each day. Through the moisture released by evapotranspiration, tropical rainforests redistribute absorbed heat. Most of the world's heat redistribution occurs through ocean currents but the atmosphere also spreads heat around the globe and tropical rainfall drives this process. The intense heat energy needed to convert water to vapour by evaporation is released back into the atmosphere when the water vapour condenses into clouds and rain. About 75% of the energy that drives atmospheric circulation comes from the heat released during tropical rainfall.

A large forest such as that in the Amazon River Basin has global importance so it is hardly surprising that the Amazon's cloud and rainfall characteristics have led to the nickname 'the green ocean.' During the past 15 years more than 494 000 km^2 of forest have been cleared from the Amazon Basin with the current rate being about 20–200 km^2 each year. Deforestation alters the exchanges of not only water but also carbon, and energy with the atmosphere, cycles that, even now, we only partially understand. Indonesia and Brazil together contribute about 10% of global greenhouse gases because of their clearance of tropical forest.

Urbanisation
More people concentrated in urban environments increases the demand for domestic and industrial water. As surface water is depleted the search for new supplies moves to underground aquifers and the transfer of surface water large distances away from its source. Following the model of large, centralised water supplies, vast multibillion dollar water re-location engineering projects are underway or planned to divert water from regions of high rainfall to those of low rainfall: for China (Yangtze diversion of water from wet south to dry north), India (dams and canals to link 14 rivers that drain the

Himalaya, transporting water to the dry south), North America (north-western Canadian Rockies to Denver and Mexico). As environmental flows decline, apart from displaced populations there is: transport of pollution, land degradation in the areas where water has been sourced, increase in soil salinity, decline in freshwater fisheries, and the destabilisation of ecosystems.

A quarter of the world now obtains its water from deep aquifers. 'Closed' aquifers contain 'fossil water' that may have been stored there for thousands or millions of years, and this water can only be used once – it is not being replaced. India's aquifers are being used faster than they can be replenished. The overuse of underground aquifers (e.g. Ogalalla Aquifer under six states in the USA) has lead to subsidence, sink holes and water intrusion. From China to Iran, and Indonesia to Pakistan, rivers are running dry through increasing extraction and these impacts are sometimes being enhanced by the effects of climate change.

Climate change

Climatologists are currently researching how climate change will alter weather patterns, rainfall and water regimes – and much remains to be done. However, a number of general trends have been established (see Info Box 5.1).

In Australia, as a result of climate change, rainfall is expected to increase in the north-west and decrease in the south-east. This will affect both ground and surface water supplies and therefore how much water can be harvested. Factors linked to rainfall include

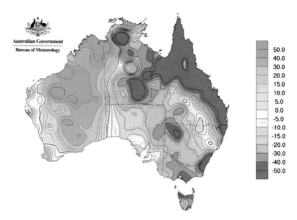

Figure 5.5 Trend in annual rainfall in Australia 1970–2005 (mm/10 yrs).[4]

surface wetness, reflectivity, also vegetation quantity and species composition, which in turn affect evapotranspiration and cloud formation, and therefore rainfall (see Figures 5.5, 5.6).

Water and trade

Globally, agriculture accounts for about two-thirds of all water use but the proportions of water allocated to agriculture and industry vary considerably from country to country.

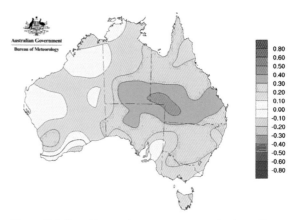

Figure 5.6 Trend in maximum temperature in Australia 1970–2005 (°C/10 yrs).[4]

> **INFO BOX 5.1: CLIMATE CHANGE AND THE WATER CYCLE**
>
> A redistribution of rainfall and water patterns:
>
> - More intense rainfall with more cyclones and floods increasing run-off while at the same time reducing infiltration.
> - When the climate is dry, small changes in temperature and rainfall could cause relatively large changes in run-off.
> - Arid and semi-arid regions will therefore be particularly sensitive to reduced rainfall and to increased evaporation and plant transpiration.
>
> Changes at the surface due to changes in the amount and timing of rainfall include:
>
> - Recharging of groundwater supplies and aquifers.
> - Changes in water quality.
> - Changes in run-off, groundwater flows and evaporation, and therefore changes to dependent natural ecosystems.
> - Altered flood regimes and water levels in lakes and streams will affect nutrients and dissolved organic oxygen, and therefore the quality and clarity of the water.
>
> Changes in water temperatures and in the thermal structure of fresh waters could affect:
>
> - The survival and growth of certain organisms.
> - Species diversity and productivity.
>
> Rising seas due to melting ice caps would result in:
>
> - Salt water invading coastal freshwater supplies.
> - Coastal aquifers may be damaged by saline intrusion as salty groundwater rises.
> - The movement of the salt-front up estuaries would affect freshwater pumping plants upriver.
> - Reduced water supplies would place additional stress on people, agriculture and the environment.
> - Regional water supplies, particularly in developing countries, will come under many stresses in the 21st century.

In Australia the proportion of total agricultural water use for food eaten *within* the country is about 21%. When the water embodied in exported agricultural products is included the total use rises to 67% (i.e. 46% of Australia's agricultural water use is embodied in exported food).

Global trade tends to distribute resources from areas of abundance to areas of shortage, so it makes sense for 'dry' countries to import water-intense products. Globally, about 16% of the world's blue and green water is embodied in exports. However, 50–70% of the world's blue (harvested) water is used to produce food for export. Most significantly for the environment, by importing food, a country is exporting the environmental problems associated with its production. Green water represents the largest share of 'virtual' (embodied) water in the international trade of agricultural commodities (mostly embodied in maize, soybean and wheat exports from the USA, Canada, Australia and Argentina) with exports going from highly productive rain-fed rich countries towards generally blue water based ones. Green water flows have rarely been estimated even though they can play a major role in ensuring global water security. However, the potential to save water by managing green water trade is limited by many factors.

Australian water use

The water footprint of a country can be defined as the volume of water needed for the production of the goods and services consumed by the inhabitants of that country.[5]

Table 5.1 shows Australia's water footprint over the period 1997–2001. (See also Figure 5.7.)

Figures like this are only estimates, but they are nevertheless an extremely useful guide to water use and its future management.

It is often pointed out that Australia is the driest continent with frequent droughts. It has the lowest percentage of rainfall as run-off of any continent, and the lowest volume of water in its rivers (on average 12%).

Australia is the driest continent *on average* only. Tully, a town in tropical Queensland, has an average annual rainfall of a staggering 4.5 m, the highest in Australia, although over the year rainfall ranges from over 700 mm in February and March to about 100 mm in September and October. Some areas have rainfall that, over part of the year, exceeds evapotranspiration, and that includes the urbanised areas of the south-west and east coast where, as we are often told, we are not short of water, we simply need to be smarter in the way we harvest and use it.

Rainfall distribution and availability

Australia's water resources are divided into 12 drainage regions, 246 river basins and 325 management areas for surface water (i.e. streams, rivers, lakes and wetlands). Regions north of the Tropic of Capricorn, together with Tasmania, receive over 50% of Australia's divertible water but contain a small proportion of the population, whereas about 65% of the population lives in coastal Victoria, New South Wales and Queensland in an area that receives about 23% of the divertible water (Table 5.2).

About 33% of usable water is diverted to human use although this proportion varies

Table 5.1 Summary of annual Australian water use for the period 1997–2001

Population	19 071 705
WATER USE	Water volume (Gm^3/yr)
National water footprint	26.56
NATIONAL WATER USE	
Internal water use	
Urban (domestic) water withdrawal	6.51
Crop water use for Australian consumption	14.03
Crop water consumption for export	68.67
Industrial water use for Australian consumption	1.229
Industrial water use for export	0.12
Imported water use	
Agricultural goods for national consumption	0.78
Industrial goods for national consumption	4.02
For re-export of imported products	4.21
Per capita water use	Water volume (kL)
Domestic water use	341
Internal water for agricultural goods	736
Imported water in agricultural goods	41
Internal water for industrial goods	64
Imported water in industrial goods	211
Total per capita water footprint	1393 kL/yr (= 3397 L/day)

Source: Hoekstra & Chapagain (2007).[6]

Table 5.2 Rainfall run-off diverted for use within drainage divisions and as a proportion of national run-off in 2000

Drainage division	Proportion of division diverted for use (%)	Proportion of total national run-off (%)
Murray–Darling	51	6.1
SE Coast	4	10.9
NE Coast	4	18.8
SA Gulf	15	0.2
SW Coast	5	1.7
Timor Sea	0.06	21.4
Gulf of Carpentaria	0.005	24.5

Source: ABS (2006).[7]

across the continent, being low in the north and high in the Murray–Darling Basin (Figure 5.3).

In many regions, the sustainable extraction level is being approached or exceeded, this being indicated by reduced water quality, algal blooms, increasing salinity and threatened biodiversity. Between 1983 and 1996, the area of irrigated land increased by 26% and use of groundwater increased by 88%. In 2000, 84 (26%) of 325 surface water management areas were either overused or close to it, and 168 (30%) of 538 groundwater management units were totally allocated or over-allocated.

Eastern states in general and the Murray–Darling Basin in particular are showing signs of stress and this is expected to worsen as climate change produces a reduction in rainfall over south-eastern Australia. The WWF lists the Murray–Darling in the world's top 10 most threatened river systems. These in order are: the Salween, La Plata, Danube, Rio Grande, Ganges, Murray–Darling, Indus, Nile, Yangtze and Mekong.

Overall, water use for the period 1985–1996/7 increased 65% and over the same period water for irrigation increased 70%.

Total surface water run-off is calculated at about 440 000 GL per annum, of which about 25 000 GL is captured for human use, about 70% used for rural irrigation, about 20% for urban and industrial use, and about 10% for other rural uses.

Table 5.3 Australian water consumption (GL), 2000–01 and 2004–05 (all states)

	2000–1	2004–5	NSW	Vic	Qld	SA	WA	Tas	NT	ACT
Agriculture	14 989	12 191	4133	3281	2916	1020	535	258	47	1
Forestry and fishing	44	51	11	8	3	1	25	4	1	–
Mining	321	413	63	32	83	19	183	16	17	–
Manufacturing	549	589	126	114	158	55	81	49	6	1
Electricity/gas	255	271	75	99	81	3	13	–	1	–
Water supply	2165	2083	631	793	426	71	128	20	8	5
Other industries	1102	1059	310	262	201	52	168	18	30	17
Household	2278	2108	572	405	493	144	362	69	31	31
Total	21 703	18 767	5922	4993	4361	1365	1495	434	141	56

Source: ABS (2006).[7]

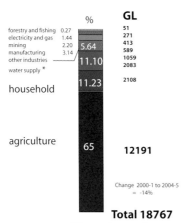

Figure 5.7 Australian water use 2004–05.[7]

From 1977–2001, the proportional use of water by the various economic sectors remained about the same, with the absolute volume steadily increasing. Agriculture accounts for around 67% (this is relatively high when compared with other industrial countries – about 33% in Europe and 49% in North America), households 9%, electricity and gas supply 7%, manufacturing 4%, mining 2%.

A simple sustainability benchmark at the national level would include a basic Environmental Water Reserve as a measure of the water required to maintain healthy ecosystems in regulated waterways.

Ideally we require an equitable distribution of water between the environment, agriculture, industry and domestic use. In practice, water use tends to be divided into urban and rural components, sectors that are sometimes seen to be in conflict.

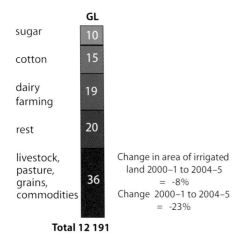

Figure 5.8 Australia's total water consumption by agriculture. GL = gigalitres = 10^9 litres.[7]

Rural

The 'pasture for grazing' sector accounts for nearly a third of Australia's total irrigation water as well as about one-third of the total irrigated area nationally (Figure 5.8). The most heavily watered crop is rice at 12.1 ML/ha, with cotton second, watered at a rate of 6.7 ML/ha.

The Murray–Darling is the nation's food basket with a catchment comprising 14% of the land area, about 1 million km² and about 6% of water run-off while supporting about 42% of all Australian farms.

The National Action Plan for Salinity and Water Quality and the 1994 and 2003 Council of Australian Governments are currently in the process of introducing substantial water reform including: careful pricing, water trading to ensure water goes to best use by an efficient market, improved resource management, allocation of water to the environment, and protection of ecosystem assets (without healthy rivers other uses are not possible),

Table 5.4 Water productivity in relation to crop

Crop	$/ML returns
Rice	179
Sugar	217
Pasture, grain	270
Cotton	420
Dairy	529
Fruit	1590
Vegetables	1817
Grapes	1859

Source: ABS (2006).[7]

encouragement of conservation and recycling in cities, and improved public consultation.

Tables 5.4, 5.5 and 5.6 indicate, for Australian agriculture, the key issues for water use: where it comes from; how much is used on different crops; how much money is made for the use of a given volume of water (crop per drop). These core concerns translate directly across to horticulture, where similar efficiencies can be made.

One management option in agriculture is to move to higher value crops, although many factors are involved in crop choices.

Urban

Approximately 60% of the water supplied to cities is used by households.

The average urban direct household water usage is continuing to fall. While the population of the state capital cities increased by 1.55% in 2004–05, the amount of water supplied to them by urban water utilities dropped by 2.1%. Over the year 2004–05, urban Australians consumed an average of 84 kL per person (average of 230.14 L/day), a reduction of some 3.65% since 2003–04 and a commendable 15% reduction in per capita consumption over the four years since 2000–01. These reductions cannot be solely attributed to restrictions but a significant culture change in our cities, with most people taking measures to become more efficient in

Table 5.5 Water consumption in Agriculture by water type and agricultural activity (ML)

Activity	Self-extracted	Distributed	Re-use	Consumption
Dairy farming	856 993	1 339 473	79 136	2 275 603
Vegetables	307 033	132 544	15 796	455 373
Sugar	404 068	858 767	6177	1 269 012
Fruit	306 978	339 315	1370	647 662
Grapes	191 363	522 029	3655	717 047
Cotton	1 697 245	122 071	2194	1 821 509
Rice	224 806	394 158	11 908	630 872
Livestock	935 396	100 078	–	1 035 474
Pasture (a)	1 000 850	887 144	39 898	1 927 892
Grains	461 815	582 098	118 356	1 162 268
Other	195 887	51 337	1436	248 659
Total	2 593 948	1 620 656	159 689	4 374 293
Grand total	6 582 435	5 329 012	279 925	12 191 372

(a) Excludes pasture for dairy farming
Source: ABS (2006).[7]

Table 5.6 Water intensities in agriculture 1996–97 derived from estimates of mean water use per land type

Land use	Water intensity $/ML	Total water use GL	% total use	Water use ML/ha
Vegetables	1295	392	2.6	3
Fruit	1276	665	4.4	7
Tobacco	985	13	0.1	4
Grapes	600	781	5.2	8
Tree nuts	507	140	0.9	6
Cotton	452	2314	15.5	7
Course grain	116	518	3.5	3
Dairy	94	5902	39.5	7
Peanuts	90	25	0.2	3
Hay	54	20	0.1	4
Rice	31	1696	11.3	1
Legumes	24	33	0.2	3
Sheep	23	13	0.1	4
Sugar cane	21	195	8.0	7
Beef	14	1080	7.2	4
Oilseeds	10	85	0.6	3
Cereals	-9	87	0.6	3
All uses	193	14 959	100	7

Source: National Heritage Trust (2002).[8]

the way they use water. At the same time, average household consumption in capital cities has fallen by 9%, from 251 kL per household per annum in 2000–01 to 225 kL per household p.a. in 2004–05 (616.44 l/day or 243.65 l/day/person with average household of 2.53). (See Table 5.7.)

What about the future? Water use in Melbourne, for example, is 59% household, 30% industry and business, and 11% other (firefighting, leaks, stolen, other). However, outside the urban area, electricity generation by the La Trobe power industry uses about 140 billion litres a year which is about 35% of Melbourne's annual consumption. Melbourne's population of 3.5 million and annual water use of 480 GL is expected to increase to 4.5 million and 659 GL p.a. respectively by 2035. Reliable yield from present catchments is an estimated 566 GL p.a.[9]

If climate change further reduces annual rainfall, then considerable savings must be made and these will be primarily through demand management and recycling. Foremost is a re-valuing of 'waste' water – treated sewage, stormwater, rainwater and greywater. Recycling strategies include: domestic greywater; re-use of large building wastewater; re-use of industry wastewater; sewer mining suitable for parks and sports grounds; retrofit of third pipe systems; harvesting roof run-off; harvesting stormwater to storage in lakes or aquifers.

Table 5.7 Melbourne city water use in 2004

Method of water use	Percentage
Sector	
Residential	60
Commercial and industrial	28
Leakage	8
Miscellaneous	4
Residential use	
Garden	35
Bathroom	26
Toilet	19
Laundry	15
Kitchen	5
Industry and commercial use	
Other commercial and industrial	61
Major industrial	29
Major commercial	10

Source: Victorian Water Industry Association (2005).[10]

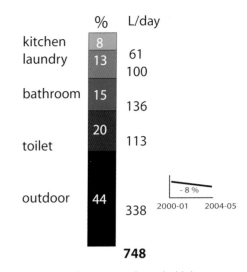

Figure 5.9 Australian average household direct water consumption per day from 2000 to 2001.[7]

Stormwater, once perceived as a health and flood risk, is now managed more sustainably by linking more closely to the whole water cycle and responding to the former role of stormwater in the environment. This translates to the retention of floodplains, construction of wetlands, retarding basins, litter traps, rain gardens and other soft engineering.

Household and individual

China, India and the United States, having large populations, are the greatest water consumers. However, water use per person is about three times higher in the United States than it is in India and China so globally there is substantial variation in individual water use.

The proportion of overall Australian water use diverted to households has remained at about 9–11% for many years.

Household mains water usage (Figures 5.9 and 5.10) is largely determined by the number of people per household, income per capita, water price, climate and attitudes to water conservation. In recent times usage has decreased as a result of water restrictions, demand management campaigns and increasing prices. Outdoor water use depends on garden size and the degree of enthusiasm the occupants have for gardening. In general the volume of water used per household increases from the more densely occupied city centre area to the more open suburbs. There

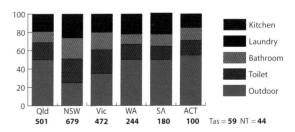

Figure 5.10 Total household water use by state and percentage allocation 2000–01 (GL/yr).[7]

Table 5.8 Embodied water used in the production of Australian commodities compared with total daily direct water use by households and individuals

Good or service	Embodied water (l)	Days of total household direct water use @ 748 l/day**	Equivalent days of garden watering @ 40% household use (299 l/day)	Equivalent days of individual water use on garden @ 2.6 persons per household (115 l/day)***
1 car (1 tonne)	47 600*			
1 kg Australian beef	17 112	23	57.00	148.8
1 pair leather shoes	8000	10.7	26.8	69.6
1 kg cheese	4544	6.1	15.12	39.5
1 cotton T-shirt (250 g)	2000	2.2	5.45	17.4
1 kg wheat	1588	2.1	5.20	13.8
1 cup coffee	140	0.187	0.46	1.2
1 glass wine	120	0.16	0.40	1.0

* 1 metric ton = 1000 kg beef = 17 112 kl/ton.
** Water Services Association of Australia (2004).[11] *** 2.6 persons/household.
Source: Hoekstra & Chapagain (2007).[6]

is little variation in indoor water use through the year but in Australian cities, as elsewhere in the world, there is a distinct seasonal cycle in outdoor use with an increase in the summer as garden watering begins.

Major savings can be made by using a rainwater tank plumbed in to the laundry and toilet systems, and recycling greywater.

However, only about 14% of our total water usage is mains water use in the home. The other 86% is embedded in the goods and services we use. For example, a water-efficient family of four with a medium-size garden in Melbourne uses on average a total of about 615 L of mains water per day (includes indoor and outdoor use), while the estimated embodied water in a hamburger is about 2400 L and a cotton T-shirt about 2000 L.

To get a very rough idea of our consumption of embodied water through eating meat, here is a very simple calculation: it is extremely approximate and over-simplifies a complex situation, but it makes an important point. We eat roughly our own body weight in meat each year (about 80 kg). One kilogram of beef embodies 17 112 L water, so 80 kg embodies 1 368 960 L. Total average Australian household direct water use is 748 L/day (= 273 020 L/yr). Therefore one person eating 80 kg of beef a year uses five times the total annual direct household water use (1 368 960 ÷ 273 020). Figures like this indicate where major water savings could be made.

We can understand these figures better when we compare the water embodied in goods and services with the direct water we use in the home (Table 5.8).

These figures show how being water-smart entails being aware of the many unlikely ways that we can conserve water. It certainly starts by being sensitive to the embodied water in consumer goods but then, since power generation often requires the use of large volumes of water in cooling towers, our weekly household electricity use brings with it an embodied water component that ranges between about 18 L and 36 L – so leaving lights on is wasting water!

Note that water conservation is important only at times and places where there is water shortage. When I buy a T-shirt made from cotton grown in an area where rainfall is plentiful, water storages are full, and environmental flows and water quality is good, then the demands made on the environment and other people is very small compared with a situation where the cotton is grown in a drought-prone area with depleted water supplies where the water used on the cotton crop takes away water from crops that are being grown as food for the local people.

Water in the garden

Most people water their gardens satisfactorily by 'feel'. With some knowledge of how your plants respond to water stress and how your watering system behaves, together with a weather forecast for the following few days and a finger poked into the soil to check dampness, it is possible to keep plants alive while holding watering to a minimum.

This may work for the canny home gardener but water is now so precious that those in charge of large irrigated parks, public gardens, golf courses and sports fields will certainly need to be familiar with the factors that contribute to water demand as it relates to the kinds of plants being grown, the climate and the soil. Home gardeners can also benefit from a working knowledge of water flows. In agriculture and production horticulture monitoring water productivity has for many years been part of routine management. Urban management of public space will need to follow suit with water sustainability accounting and audits. This means thinking in terms of megalitres (ML) of water and developing the ability to calculate the likely volume of water needed to irrigate sports grounds and urban trees, and to maintain a specific vegetation quality according to the particular landscapes and types of plants used.

The better we understand exactly what happens as water flows through our parks and gardens, the more effective will be our management, and the smarter our water budgets. A quick overview of how water flows through the garden will set the scene before we examine each step in more detail (see Figure 5.11 and Table 5.9).

Water flows

Rain and irrigation water falling on a garden lands on plant leaves and the soil surface where some will be lost by evaporation before the remainder infiltrates the soil. If the rate of water supply exceeds the rate of infiltration, as in a heavy rainstorm, there will be surface pooling or run-off (including that from hard surfaces) into surface channels, gutters and drains. In most of our established urban areas, surface run-off flows into the neighbourhood stormwater drains, and from there into larger water bodies like rivers or the sea.

Figure 5.11 The major water flows in a garden.

The rate of infiltration and the water-holding capacity of the soil depends largely on the soil structure and texture (particle size), so sandy soils drain quickly and hold less water than slow-draining clay. If the volume of infiltrating water is sufficiently large, it will percolate into the soil and may pass into subsurface drains or seep even further into an aquifer (a band of soil or rock containing water). This groundwater is referred to as the *water table*. Water tables rise in wet seasons and fall in dry ones and may be close to the surface near rivers, lakes and other water bodies. The majority of plant roots are generally in the top 30–50 cm of soil, depending on soil and plant type, and are usually well above the water table, so in extended dry periods plants may require watering.

If rain or watering is heavy over a prolonged period then all the air spaces in the soil fill with water, the soil becomes saturated and the ground becomes waterlogged. This might happen quite quickly if there is a hard layer of rock or clay near the surface. Additional water then forms pools or flows away as surface run-off. If the soil remains waterlogged, plants may die because their roots cannot obtain oxygen. Excess water

Table 5.9 Water flows in the garden

Water inflows (sources)
Rainfall
Irrigation
Mains water
Rainwater from tank
Re-used or recycled water (greywater)
Water movement and storage in the soil
Infiltration – flow into soil
Percolation – flow through (saturated) soil
Water storage
Water outflows (losses)
Evaporation
Transpiration
Drainage
Surface run-off
Subsurface drainage
Deep percolation

may be removed by installing subsurface drains. When watering or rain stops, the water gradually drains through the soil, but not all of it will pass through. At a certain point the soil particles 'hold on' to what remains. This bank of water in the soil is drawn on by the capillary action of moisture evaporating from the soil surface and absorption by the plant roots as the leaves transpire, the combined process of evaporation and transpiration being known as *evapotranspiration*. The more water that is taken out of the soil, the harder it is for the plant roots to absorb what is left.

Assuming there is no more rain or watering, evapotranspiration steadily removes more and more water from the soil. With lack of water, plants progressively 'shut down' by closing stomata (openings in the leaves allowing gas exchange), diverting sugars to the roots for increased water suction, and shedding leaves so growth slows down. Those plants with softer foliage will wilt in the heat of the day, recovering at night. Eventually, after continuing and prolonged loss of water from the soil, wilting is permanent and the plant dies.

This process can be divided into three phases of water flow: inflows, storage and outflows. We have divided these into further parts to help explain what is happening and indicate strategies for water conservation.

It helps to think of the garden soil as a large, leaky water tank with inflows and outflows. Water inflows for this leaky tank are direct rainfall and irrigation. Irrigation water can come from rainwater harvested and stored in a tank, from the water mains, or possibly from recycled domestic water, or greywater. Outflows are through evaporation, transpiration and drainage. Drainage occurs as surface run-off, subsurface drainage and deep percolation. Water is stored in the soil 'tank' itself but has the potential to move in any direction, although this is mostly downwards by gravity when it is saturated, and upwards due to evapotranspiration when the soil has drained (see Figure 5.12).

There is now an established colour coding system for different water uses and we shall be using this from now on (see Info Box 5.2).

We can now examine the garden water flows in more detail.

Water inflows

Rainfall

Eighty-eight per cent of the Australian population lives in seaboard cities with moderate to high annual rainfall. In southern

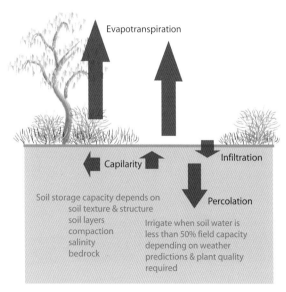

Figure 5.12 Water movement in the soil.

states, there is certainly a period of high water demand in summer, but there is the opportunity at other times of the year to harvest urban rainfall in preparation for times of need. Because Australia's major dams rarely fill completely, saving water in tanks to reduce demands on the dams is important at all times of the year throughout the country.

Weather and rainfall statistics for your area can be obtained from the nearest Bureau of Meteorology weather station.

Irrigation

Using mains water In 2004–05, a summary of water supply and use in the Australian economy noted that 33% of the 242 779 GL of run-off from natural rainfall was diverted to human use, leaving only two-thirds to maintain all Australia's natural ecosystems.[7] Even with this massive diversion, after prolonged drought beginning in the 1990s, water storages in mid 2007 were reaching critically low levels for most of Australia's larger cities (Figure 5.13).

Greywater The second source of irrigation water is greywater (or recycled water). Greywater is non-toilet domestic wastewater. It makes up 50–80% of indoor domestic wastewater. Using greywater has the potential to reduce mains water use by 30% or more and a greywater harvest of about 210 L per day (about 76 000 L/yr) could be expected. It can be used for watering the garden or flushing toilets. Using greywater reduces water bills and the demands on sewage treatment plants. More than a half of all Australian households now report using greywater, with nearly a quarter (24%) using it as their main source of water for the garden.[12] However, reduced flows from domestic greywater may result in less reclaimed water for municipal use.

Recycled blackwater In the future we can expect much greater use of water recycled from sewage. This water is treated to different quality levels by removing

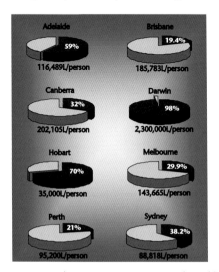

Figure 5.13 Capital city water storages. Levels and litres available per person in May 2007.[13]

INFO BOX 5.2: WATER COLOUR CODING

In managing water we need simple terms to distinguish between fresh potable water, recycled water, sewage and so on, and in nature we need to denote particular components of the water cycle. The following colour terms are gaining currency:

Greywater (sullage)

This is household wastewater that has not come into contact with toilet waste; it includes water from baths, showers, bathrooms, washing machines, laundry and kitchen sinks (fat free).

Blackwater

Household wastewater that contains solid wastes, i.e. toilet discharge.

Green water

This is water in the root zone that is available for plant growth. Because green water cannot be piped or drunk, and cannot be sold, it is generally ignored by water management authorities; but it is crucial to plants in both nature and agriculture and needs careful management as an important part of the global water cycle.

In nature, the global average amount of rainfall becoming green water is about 60% – 55% of this total falls on forests, 25% on grasslands and about 20% on crops. We can increase green water productivity by rainwater harvesting, and improving soil infiltration of run-off. Effective management of green water has large implications for global food production.

White water

Water in the atmosphere that has vapourised through evapotranspiration.

Blue water

Collectable water from rainfall; in nature this is the managed water of rivers, dams and aquifers making up about 40% of total rainfall. Water budgets based on nature have concentrated on blue water but improvement in efficiencies in other areas, like green water, could reap major rewards. In gardens, blue water is the harvestable water that falls on roofs, hard surfaces and the garden, much of this usually flowing through the stormwater system into rivers and the sea or recharging groundwater and aquifers. The greatest water savings in gardens can be made by harvesting and using blue water for both house and garden, although improvements in green water efficiency and reductions in white water production also play their part.

Lilac water

Recycled water. Non-potable recycled water sources have lilac pipes and lilac 'not suitable for drinking' signs. State government recycled water use guidelines are available online.

---

disease organisms and so may be suitable for use on turf, crops and vegetables, or even for drinking.

The use of greywater and blackwater in gardens and landscape, and recommended precautions in their use, is discussed in more detail in the chapter on design.

Water flows and storage in the soil

Having looked at water sources and their use, it is now time to see how water behaves in the soil. There are many important questions to consider here. For example, how deeply does a given rainfall penetrate the soil? How rapidly does water move through the soil and what controls that movement? How much water does a given volume of soil hold and what methods can we use to retain it as long as possible? How much water should we apply to avoid wastage through deep percolation? How much water is drawn out of the soil by transpiring plants, and how can we estimate the rate at which this will occur? How much water do drought-tolerant plants save and how is this done? We need answers to these sorts of questions before confronting more direct questions about irrigation; how often, how much, and how long should we water?

These questions are dealt with in detail in the chapter on maintenance, but we provide a brief overview of general principles here.

Water movement and storage in the soil is largely dependent on the physical properties of the soil and water and these can be measured quite accurately. The downward movement of excess water in the soil is produced by gravity. The lateral and upward movement is induced largely by evaporation from the soil surface and transpiration by plants, which creates a pressure (water potential) that is sufficient to draw the water along. Saturated soils will generally drain for 2–3 days, sometimes more, but when draining stops there will be water stored in the soil 'tank' which is then said to be at *field capacity*. The amount of water now available to plants will depend primarily on the extent of the plant root system and the storage capacity of the particular soil. Soil storage capacity, in turn, depends mostly on soil texture and structure but also salinity (which decreases water availability), the presence of a bedrock (which reduces percolation), soil layering/stratification, and compaction. Fine textured soils like clay hold more water than coarse textured soils like sand, but infiltration and percolation are slower and less amenable for plant growth in clay soils. Medium-textured loamy soils with high silt content hold the most available plant water. Water-holding capacity and healthy plant growth can be improved by a carefully considered program of soil improvement and, in particular, the addition of organic matter.

Water outflows

Water flows out of the garden by evapotranspiration and drainage. Evapotranspiration includes plant transpiration and surface evaporation from soil, foliage and hard surfaces. Drainage includes: run-off into the surface drainage system; subsurface drainage through the soil and drainage pipes; and deep percolation into the water table. We can calculate evaporation rates with some precision, but calculating transpiration is much less precise as there is still much to learn about plant–water relations. It is nevertheless possible to have a clear understanding of how water is taken up and released by our garden plants (Figure 5.14).

Evapotranspiration

If there is harvesting of rainwater on-site then, in most situations, the greatest water loss is by evapotranspiration.

This is the most difficult area of water outflow to measure and understand because plant transpiration is the result of complex interactions between the plant, the soil and

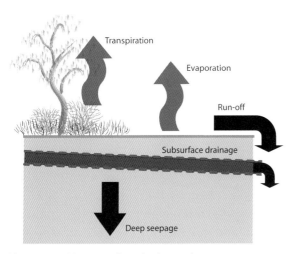

Figure 5.14 Water outflows in the garden.

the atmosphere. This is still an active area of research known as the *soil-plant-atmosphere-continuum*. (See Figure 5.15.)

We shall be looking at many ways of reducing evapotranspiration ranging from general garden design to plant selection, mulching, use of shade and shelter, soil improvement, and more, at a later stage.

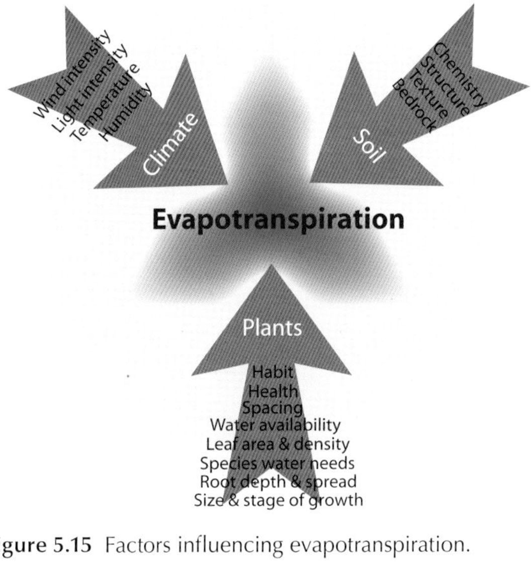

Drainage

If water is added to the garden faster than it can infiltrate the soil, then it will either pool on the surface or flow down slopes. *Surface run-off* usually passes directly into the stormwater system but it can be directed onto the garden (if waterlogging is not a problem), or attempts can be made to move it into soil storage. Often water will pass through the upper layer of soil quite quickly but then meet a less permeable layer in the soil profile. In heavy rain this can lead to waterlogging that can in some cases be remedied using subsurface drainage pipes. Gardeners will be familiar with a layer of clay lying within a metre of the soil surface in many areas. If this layer is shallow it may be broken up to encourage deeper percolation of water.

Surface run-off

Excess surface water in nature flows into rivers, streams, creeks, swamps and wetlands. In urban areas we use surface drains (gutters, stormwater drains) or subsurface drains (aggie pipes) to keep the groundwater from rising above a certain level. The impervious surfaces of roofs, driveways, roads, car parks and paths generate torrents of stormwater. This results in flash flooding and erosion because the water has failed to infiltrate the ground and drains are suddenly overloaded. At the neighbourhood level it makes ecological sense to design landscapes that minimise run-off and flooding and maximise infiltration, helping to maintain the health of the trees and other plants in the landscape.

Figure 5.15 Factors influencing evapotranspiration.

This strategy will apply from the top of the watershed to the bottom. Stormwater retention basins and wetlands are now widely used to slow the flow of stormwater, trap silt and catch rubbish. Chapter 10, 'Sustainability in the broader landscape', discusses this in more detail.

In general, water run-off is an opportunity not to be missed. It can be harnessed in several ways: by storing it in some way, by directing it onto the garden, and by making the soil more permeable so that it soaks in.

Subsurface drainage

With subsurface drains there is a need to balance the advantage of clearing waterlogged soils with the disadvantage of drying soils too quickly. At field capacity, water will not be flowing into the drains (see Chapter 12 on landscape maintenance).

Deep percolation

In some situations where soil is deep and porous, it may be possible to water continuously without the soil becoming waterlogged. In cases like this it is important to know how much water is needed to bring the soil in the root zone to field capacity, and not to water beyond the root zone.

This background on water flows in the garden is the foundation for the water-conserving strategies discussed in the chapters on garden design, construction and maintenance.

Environmental impacts of garden water use

In times of drought, we protect the mains water supply primarily because of its significance for agriculture, domestic and industrial use but, of necessity, we are now becoming more sensitive to the environmental impacts of reduced environmental flows, as freshwater ecosystems are the most vulnerable on the planet.

The environmental effects of garden water use can be divided into those that occur upstream (on the way *to* the garden, including collection from rivers and storage in dams before passing through treatment plants and pipes), and those that occur downstream (the water that flows *from* the garden into the stormwater and sewerage system). Infrastructure, both upstream and downstream, incurs the usual environmental costs in land, energy and other resource use, together with ongoing maintenance costs.

Upstream of households, water withdrawn for human consumption (for crop irrigation and storage in dams) reduces environmental flows which can lead to a reduction in ecosystem vigour, degrading of water quality (pollution), salination and algal blooms, while also interfering with natural processes such as the migration of spawning fish, and in some cases loss of life and species or, in extreme cases, species extinction. If water supplies in dams run low then there are calls for new dams or desalination plants with further habitat interference, biodiversity loss, social disruption and, in the case of desalination plants, high energy use. Households and local water users have little control of the upstream mains water supply but can minimise demand by good on-site water management.

Downstream management of stormwater is largely the responsibility of local government and water utilities who strive to improve water quality, but the gardener can be mindful of what happens to water after it leaves the garden. Run-off may contain nutrients, especially nitrates, phosphates or other chemicals (although this is more a problem of fertiliser use in agriculture than for domestic run-off) causing excessive plant growth (known as *eutrophication*). It may also be contaminated by pet faeces and leaky septic systems and so contain disease organisms that may be a health threat. The water flow from a house block will vary largely in proportion to the area of impermeable surface (roofs, sheds, driveways, paths, etc.); the greater the hard surface area, the greater the run-off (unless the run-off is harvested). The overall volumes of this outflow water are environmentally significant as flash floods can cause damage and erosion, and extreme water harvesting may leave little water left for natural ecosystems. Water from roofs and hard surfaces passing directly into drains and channels leaves insufficient time to recharge soil water. Table 5.10 lists the various threats to water quality and biodiversity from stormwater pollutants.

Flows into the sewerage system entail sewage treatment with its associated infrastructure (treatment plants, pipes, drains, etc.) and any impacts when this is released to the environment.

The *Australian Water Resources Assessment 2000* studied about 30% of Australia's 246 river basins and found major nutrient excess, mainly nitrogen and phosphorus, in about two-thirds of these. High salinity was also a factor. Australia's first comprehensive assessment of catchments, rivers and estuaries, the *National Land and Water Resources Assessment* used an aquatic biota index to monitor the response of macro-invertebrates (e.g. worms, snails, insects) to environmental change.[14] It showed that about a third of the total river length assessed (21 909 km) had lost between 20 and 100% of the various macro-invertebrates that could be expected to live there. The study also used an environment index (including catchment and hydrological disturbance, habitat, nutrient and suspended sediment) to assess river water quality. Apart from Tasmania, more than 80% of river length tested was assessed as moderately to substantially modified.

Monitoring water quality is an expensive and time-consuming task that has been taken up by a small army of volunteers. Waterwatch Australia, established in 1993, is a national community water-quality monitoring network for the protection and management of waterways and catchments. Waterwatch groups conduct biological and habitat assessments via physical and chemical water tests. There are now about 3000 groups Australia-wide. At over 7000 sites and covering over 200 catchments, they are assessing the health of their waterways and catchments and the quality of the water. Their role includes fencing areas of riverbanks, eradicating weeds and invasive species, and reducing the use of pesticides and other pollutants.

Sustainable water use

Water use on the garden as a proportion of our total individual water use (direct and embodied) is extremely small, probably only about 1–2%. However, in urban situations residential direct water use averages about 60% of all urban water use of mains water

Table 5.10 Stormwater pollutants

Pollutant	Effect	Urban source
Sediment	• Reduces light and inhibits plant growth and therefore the food supply • Suffocates organisms; clogs fish gills	• Land surface erosion • Pavement and vehicle wear • Building and construction sites • Spillage; illegal discharge • Organic matter (leaves, grass, etc.) • Car washing • Atmospheric deposition
Nutrients	• Stimulation of growth of weeds and algal blooms	• Organic matter • Fertiliser • Sewer overflows • Animal faeces • Detergents • Atmospheric deposition • Spillage, illegal discharge
Oxygen demanding substances	• Oxygen displaced from water more rapidly than it can be replaced from the atmosphere • Fish kills; unpleasant smells	• Organic matter decay • Atmospheric deposition • Sewer overflows, septic tank leaks • Animal faeces • Spillage, illegal discharges
Toxic organics	• Kill or threaten the health of organisms	• Pesticides • Herbicides • Spillage, illegal discharge • Sewer overflows
Heavy metals	• Kill or threaten the health of organisms • Persist in the environment	• Atmospheric deposition • Vehicle wear • Sewer overflows • Weathering of buildings • Spillage, illegal discharges
Gross pollutants	• Unsightly litter and debris that can also be a health threat	• Pedestrians and vehicles • Waste collection systems • Leaves, lawn clippings • Spillage, illegal discharges
Oils, detergents and shampoos (surfactants)	• Toxic poison to fish and other organisms	• Spillage, illegal discharges • Vehicle leakage • Car washing
Micro-organisms	• Bacteria and viruses that cause disease e.g. hepatitis and enteritis	• Animal faeces • Sewer overflows • Organic matter decay

Source: Adapted from Table 1.1 Urban Stormwater. Best practice environmental management guidelines, CSIRO, 1999.

and, since the average outdoor water use in Australia is 40% of household use, then this means that on average about 24% of all mains urban water use is for gardens; sufficient to be of concern.

Water saving in the garden is best considered as part of a general household strategy of water conservation including the use of rainwater tanks linked to both garden and house (garage, sheds etc.), the possible use of recycled greywater, and a strategy for dealing with embodied water consumption. Changes in consumption habits can make much larger savings than changes in garden water management. The precise connections between the quantities of water used in agriculture that are embodied in our food are unclear. Nevertheless it is evident that changing our consumption patterns, altering our eating habits and/or producing our own food is likely to have a far greater impact on our overall water use than all of our household and garden water-saving strategies combined.

We recommend three major strategies for reducing garden water use:

1. minimising the use of mains water by installing a rainwater tank and considering the use of recycled greywater (aiming at self-sufficiency in garden water use)
2. using water-conserving garden design, and
3. maximising irrigation efficiency by careful management of all water flows.

If you are self-sufficient in your garden water supply then your water demands on the environment are minimal and water sustainability has been achieved – providing there are no downstream environmental consequences.

Materials used for horticulture include those for glasshouses and other garden structures, tools, hard landscaping, products such as pots and chemicals.

MATERIALS

Don't buy rubbish   Swap 'til you drop

Buy things that don't cost the Earth

We turn goods into Gods   Don't shop 'til you drop: swap shopping for swapping

> **KEY POINTS**
> Materials and chemicals have become more diverse, toxic and 'global'.
> We need dematerialisation and cyclical materials metabolism.
> Large material savings can be made during design, engineering and construction of goods.
> Individuals can contribute by using green consumption.

Each year we move more and more 'stuff' around the planet – everything from food, raw materials and commercial products to animals and plants. This is all part of global industrial metabolism as 'things' are sourced, produced, distributed, used and discarded. Resources are consumed at every step of each item's lifecycle. By the time goods are sold, many have generated several times their own weight in waste.

In general, with greater technological sophistication and increasing wealth, the quantity of waste has not only increased, it has become more diverse, toxic and difficult to manage, and the massive environmental costs are rarely included in the market prices of products.

Much of the problem can be resolved by careful treatment of waste. Sustainable materials management must follow, in strict order of priority, the well-known path of 'reduce, re-use and recycle'. To-date, emphasis has been placed on the last of these recommendations through 'end of pipe' waste management. However, the more desirable preventative 'start of pipe' production sustainability is now gathering momentum in the design, engineering and construction industries.

Dematerialisation, or *zero waste* as it is also known, is the single guiding idea behind materials sustainability. Material flows in human communities would be sustainable if they occurred in 'closed' cycles as they do in nature. Dematerialisation attempts to convert the linear path of a product (from extraction, to manufacture, to use, to disposal in landfill) to a cyclical path where materials are re-used and recycled as they are in natural ecological metabolism; turning waste into resource. Two formal procedures are being set in place to deal with this situation:

1. *Product stewardship* recognises that manufacturers, importers, governments and consumers all have a shared responsibility for the environmental impacts of a product through its life cycle. Product stewardship schemes may operate with various degrees of regulation and have been adopted in the EU, parts of Asia and North America and are also under consideration in Australia.
2. *Integrated Product Policy* addresses the life cycle impacts of products. In a review of 11 studies analysing the life cycle impacts of total societal consumption and the relative importance of different final consumption categories in the EU it was found that housing (building materials, heating, electrical appliances), transport (mostly cars and air travel) and food (mostly meat and dairy) are responsible for 70% of the environmental impacts in most categories although representing only 55% of the final expenditure in the 25 countries making up the EU.[1]

In Australia the *National Packaging Covenant* brings companies from all sectors of the packaging supply chain – raw material suppliers, packaging manufacturers, packaging users and retailers – together with the Commonwealth, state, territory and local governments to reduce packaging waste. In June 2002, there were 500 signatories.

High on the list of eco-efficiency targets for material resource use are minimisation of product toxicity, and creation of economies based more on services and less on products.

Changing consumption patterns involves all levels of the sustainability action hierarchy as economies become more interdependent through world trade. Much is to be gained from Life Cycle Analysis (LCA) of products and a new consumer purchasing awareness.

Materials and trade

Fossil fuels have facilitated transport of raw materials to factories and markets. At the same time mass production has made possible a vast increase in output at much reduced cost.

In the early 20th century plastics were introduced – at first plant-based celluloid but later more complex petroleum-based chemicals – and these often replaced metals. At this time there was a surge of synthetic chemicals such as pesticides, solvents, paints and refrigerants. Between 1940 and 1960, energy intensive fertiliser production and aluminium smelting boomed, and this was followed by a massive increase in car production that both depended on and encouraged the development of roads, buildings and other infrastructure.

Most of these changes improved the convenience and comfort of peoples' lives but it was at the cost of increased wastage of all materials, environmental damage and resource depletion which was initially ignored. In the 1960s there were the first stirrings of a change in perception as recycling gradually became accepted as normal practice. In industrial countries between 1980 and 1990, paper and cardboard recycling increased from 30% to 40%, glass from 20% to 50%, and metal recycling was also on the increase. However, the amount of material extracted and produced doubled between 1963 and 1995, metal extraction increasing four times from 1950 to 2000 and products based on fossil fuels, like plastics, increasing even more rapidly.

In the 1990s, in the face of gathering environmental problems, scientists recommended a dematerialisation of industrial economies, calling for a reduction in material flows of 50% to 90%. Some countries, like Austria, produced national environmental plans.

As production of infrastructure decreased in industrial economies, economic emphasis shifted from goods to services: material intensities (material needed to generate a given cost of output) declined by 20–30%. However, total consumption increased by well over 50%, emphasising the need to reduce absolute levels of material use. There was the additional problem that gains in efficiency generally only encouraged consumption in other areas.

Globally, we are still in a phase of escalating levels of material use with its associated increase in environmental pressure. Global mineral reserves are shown in Table 6.1.

True globalisation of world trade is generally assumed to have occurred in the 1990s as the monetary world trade (exports and imports) almost tripled between 1990 and 1998 affecting mostly newly industrialised countries.

One important ecological dimension of this world trade globalisation is the production of a 'north-south' divide. Material flows analysis shows a transfer in environmental pressures from industrial countries to newly industrialised countries and non-industrialised countries. As always there is a need for an effective accounting system for the movement of materials around the world. Material Flow Accounting appeared in the 1960s but only began to gather momentum in the 1990s.

Waste

The average human requires 45–85 tonnes of material (natural resources) each year.[2]

Eighty per cent of what we make is thrown away within six months of its production. Products contain more components and these are more difficult to biodegrade than before. Many items are deliberately manufactured to be thrown away: tissues, plastic razors, nappies, plastic bags, toner cartridges, e-waste and so on. Waste is something we discard because it is no longer needed, but it is very difficult to classify because it is categorised in many different ways: by its origin (how it was made), by its danger, what it is made of, and the way it is managed. This makes the compilation and comparison of statistics difficult. However, at the municipal tip, in Australia at least, it is divided into three types: municipal (kerbside, hard waste); commercial and industrial (metals, plastics,

Table 6.1 Mineral reserves, recycling rate and time to run out

Mineral	Known reserves (million tonnes)	Time to run out (yrs)*	Proportion (%) of consumption met by recycling	Estimated amount used in average American lifetime (77.8 yrs in 2007)
Aluminium	32 350	510	49	1576 tonnes
Antimony	3.86	13	n.a.	7.13 kg
Chromium	779	40	25	131 kg
Copper	937	38	31	630 kg
Gold	89 700 tonnes	36	43	48 g
Hafnium	124 tonnes	~10	n.a.	n.a.
Indium	6000 tonnes	4	0	32 g
Lead	144	8	72	410 kg
Nickel	143	57	35	58.4 kg
Phosphorus	49 750	142	0	8 322 tonnes
Platinum/Rhodium	79 840 tonnes	15	0	45 g
Silver	569 000 tonnes	9	16	1.58 kg
Tantalum	153 000 tonnes	20	20	180 g
Tin	11.2	17	26	15 kg
Uranium	3.3	19	0	5.95 kg
Zinc	460	34	26	349 kg

Source: Cohen D (2007).[9]
* assuming no recycling and each used at half the rate currently used by an average US resident in 2007

timber); construction and demolition (building waste, rubble, earth).

In industrial countries industry itself is the top producer of waste, but each day every one of us uses and discards many kilos of materials ranging from plastics, glass, wood, cement, metals and chemicals. In non-industrial countries almost all this waste can be composted but the compost-ready waste of wealthy nations makes up only about one-third.

Depending on what it is, solid waste may be put in landfill, incinerated, composted or exported. Landfill is expensive and produces methane gas (21 times more potent as a greenhouse gas than CO_2), although about half of this is captured for power generation. The siting of tips is problematic although modern ones are designed to control leachate (water that has filtered through the waste) and gas emissions while taking account of local hydrology, climate, haulage and other issues.

Knowing a little about waste stream categories can help in making environmentally responsible choices about the goods and materials we purchase and use.

Jumbo waste includes cars (more produced each year), boats and planes down to whitegoods. All benefit from green design and clever construction that allows for dismantling and recycling.

Electronic waste (e-waste), especially refrigerators, washing machines and other household electrical appliances, mobile phones, TVs and PCs, DVD and video players, radios, lighting, electronic tools and ticket machines, is currently the fastest growing global waste stream accounting for 5% of all solid municipal waste.

Batteries contain heavy metals that can contaminate leachate.

Mining waste – Australia is the world's largest miner of bauxite ore for aluminium at 50 M tonnes/yr with heavy mining impact and an energy hungry smelting process (production of 1 tonne aluminium requires ~ 15 000 kWh energy). Recycling aluminium reduces its greenhouse gas production by 95%.

Industrial waste is the major contributor to waste and a large proportion of this waste is the hazardous chemicals which must be monitored from source to final disposal.

Wood, paper, pulp and cardboard production is one of the worst polluters and uses more than 40% of all the timber felled worldwide (330 M tonnes in 2006). State-of-the-art mills use recycled waste with zero effluents, cleaning and re-using wastewater and effluent gases, or waste materials such as bark and sawdust are used to generate electricity.

Organics like food waste and garden waste also contribute to methane emissions.

Packaging protects goods from dirt and vibration, and supplies marketing appeal. Around 50% of goods are in plastics of various kinds and each requires its own recycling method. We tend to produce things in individual portions (tea and coffee bags).

Mixing recyclable materials in packaging can be a major problem.

Household hazardous waste includes oils, paints and pesticides, any materials that contain corrosive, toxic, ignitable or reactive ingredients.

Used oils and waste tyres litter landscapes and waterways and take up landfill. Up to 100 million litres of motor oil goes 'missing' in Australia each year and is a serious environmental problem.

Glass – crushed glass, called cullet, is now the major raw material for glass manufacturing in Australia. Each tonne of cullet saves 1.1 tonnes of raw materials.

Plastic – first invented in 1860 and becoming widely used in the last 40 years or so are polymers manufactured largely from crude oil, gas and coal (mostly waste from petroleum refining). An enormous amount of waste can be directly attributed to this material (Figure 6.2).

Australia

Australians throw out about 21.2 million tonnes of rubbish each year and over the same period the average Australian family produces 800–900 kg of waste. At present over 40% of household garbage by weight is made up of food and garden waste, so most of this grand total could be composted or mulched (Figure 6.4). Waste generation per person increased from 1.23 tonnes in 1996–97 to 1.62 tonnes in 2002–03 (Figure 6.3).

The *State of the Environment Report* (2006) draws attention to the major waste issues in Australia. It indicates that the amount of

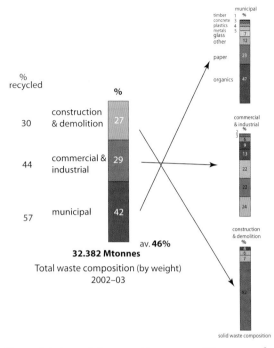

Figure 6.2 Plastic waste in Australia.[4]

Figure 6.1 Australia's waste. Mtonnes = million tonnes.[3]

waste generated has increased along with the population.

Most households in Australia's cities recycle some waste, and waste recycling and re-use rates in general have increased to an average of 36% across Australia (Figure 6.1).[5]

In areas outside the capital cities, the recycling rate is lower because of the logistics and costs associated with collection and transport from rural areas to processing plants. Re-use of waste, in contrast, is higher among rural and regional households. Most states and territories have implemented waste reduction policies with a view to reducing the amount of waste to municipal landfills. These have been broadly successful. For example, of the annual volume of waste generated from building activity in the Sydney region, approximately 10 million tonnes are now recycled or re-used, 2.5 million tonnes are reprocessed into building materials off-site, and 1 million tonnes are disposed of annually to landfill.

At present, growth in waste generation is related to household incomes and corporate earnings so the challenge of decoupling this connection is clear. Product stewardship schemes are now being used by industry and governments to bring the key players together

Figure 6.3 The 800–900 kg of waste produced annually by the average Australian family. (Photo courtesy of VISY Recycling)

104 SUSTAINABLE GARDENS

to understand and correct market failures in the life cycle of products and materials. Schemes are also being developed for tyres, televisions and computers. Many of Australia's landfill waste disposal sites do not incorporate measures for the collection and treatment of landfill gas. An estimated 80% of Australia's municipal solid waste is available for this purpose, representing a source of approximately 50 GJ of biogas annually excluding the total potential from existing landfill sites.

Despite some effort being made a large amount of Australia's waste ends up in landfill. The Productivity Commission (2006a) released its draft report on *Waste Generation and Resource Efficiency* in May 2006. The Report focuses solely on the downstream environmental impacts of waste disposal and does not investigate outcomes from implementation of more resource efficient practices or cleaner technologies.

Government agencies are now beginning to focus on the task at hand. In Victoria, for example, the agency Ecorecycle provides extensive information on sustainable waste management and waste minimisation including information sheets on the various components of our waste. Through its *Waste-wise* program it advises how businesses and government organisations can do an audit of their waste procedures and develop a 12-month action plan to measure performance.

Key waste statistics for Australia are given in Tables 6.2 to 6.7.

Community, household, individual
Changing behaviour

Historically, we have depended much more on organic materials than we do nowadays. Wood was a major component of the structures we used from day to day and when metal objects became more widely used they tended to be used for long periods before being melted and re-used. Only a few decades ago, partly due to the deprivations of war and the Depression, but also as part of a general philosophy of frugality, products were saved and re-used, or modified for other purposes. People were proud to be self-reliant. Socks were darned, clothes were passed on through the family, rusty nails were stored in the shed for when they might 'come in handy', vegetables and fruit were grown in the garden and excess preserved for leaner times. This was done, not so much to protect the environment, more to 'make ends meet'. It is only in recent times that 'disposable' goods have become accepted.

In very simple terms our environmental impact relates to our consumption level, which relates to our expenditure. We want to know that when we spend our hard-earned

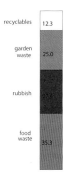

Figure 6.4 Garbage bin audit 2007 City of Whitehorse Victoria prior to the introduction of green waste collection service.[6]

Table 6.2 Solid waste composition by material type, Australia 2002–03

Component	Municipal %	Commercial and industrial %	Construction and demolition %
Organics (food/garden)	47	13	1
Paper	23	22	–
Plastics	4	6	–
Glass	7	2	–
Metals	5	22	7
Concrete	3	3	82
Timber	1	9	4
Other	12	24	6

Source: ABS (2006).[3]

money we are helping the environment (breaking the consumption/environmental damage link) while at the same time maintaining our quality of life.

This can be done by first making a general assessment of our lifestyle and consumption patterns and thinking about how our environmental impacts relate to life-choices: the forms of transport we use; the number of children we have; the kinds of food we eat; how we spend our money on hobbies and interests; the kind of housing we choose, and so on.

Then we can exercise our consumer choice when buying goods, being careful in the way we use and dispose of them.

Green purchasing

Our economy is based on a choice of competing products and, as often as not, the product that wins out is the one that costs less. Most of us are convenience shoppers, going to nearby shops and buying the first goods we see that do the job we require at a price we feel is acceptable. Consumers, through their purchases, send strong messages to producers, retailers, distributors and other shoppers. It may well be that the most important ethical decisions affecting our society and the environment are made unwittingly as we do our shopping.

Green purchasing means finding out about the sources of products, the critical

Table 6.3 Waste generation, selected indicators in Australia

	1996–97 tonnes	2002–03 tonnes	% change 1996–97 to 2002–03
Waste to landfill	21 220 500	17 423 000	-18
Waste recycled	1 528 000	14 959 000	879
Total waste generation	22 748 500	32 382 000	42
Waste to landfill/person	1.15	0.87	-24
Waste to landfill/M$ GDP	41.76	23.47	-44
Waste generation/person	1.23	1.62	32
Waste generation/M$ GDP	44.77	44.07	-2
Recycling/person	0.08	0.75	838
Recycling/M$ GDP	3	20.37	579

Source: DEH cited in ABS (2006).[3]

Table 6.4 Per capita waste generation and change over time in Australia

State	Waste/person (kg) 2002–03	Total recycled %
NSW	1820	48
Vic	1751	51
Qld	1046	31
WA	1804	23
SA	2248	63
ACT	2087	69
Average	1629	46
Increase 1996/7–2002/3 = 32%		

Source: ABS (2006).[3]

examination of the methods used in their production and distribution, and seeing if these measure up socially and environmentally. Environmentally friendly products use materials, work practices and processes that minimise negative environmental impacts. We cannot assume that all the goods in shops have been produced using environmentally and socially acceptable standards. Convenience foods are often heavily packaged (single-serve yoghurt, take-away foods, etc.).

Consumers can reasonably expect well-researched factual information on the consumer choices they make, but this is difficult to find. Green purchasing can be very effective at the institutional level when purchasing office equipment, vehicles, power, cleaning services and catering. We can also be guided by ecolabelling. An ecolabel is a label that indicates an overall environmentally preferred product or service, or an assessment of environmental performance, based on an independent third-party assessment. The third-party assessment is important as it distinguishes this kind of labelling from various 'green' symbols or claims made by the manufacturers or service providers themselves.

Already, in early 2007, Britain's biggest retailer and largest supermarket chain, Tesco, in association with the British Government and the UK's Carbon Trust, has committed to 'carbon labels' showing greenhouse emissions associated with its 70 000 products. The Carbon Trust, is a private company set up by Government in response to the threat of climate change, to accelerate the transition to a low carbon economy. The Carbon Trust works with UK business and the public sector to create practical business-focused solutions to explain, deliver, develop, create and finance low carbon enterprise. In the USA, corporations like PepsiCo and Wal-Mart are conducting inventories of their products' emissions.

Standardisation of methods for assessing emissions are also under investigation because of the many difficulties involved. Circumstances may vary significantly over the supply chain of a particular product and many factors can change rapidly. A carbon footprint methodology has been developed by the Carbon Trust. Products will display labels

Table 6.5 Major categories of waste, amount recycled and going to landfill in Australia

	Municipal	Commercial and industrial	Construction and demolition
Proportion of total waste 2003	27	29	42
Proportion to landfill 2003	70	56	43
% recycled (overall = 46%)	30	44	57

Source: ABS (2006).[3]

Table 6.6 Recycling rate of waste components in 2006[3]

	%	Material recycled
Highest	>70	Paper (92%), glass (90%), plastic bottles (90%), plastic bags (89%), metals (82%), concrete (74%)
High	>50	Drink packaging, vehicle batteries, cars, cables, roofing iron
Medium	20–50	Hot water systems, appliances, clothing, gas cylinders, flexible plastic freight packaging, bricks, roof tiles
Low	<20	Mobile phones and other e-waste, power tools, footwear, mattresses, n-cad batteries, grocery packaging, retail carry bags, tyres, asphalt, office fittings, paint, piping, window glass
None	0	Treated timber, fixed line phones, TVs, CDs, DVDs, toys, videos, personal batteries (some exporting)

showing the greenhouse gas emissions created by their production, transport and eventual disposal, similar to the calorie or salt content figures on food packaging.

In Australia, all state governments have endorsed the 1996 *National Government Waste Reduction and Purchasing Guidelines* which encourage the purchase of recycled products.

Recycling

Recycling is viewed as a procedure performed at the end of a product's life but it can begin with product design and use of materials that, instead of having built-in obsolescence, have built-in durability. This means that they are deliberately constructed for easy disassembly, upgrading, repairing, re-use and recycling. Currently, recycling is based on materials, like paper and glass, which can be easily separated and re-processed. However, separation and recycling remain uneconomical for many products so they end up in landfill, and this is encouraged because it is often cheaper to buy a new product than to get the old one fixed. Much remains to be done.

The Freecycle Network™ is made up of 4117 groups with 3 893 000 members across the globe. It is a grassroots and entirely non-profit movement of people who are exchanging goods for nothing in their own towns. It is all about re-use and keeping valuable materials out of landfills.

Chemicals

Chemicals are an important component of the materials category. We now live in an exceedingly complex environment of natural and synthetic chemicals. The trend since World War 2 has been towards a rapidly increasing number of synthetic chemicals, in increasing quantities, used and dispersed over ever-wider areas. It is, as usual, the scale of this activity involving so many people and such advanced technology that has transformed concerns about chemicals from local issues about spills, pollution and the like, to global environmental threats.

Table 6.7 Estimated e-waste[3]

	E-waste (total Australia)	
Item	Total number bought/yr (mill.)	Current millions
PCs	2.4	9
TVs	1	–
Printers	–	5
Scanners	–	2

We need a clear picture of how the range and quantities of natural and artificial chemicals is affecting the sustainability of the planet. Altogether about 10 million chemical compounds have been synthesised by humans. Of these, about 110 000 are listed as commercially available on the *European Inventory of Existing Commercial Substances* and 1–2000 new ones are added each year (see Figure 6.4).

The momentum of new chemical production has gathered over the last few decades, and global production of chemicals has increased from about 1 million tonnes in 1930 to about 400 million tonnes in 2001.

The amounts of life-important chemicals that are part of the Earth's major geochemical cycles are being altered by human activity. Especially important here are water, carbon, nitrogen and phosphorus as they cycle through living and dead matter and between the land, atmosphere and oceans. Water is vital to all living organisms; carbon is an important element in organic matter and closely linked to plant photosynthesis and the carbon dioxide of fossil fuels which is related to climate change; nitrogen and phosphorus are valuable plant nutrients that can be critical in limiting or encouraging growth and are therefore of particular significance for agriculture. Living organisms are greatly influenced by their chemical environment. In the soil there are natural minerals, nutrients, moisture and organic matter while the atmosphere has carbon dioxide, nitrogen and oxygen. Changes in the balance and concentrations of these chemicals can have dramatic effects on the Earth's life-support systems.

One example of human impact on biogeochemical cycles is the effects of our demand for nitrogenous fertilisers for food crops. Artificially synthesised nitrogen compounds for fertiliser have been produced in such quantity that the amount of nitrogen available for biological uptake has doubled since 1940, and human nitrogen production now contributes more to the global supply of biologically usable nitrogen than natural processes (see Info Box 6.1). Global fertiliser consumption has increased substantially since 1950, while the world's population has grown from 2.5 to 6 billion.

Fixed nitrogen from human activity is created by agriculture, energy production and transport with agriculture accounting for about 86% of human-generated nitrogen. Inorganic fertilisers bypass all the microbial activity needed to break down organic matter into a form appropriate for the plant, instead feeding them directly with the final product. So organic composts feed the soil (micro-organisms), inorganic fertilisers feed the plants. (See Figure 6.5.)

Negative consequences of increased nitrogen availability include: increased atmospheric nitrous oxide and nitrogen components of acid rain and smog; acidified soils with locked up calcium and potassium essential for long-term soil fertility; acidified water with local algal blooms, probably declines in fisheries, and losses in plant and animal biodiversity. In short, detrimental changes to the ecological functioning of water and land systems and the living communities they support.

Phosphorus does not leach as readily as nitrate because it is more tightly bound to soil

> ## INFO BOX 6.1: NITROGEN FERTILISER
>
> The major reservoir of nitrogen on Earth is pure nitrogen gas in the atmosphere (which is 79% N) but organisms cannot tap this vast reservoir of free nitrogen until it has been 'fixed' (changed) to a form such as nitrate, ammonia or urea.
>
> In nature, nitrogen cycles through the biosphere in four basic processes:
>
> 1. *Nitrogen fixation* – the combination of nitrogen with other atoms which requires considerable energy derived by three methods (*atmospheric fixation* by lightning (5–8%); *biological fixation* by microbes/bacteria firstly as ammonia quickly converted to protein etc; *industrial fixation* to form ammonia (then used in various forms as an agricultural fertiliser).
>
> 2. *Decay* – nitrogen returned to the environment from organisms excretion and decomposition, micro-organisms converting this matter into ammonia.
>
> 3. *Nitrification* – ammonia produced by decay is converted to nitrates (available to plants) by nitrifying bacteria, first, a transformation to nitrites by *Nitrosomonas*, then to nitrates by *Nitrobacter*. Processes 1–3 remove N from the atmosphere. N is replaced by denitrification.
>
> 4. *Denitrification* – is anaerobic (oxygen depleted). Bacteria reduce nitrates to N, thus replenishing the N gas in the atmosphere.

particles. However, it is carried with eroded soils into surface water bodies, where it may cause excessive growth of aquatic plants. In extreme cases, lakes and reservoirs become choked with decaying mats of algae which have offensive smells and cause fish kills from the resulting lack of dissolved oxygen. Potassium does not cause water quality problems because it is not hazardous in drinking water and is not a limiting nutrient for growth of aquatic plants.

A study in the *New Scientist* made crude estimates of the known reserves of a variety of minerals remaining for exploration and how long it would take for them to run out assuming that each were used at just half the rate that they are currently used by an average US resident in 2007, also assuming no recycling (Table 6.1).

Australia

The National Chemicals Taskforce concluded, in 2002, that while Australia has made significant progress towards ecologically sustainable chemicals management, more work is needed.[8]

The Australian National Pollutant Inventory is a publicly accessible database containing information on emissions of 90 substances from more than 3800 facilities around the country. Data for 2005–06 on emissions to Australia's air, land and water during 2005–06 showed that 47 of the 90 NPI substances reported by industrial facilities decreased compared to the previous year.

In addressing the problems posed by fertiliser use in Australia, environmental precautions

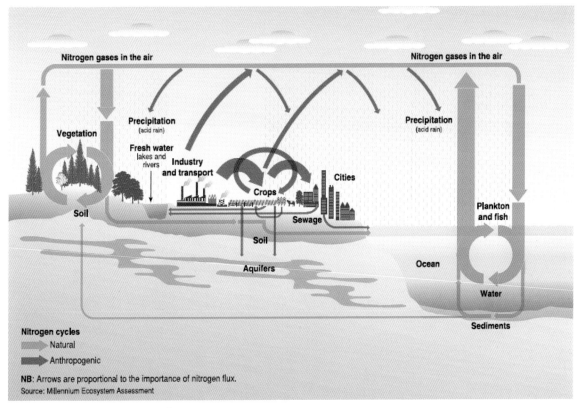

Figure 6.5 Global nitrogen cycle.[7]

are gradually being put in place. National Land and Water Resources Audits assess the loss of nutrients from farming systems. Farming industries now have a Nutrient Management Code of Practice. In 2002 the Fertiliser Industry Federation of Australia Inc. (FIFA) endorsed an eco-efficiency agreement with Environment Australia. This was in line with the World Business Council for Sustainable Development's aim at achieving eco-efficiency.

Waste is still mostly a state responsibility but change to procedures, as with e-waste, often requires a cooperative effort involving at least all state governments and manufacturers.

Garden

The garden products we use, like all products, have sourcing, production, energy, water and waste costs. In most cases, at present, we have no idea what these are. The environmental and social costs of goods that are manufactured overseas are almost impossible to obtain.

Our aim is to reduce, re-use and recycle, targeting zero waste, minimising the tendency for a linear flow of materials and chemicals into and out of the garden and maximising on-site cycling (see Figure 6.6).

- Only use products with low resource intensity.

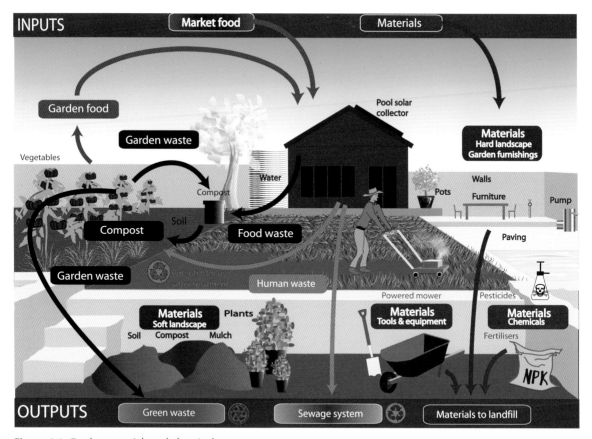

Figure 6.6 Garden materials and chemicals.

- Recycle as much as possible – this includes both organic and inorganic materials.
- Minimise use of pesticides, fertilisers and other synthetic chemicals.
- Link to the house by composting as much household organic waste as possible.
- Use as much green waste as possible on-site.

FOOD

*Food security exists when all people, at all times,
have physical and economic access to sufficient,
safe and nutritious food to meet their dietary needs and
food preferences for an active and healthy life.[1]*

> **KEY POINTS**
>
> Agriculture occupies about 40% of the global land surface and consumes about 20% of its plant productivity.
>
> The environmental demands of agriculture are a major driver of biodiversity loss and land degradation.
>
> Food security can be threatened by climate change, water shortages, diminishing grain stockpiles, increasing costs of production, and use of arable land for biofuels.
>
> A low meat, low dairy diet together with a strong community and home garden based urban agriculture can make major sustainability savings.

In populated areas, a plane flight on a clear day will most likely reveal below you a neat mosaic of cleared agricultural land. This is the vast 'vegetable patch, orchard and livestock pasture' that is supplying food to our cities: the urban back garden (Figure 7.1). It is also land that once supported natural plant and animal communities. Almost 40% of the Earth's land surface is now used for agriculture and in Australia, with its large area of cattle station pasture, this figure is 57% (see Table 7.1).

Global net primary production (NPP) is the rate at which all the plants in the world store energy as biomass; a measure of the total plant matter produced over a particular time period. Estimates of global NPP are becoming more precise through the use of satellite photography and other methods. A recent study has shown that about 20% of the total vegetation produced on Earth is for human use (crops, pasture, timber, etc.). Tables 7.2 and 7.3 give a breakdown of this human land

Figure 7.1 Land for food: a vegie patch for the city. (Photo: Gregory Heath, CSIRO Land and Water)

use. In terms of both land space and primary production, a large proportion of the planet is now dedicated to plants in our service.

Globally there are about 3000 known food plants of which about 150 have been extensively cultivated and traded. In spite of this apparent variety about 90% of the human diet consists of only about 15 species and, of these, only four (wheat, rice, corn and potatoes) make up over 60% of the world's food supply. We now rarely eat wild food, possibly a few berries, greens and field mushrooms. Regional agriculture evolved based on local plants. In East Asia this was rice, in the Middle East, wheat and barley, and in Central and South America, maize. All staple foods were domesticated in prehistoric times and the full range of cereals, vegetables and fruits have been altered from wild ancestors.

Our basic need for food is the single most environmentally demanding aspect of our lives. A flourishing global agricultural industry is critical for food security but it is also closely linked to a range of environmental difficulties, world health, water supply and poverty; it is also highly sensitive to the effects of climate change. Food security is dependent on water security, especially in areas where people depend on local agriculture. In these areas it is common to find erratic rainfall and seasonal differences in water availability that produce temporary food shortages (although it is floods and droughts that cause most food emergencies).

We have seen that agriculture soaks up about 70% (to 95% in non-industrialised countries) of total human water use, diverting huge volumes away from natural systems, stretching planetary reserves of freshwater to their limits.

In the 20th century, agriculture in the industrialised world has passed through three major phases: 1900–1920 rapid clearing of land for food; 1920–1970 use of technology and fossil fuels to vastly increase production; 1970 on, diminishing returns in the face of increasing population, gene technology, concern over peak oil and food shortages.

Global

Many of agriculture's environmental impacts are global in scale and include: water pollution, species extinction through land

Table 7.1 Land use and agriculture in 2003

Area	World (1000 ha)	Australia (1000 ha)
Total land area	13 004 397	768 230
Agriculture	4 973 406 (38.24%)	439 500 (57.2%)
Arable land	1 402 317 (10.78%)	47 600 (6.20%)
Permanent pasture	3 432 834 (26.4%)	391 565 (51%)
Permanent crops	138 255 (1.06%)	335 (0.04%)

Source: FAOSTAT.[2]

Table 7.2 Human use of land net primary production for selected regions (HA = human appropriation)

Region	Area million km²	Popn millions	Per cap. (tonnes per capita NPP)	Total NPP (pg)	HA % NPP
Africa	31.1	742	2.08	12.50	12.40
E Asia	11.9	1400	1.37	3.02	63.25
SC Asia	10.9	1360	1.21	2.04	80.39
W Europe	1.20	181	2.86	0.72	72.22
N America	19.7	293	5.40	6.67	23.69
S America	18.4	316	3.11	16.10	6.09

Source: Imhoff et al. (2004).[3]

clearance, alteration of global biochemical cycles especially nitrogen (used in fertilisers), soil erosion, salinisation, soil degradation, livestock methane production and more.

Although the 20th century has seen a successful expansion of food production the FAO estimates that by 2030 about 50% more food will be needed than was produced in 1998. The resource and environmental impacts of food production are increasingly reflected in shop prices and it looks as though costs will increase for some time due to structural changes in the world economy that are the result of a decrease in stockpiles, the demands of a new middle-class in China and India, competition for land from biofuels, damage by extreme weather events, increasing production costs (rising energy costs flowing through to cost of fodder, chemicals, fertiliser etc.), even speculative hoarding.

Nations affected by food shortage include India, Indonesia, Philippines, Madagascar, Thailand, Vietnam, Haiti, Bangladesh, Pakistan and African nations. Although poor nations are often the first affected by food insecurity, in many African countries millions live as subsistence farmers so global food prices have no meaning; here people do not buy food, they simply eat what they grow.

Table 7.3 Annual global human appropriation of terrestrial primary production in billion tonnes of carbon

Product	Low estimate	Medium estimate	High estimate
Vegetable	0.90	1.73	2.95
Meat	1.69	1.92	2.22
Milk	0.15	0.27	0.43
Eggs	0.09	0.17	0.26
Food subtotal	2.83	4.09	5.86
Paper	0.20	0.28	0.38
Fibre	0.32	0.36	0.42
Wood (fuel)	2.68	4.31	4.71
Wood (construction)	1.97	2.50	3.44
Wood/fibre subtotal	5.17	7.45	8.95
Grand total	8.00	11.54	14.81
% annual NPP (56.8)	14.10	20.32	26.07

Source: Imhoff et al. (2004).[3]

The total number of chronically undernourished people in the world is estimated by the FAO at 854 million in the period 2001–2003, of which about 820 million live in developing countries, 25 million in transitional countries and 9 million in developed market economies. Of the undernourished, 61% live in Asia and the Pacific and 24% in sub-Saharan Africa.[4]

Although food aid has saved millions of lives it has also been criticised as creating dependency of recipients while undermining local agricultural producers and traders upon whom sustainable food security depends. Although emergency food aid is essential, the timing and targeting of other forms of food aid needs careful management.

Food production must be based on sustainable practices, and this enterprise is already well underway. The Independent Science Panel (ISP) and the Institute of Science in Society have launched a *Sustainable World* initiative to encourage sustainable food production systems. A special ISP group for Sustainable Agriculture is compiling a comprehensive report including a series of recommendations for government and inter-governmental agencies on the social, economic and political policy, and the structural changes needed to implement a sustainable food production system with regard especially to the oil and water demands of industrial agriculture.

As we rely on relatively few plants for food, pest and disease threats to crops growing as monocultures can have drastic consequences. In natural ecosystems the variety of plants provides a natural resistance. The environment of an agro-ecosystem has been adjusted to favour the crop by manipulating competition, allowing favoured species (the crops) to thrive, and removing species which might otherwise crowd them out (weeds), or threaten them in any way (pests and diseases). This is a very simple ecosystem with a maximum of three trophic (feeding) levels: producers (crops), primary consumers (livestock, humans) and secondary consumers (humans). Relatively little energy is lost between these trophic levels compared with natural systems but there is a large energy input to keep it this way.

Plants need inorganic nutrients to grow, but in cultivated land these nutrients are harvested along with the crop and taken away from the site to end up as human waste passing into rivers and the ocean. With removal of nutrients, crops need chemical fertilisers and these are expensive in embodied fossil fuel energy through mining, transportation, run-off, etc. Most terrestrial herbaceous ecosystems are nutrient-limited, firstly by nitrogen, often followed by phosphorus. In nature, varying a major limiting factor such as nutrient levels can change both the species composition and structure of a plant community.

Soil cultivation alone may lead to nutrient leaching, loss of organic matter and breaking down of soil structure. The fossil fuel cost is high. It has been estimated that to double the yield of crops requires roughly a ten-fold increase in the use of fertilisers and chemicals.

The chapter on the origins of sustainable horticulture described some of the links between agriculture and horticulture, and the significant contribution to sustainability and food security that can be made in our parks and gardens.

In many areas of the world, the amount of arable land remaining for agriculture is very small. Also, when settlers established the first towns they did so mostly in fertile areas. It is ironic that, as time passed, this prime agricultural land has been covered with concrete and bitumen as the towns expanded.

Table 7.4 shows the impacts for various regions of the world and it can be seen how wealthy industrialised regions consume more per capita while highly populated regions, though consuming less per capita, nevertheless use a far greater proportion of their net primary productivity.

In 2005 total global food exports, valued at about US$612 billion, increased by about 5% over the previous year. Over 66% consisted of transformed or processed products, mostly alcoholic drinks (~11%), fresh, chilled or frozen meats (~9%) and animal foodstuffs (~7%). Generally the world's largest food exporters (by value) are also the world's largest importers: the US, Japan and high-income industrialised countries of Western Europe.

Food production and climate change

Rising temperatures due to climate change can affect agriculture in many ways: through rising sea levels inundating coastal areas, but also by changing optimal crop growing areas through altered rainfall and evapotranspiration conditions, as well as a redistribution of pests and diseases. Poorest countries are generally expected to be the most affected. Redistribution of rainfall patterns is likely to result in decreased rainfall in lower latitudes, and increases in rainfall in high latitudes so the most productive crop zones will move pole-wards to higher elevations with a redistribution of food capacity that is likely to impact on sub-Saharan African nations.

Food production, because it now makes up such a large part of global primary productivity, has the capacity to influence

Table 7.4 Food security and nutrition

Region	No. undernourished (M)		Proportion undernourished in total population (%)		Dietary energy supply		
	1990–1992	2001–2003	1990–1992	2001–2003	1990–1992 (kcal/person/day)	2001–2003 (kcal/person/day)	Av. annual % increase 1990/1992 to 2001/2003
World					2640	2790	0.50
Developing countries	823.1	820.2	20	17	2520	2660	0.49
Australia					3170	3120	-0.14
UK					3270	3440	0.46
USA					3500	3770	0.68
India	214.8	212	25	20	2370	2440	0.26
China	193.6	150	16	12	2710	2940	0.74

Source: FAO (2006).[5]

CO_2 levels and climate change. A recent report suggests that organic, sustainable agriculture that localises food systems has the potential to mitigate nearly 30% of global greenhouse gas emissions and save a sixth of global energy use.[6]

Australian context

The Murray–Darling Basin is known as the Australian food-basket because it produces 41% of the gross value of the country's agricultural production and occupies 71% of the nation's total area of irrigated crops (Table 7.5).

Australia's most important food and agricultural commodities are shown in Table 7.6.

Table 7.7 shows the high levels of ruminant emissions as a component of overall agricultural emissions. Methane is the second most significant greenhouse gas and cows are one of the greatest methane emitters. Their grassy diet causes them to produce methane, which they exhale with every breath.

The food and grocery products industry is Australia's largest manufacturing sector, employing more than 200 000 people and contributing 2.5% of gross domestic product. The Australian Food and Grocery Council (AFGC) is the national body representing the nation's food and grocery products manufacturers. The Council's *Environment Report 2003* flagged the need for improved processing packaging and recycling although this has a relatively minor impact. It estimated that Australians waste 2.2 million tonnes of food per year. Primary production is the most water-intensive process in the supply chain followed by consumption in the home. Emissions are fairly evenly spread across the production and consumption chain. The rate of waste generation increases towards the consumer end of the supply chain. The industry is improving its reporting on key environmental performance indicators including water and energy intensities per kilo of product as well as waste generation and recovery.

Figure 7.2 illustrates the Australian food value chain for 2006, showing the dollar value of basic food materials and change in value as they pass through the wholesale and retail sectors of the industry. It also shows the dollar value of the various food exports and imports.

Table 7.5 Murray–Darling basic statistics: Australia's vegetable garden, orchard and pasture

Murray–Darling Basin	
Total area	1 061 469 km²
Extent	1250 km E-W, 1365 km N-S
Proportions of state land	75% NSW, 60% Vic, 100% ACT, 7% SA
Percentage of total national land area	14%
Average water flow into the basin	14 000 Gl
Percentage of total water for agriculture	95%
Percentage of population within its bounds	11%
Amount of food produced	41% gross value of agricultural production
Area of basin under irrigated crops and pasture	1 472 241 ha 71% nation's total area of irrigated crops
Length of rivers	Darling – 2740 km Murray – 2530 km Murrumbidgee – 1690 km

Source: Erlich R, Wahlqvist M & Riddell R (2007).[7]

Table 7.6 The 20 most important food and agricultural commodities (ranked by quantity produced) in Australia for 2003

Rank	Commodity	Production (mt)
1	Sugar cane	36 892 000
2	Wheat	22 500 000
3	Cow milk, whole, fresh	10 377 000
4	Barley	7 792 000
5	Australian-produced cattle meat	2 130 000
6	Sorghum	1 900 000
7	Grapes	1 800 000
8	Rapeseed	1 549 000
9	Oats	1 408 000
10	Potatoes	1 200 000
11	Lupins	996 000
12	Cottonseed	693 000
13	Australian chicken meat	680 740
14	Australian sheep meat	647 000
15	Triticale	641 000
16	Rice, paddy	535 000
17	Wool, greasy	520 000
18	Oranges	490 000
19	Cotton lint	490 000
20	Peas, dry	466 000

Source: FAOSTAT database.

Table 7.7 Australian agricultural sector emissions in 2003

Source	Mt CO_2-e
Enteric fermentation (ruminant emissions)	62.7
Agricultural soils (disturbance, cropping, improved pasture, addition of fertilisers and manures)	18.7
Savanna burning (prescribed)	11.8
Manure management (emissions when stored)	3.3
Rice cultivation (methane from flood fields)	0.4
Field residue burning (sugar cane and cereal stubble)	0.3
Total	97.3

Source: National Greenhouse Gas Inventory (2003).

Horticulture for Tomorrow (H4T) is a program that aims to help growers link production targets to their care for the environment as an integral part of their daily business management. An introductory guide to environmental management systems isolates the following key issues:

- Soil/growing media.
- Water – increasing efficiency through irrigation scheduling and appropriate irrigation systems.
- Nutrition – feeding by calculation of nutrient requirement.
- Pest, disease and weed management – careful management of chemicals.
- Air quality – care with spray, dust, odours.
- Smoke.
- Energy – care in selection and management of power machinery and fuel use.
- Waste management – care with containers, paper, plastics, steel and timber (re-use, return, recycle).

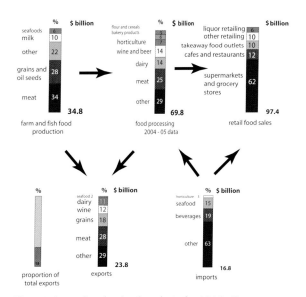

Fig. 7.2 Australian food value chain for 2006. (Source: Australian Food Statistics 2006)

7 – FOOD 119

> **INFO BOX 7.1: CURRENT MAJOR TRENDS AND CHALLENGES IN FOOD PRODUCTION THAT AFFECT SUSTAINABILITY**
>
> **Trends**
> - Increased demand for convenience shopping.
> - People shopping more frequently.
> - More people eating away from home and 'on the go'.
> - Lower proportion of income spent on food.
> - A desire to know the story of the food.
> - Greater emphasis on general health, especially obesity.
> - Indifference to local origin/source of food in relation to low price imported goods.
> - Preference for reduced meal preparation time.
> - Greater use of snacking products and 'ready meals'.
>
> **Sustainability challenges**
> - Apply sustainability principles across the production phase.
> - Emissions are fairly evenly spread across the production and consumption chain.
> - The rate of waste generation increases towards the consumer end of the supply chain.
> - Plan for a changing climate.
> - Manage supply and cost of water more carefully.
> - Manage competition from quality imported processed foods.
> - Maintain disease-free status.
> - Improve processing, packaging and recycling.
> - Encourage reduced waste (Australians waste 2.2 million tonnes of food per year).
> - Improve reporting on key environmental performance indicators including water and energy intensities per kilo of product as well as waste generation and recovery.

- Biodiversity – protecting remnant vegetation and revegetating areas unsuitable for production. Encourage biodiversity along crop lines, fence lines, roadsides, etc.

'Enviroveg' is a simple set of tools for growers to assist them to achieve, measure and demonstrate good environmental practices on farm. The program is grower-owned and developed, and outlines the principles and practices required for maintaining or improving environmental management. Using the Enviroveg program gives growers a way of measuring environmental improvements each year and recognition for improvements. Practices can be introduced and managed at a pace to suit the grower and their circumstances.

Food and energy

Simple traditional farming methods (without modern technology and fossil fuels) would have produced a solar energy surplus by yielding more energy in the crops produced than that used by the human and animal labour required in their production. As a plant grows and metabolises in the wild, there is a net energy gain because the energy stored during photosynthesis exceeds that lost by the plant's respiration. In agriculture, yields are high but this is because there has been a large fossil fuel subsidy from fertilisers, chemicals to control pests and diseases, irrigation systems, use of machinery for cultivation, the 'food miles' of distribution, and so on. This means that overall, there is net energy loss. This is one reason why in the non-industrialised world, the vast majority of people are still employed in agriculture. There is also a substantially higher use of energy in the production of meat than in crops, of greenhouse-produced crops relative to open-air cultivation, and of canned, frozen or processed food produce relative to fresh.

All these considerations lend weight to the idea of growing more vegetables in the home garden as a major means of reducing energy use.

Nowadays the energy needed to provide the food on our tables is about nine times greater than the calorific value of the food itself. This has been achieved through the use of cheap emission-producing fossil fuels.

Food miles

The expression 'food miles' came into use in 2006 and has already become the accepted international term referring to the distance travelled by food products between production and consumption. With less costly shipping and trade globalisation, large companies are now able to source the cheapest food on the international market. Food now travels further than ever before and therefore requires longer storage (often in refrigerated trucks and cargo holds), more fossil fuels, and often several layers of packaging which is sometimes carried out at sites remote from production (Table 7.8). In the US, the average food item travels between 2500 and 4000 km.[8]

Australia's first formal study of food miles was done in Melbourne and carried out by the organisation CERES as the basis for a new Food Education Program. The work, published in July 2007 (see Table 7.9), found that for a 'typical' food basket of 'commonly available and popular foods', road travel of the component food items was 23 073 km, while the total for all transport was 70 803 km. As the study points out, ideally food miles are only one aspect of total product LCA, and it is possible for products transported over long distances to be more energy efficient than the same product produced locally. However, without baseline data informed decisions cannot be made.

Perishables can be given extended lives by processing, packaging, canning, refrigerated storage, the use of additives and preservatives, controlled ripening and so on. The collective effect of packaging, processing and preservation together with long-distance travel is to sanitise food, distancing us literally and metaphorically from the fields, orchards and pastures where it was produced. It is

Table 7.8 'Food miles' for a Melbourne meal in May

Food	Origin	Distance (km)
Pre-cooked potato chips	A third mostly from NZ, but also Canada and US	Up to 15 100
Zucchini	Bundaberg or Gympie	1600
Frozen peas	NZ, Belgium	Up to 16 700
Fresh peas	S Qld	1600
Carrots	WA, Tas, Qld	Up to 3000
Broccoli	Orbost (some from Lockyer Valley SE Qld)	Up to 1300

Source: *The Sunday Age*, 7 May 2006 (P Weekes).

Table 7.9 Food kilometres* and emissions for Melbourne

Product	Distance travelled (km)	Emissions (g)
White bread	486	107
Chips	2024	446
Savoury biscuits	1802	397
Apples	112	25
Oranges (Australia)	567	125
Orange juice	2024	446
Pumpkin	361	80
Full-cream milk	348	77
White sugar	2315	510
Cheese	688	152
Jasmine rice (Thai)	9709	n/a
Onions	782	172
Beef	298	66
Carrots	311	69
Tomatoes	1618	357
Potatoes	155	34
Bananas	2746	605

* From point of production to Melbourne CBD – for a 'typical' food basket averaged over a year.
Source: Gaballa S & Abraham AB (2007).[9]

sometimes pointed out that long-distance transport has the potential to spread pests and diseases long distances too.

A long-distance centralised food distribution system is vulnerable to disruptions such as strikes, international incidents, transport problems, application of international trade agreements, changes in international market conditions, currency fluctuations and so on. There is also some uncertainty and concern about the standards applied in other countries in relation to hygiene, use of chemicals, and social justice – or the degree of choice we have in relation to the consumption of genetically modified food. The demand by large retailers for 'commercially desirable' crop varieties has in some cases resulted in a reduction of choice of local genetic diversity in favour of standardisation. The most profitable markets are often in highly populated accessible areas so poorer more isolated communities may be poorly served. There is an unbridgeable gap between grower and eater.

Of course care is needed in making assertions about the environmental impact of food transport. For instance, it could be that a typical shopping trip by car uses more energy per kilo of food than transporting the same average shopping load as part of a large cargo over a long sea journey. The growing of tomatoes under cover in a cold Australian climate may be much more energy intensive than importing them from a warmer region. However, short-haul air transport is claimed as 150 times more energy intensive than bulk sea transport so perhaps further technological development is needed to maximise the use of sea freight. Clearly there is some detailed work here for sustainability accountants.

Household and individual context
Diet

Humans are natural omnivores with canine teeth for tearing meat. Nevertheless increasing numbers of people are choosing to eat less (or no) meat and dairy products. Some would argue that this is more healthy, decreasing the likelihood of obesity, some cancers and other complaints. It also assuages concerns about animal welfare, the tens of billions of sentient animals that are slaughtered annually, many inhumanely.

But there are also the environmental gains that result from eating less meat. Intensive agriculture makes numerous demands on the environment apart from the land itself, all of which can contribute to ecological problems. Eating lower on the food chain undoubtedly helps the Earth as it reduces the use of the world's limited resources. But if we eat less meat, will we need more land for crops? Yes, but much of the world's cultivated land is used as pasture for livestock. In fact, about 40% of the world's cereal crop yield is used to feed animals that are bred for our food (70% in USA). In a world where the large-scale poverty and starvation of the Third World exists side-by-side with the obesity and unhealthy excess of the rest there is a clear need to establish healthy eating habits that have minimal environmental impact.

The World Health Organization (WHO) has published a Global Strategy on *Diet, Physical Activity and Health* which was endorsed by the May 2004 World Health Assembly. It addresses the significant change in diet and physical activity levels that have occurred worldwide as a result of industrialisation, urbanisation, economic development and increasing food market globalisation.

A plant-based Mediterranean-style diet has been suggested as a means of meeting both health and environment expectations and a sustainable diet (defined as one which is protective of health as well as having relatively low environmental impact). Here a model is presented that identifies patterns of production and trade and quantifies physical inputs of land, water and energy as well as the costs and prices associated with different assumptions about diets, production techniques and policies. This is a valuable tool as meals with similar nutritional characteristics may have different environmental impacts.

The Mediterranean diet is associated with health and longevity and is low in meat, rich in fruits and vegetables, low in added sugar and limited salt, and low in saturated fatty acids; the traditional source of fat in the Mediterranean is olive oil, rich in monounsaturated fat. The healthy rice-based Japanese diet is high in carbohydrates and low in fat. Both diets are low in meat and saturated fats and high in legumes and other vegetables, and they are associated with a low incidence of many of the ailments common in industrialised countries.

A diet of fresh, unprocessed, locally produced fruit and vegetables with no or low meat is not only more healthy for people but also for the environment than a more resource-hungry and unhealthy meat-based diet with processed foods rich in sugar and salt.

At least 8% of Australia's food goes to waste.[10] Apart from the wasted embodied resources this represents, food waste can be used to recover biogas used as fuel for cogeneration engines which produce electricity. This electricity can be sold into the grid and classified as 'Renewable Energy'. The anaerobic digestion process also produces a high nutrient (NPK) fertiliser sludge, which can be dried and granulated for sale into the agriculture and horticulture market as organic fertiliser.

Community context
Food shopping
A few decades ago fruit and vegetables grown at nearby farms and market gardens

were bought from the local greengrocer. Nowadays many people buy their produce from a supermarket. Critics point to some of the perceived deficiencies of this system. At one time the local farmer, like the greengrocer, was an independent business unit. Now he is often beholden to the demands of a supermarket chain. Of course this can be lucrative, but if he is not one of those chosen as a supplier, then prospects may be limited. Processing has been moved to large concerns and the patenting of plant varieties and products can exclude people from the benefits of new technology and lock them into unwelcome dependencies. Food characteristics can be chosen that suit the industry rather than the customer. And the local greengrocer now has to compete with the marketing power of international food branding.

Supermarkets world-wide look similar, not only because of their shopping model but also because of the products that are sourced globally. A mass market demands products that look good and allow for transport time so they must survive long, often refrigerated, storage. The global trend is towards a culinary uniformity and a mass production which does not explore the variety and flavours of local produce and culinary traditions.

To ease the many environmental pressures from agriculture, including the perceived negative side of agribusiness, there has emerged a movement towards self-reliance, food self-sufficiency (with potentially vast environmental benefits), chemical-free, healthy, locally produced seasonal food, together with a greater awareness of the environmental demands of food production and sustainable agriculture.[1] This has taken many forms.

Urban agriculture

As cities are often built on top-grade arable land it is appropriate that they take on whatever 'farming' they can.

Urban agriculture is a convenient term for a whole host of new and reinvigorated urban movements including: vegetable allotments, community gardens, farmers markets, conversion of amenity landscapes to food systems, the extended home-garden vegetable patch, roof-top gardens, mushroom farms, wetlands for food production, slow food, organic and biodynamic food, hydroponics, farmers cooperatives, local food policy councils, fresh-food-to-schools programs, and the linking of local food sources with restaurants, hotels, cafeterias, caterers and the like.

Advantages of this approach include: ruralisation of suburban landscapes; seasonal healthy eating; rural resettlement; and local control of production methods with money paid to local people and the local economy. Greater community involvement in food production means healthy light exercise where you can socialise and develop a sense of community – a form of horticultural therapy. Small-scale methods cannot compete with the economy of scale of mass-production, but the additional costs of small-scale methods are offset by not having to meet the expense of storage, transport, processing and marketing.

It has been suggested that the present-day disconnection between ourselves, food, nature and land might account for our apparent environmental indifference.

Proponents encourage 'ethical eating', an acknowledgement that our consumption patterns have social and environmental consequences.

The desire is for local food that has travelled little distance, has a short shelf-life, is tastier, especially as it has been harvested at maturity, and has not needed to be fumigated, refrigerated, processed or packaged. There is also the wish for much greater local and individual control over what we eat. The 'Slow Food' movement espouses many of these goals. Our shopping habits are an extremely powerful way of encouraging change. Supermarkets and other food shops can be encouraged to stock more organic foods, some local produce and so on. The ideal for the 'locavore' is to deliberately try not to export food until local demand is met, and conversely to not import food that can be readily produced locally.

Over the last decade, and over much of the Western world, small companies have set up dealing in heirloom seeds and old flavoursome plant varieties.

Another alternative is the food co-op, an example being Alfalfa House in Sydney's inner west and the Friends of the Earth Co-op in Collingwood, Melbourne. These are not-for-profit organisations selling organic food that is 'ethically produced'. Packaging is minimal and food is sold in bulk from large containers. Shares are bought for a nominal sum.

Community gardens are thriving. They pride themselves in the polyculture that distinguishes their production approach from the monocultures of industrial agriculture, and they represent perhaps a more strategic approach to the provision of nutrition for the surrounding neighbourhood along with surveys of vacant land and ideas like backyard sharing and the planting of hardy fruit trees in nature strips and in public parks.

It may be possible to establish neighbourhood dairy and chicken farms; reduce restrictions on animals in homes; encourage barter; dismantle intensive livestock farming; have tax breaks for self-reliance and certification for embodied energy and water.

INFO BOX 7.2: SOME STEPS TO IMPROVE FOOD SUSTAINABILITY

- Find out, as far as possible, where your food comes from – country, location, farmer: this should be public knowledge as it relates strongly to social, environmental and other issues.
- Become a 'locavore', eating fresh local and seasonal food produced within a radius of about 150 km of your house (the '100-mile diet') as much as possible.
- Grow your own food.
- Eat less generally, and eat less (or no) meat and dairy products; explore vegetarian recipes, products and meat alternatives.
- Buy organically certified products when possible.
- Buy products from companies that are open about their production methods and have clearly tried to reduce carbon emissions from travel and packaging.
- Buy Fairtrade.
- Buy free-range eggs.
- Avoid processed foods and those that need refrigeration and storage.
- Avoid wastage and compost all organics.

INFO BOX 7.3: SOME SUGGESTIONS FOR SUSTAINABLE HOME FOOD PRODUCTION

- Consider mixing your ornamental plants with vegetables like beans, rhubarb and silver beet.
- Grow from your own saved seed where possible and therefore save the water, energy, materials and chemicals needed to raise seedlings at another site.
- Use crop rotation.
- Grow crops that provide high yield over a long period.
- Plan for consecutive harvests with ground never fallow in the growing season.
- Mix root, leaf and fruit crops.
- Above-ground vegetables such as cabbage, broccoli and cauliflower will need chemical treatment to avoid attack by pests but root vegetables like parsnips, carrots and potatoes are easier to grow.
- Blanche excess vegetables and freeze them on a tray before putting them in a plastic bag and into the freezer where they can last for up to a year.
- Mulch in summer to save water and stop weeds.
- Know your site and climate to plan sowing and harvesting dates that allow for two crops and minimise shading by larger plants – this can be calculated at first by season.
- August – sow (spinach, lettuce, beetroot, peas, potato, garlic).
- September – leeks, garlic.
- Late October – first crop of spinach, carrots, beetroot (cropped for several weeks).
- November – onions.
- January – second crop.
- Record on a calendar sowings and harvesting periods noting which seem to work best and how late and early plants can be sown or planted out.
- The same vegetable can be sown two to three times 2 weeks apart to allow for extended cropping.
- High-yielding crops include: silver beet, cherry tomatoes, potatoes, carrots, onions, choko, zucchini. Long-lived perennials include: citrus, guava, feijoa.
- Sow crops at intervals to extend the cropping season as much as possible.
- Grow hardy productive fruits like lemon, grapefruit, mandarin, plum, figs.
- Set up a simple 'cage' or net to protect fruits like currants, raspberries and strawberries.
- Consider the use of long-cropping, high flavour heirloom cultivars like Tomato 'Tommy Toe' and Strawberry 'Cambridge Rival' and 'Chandler'.

Garden context

The more food we can grow locally or in our home gardens the greater the environmental benefits. Horticulture can play a vital role here. This is done in two main ways: firstly, by relieving agricultural land of the space needed to grow a proportion of our food supply; secondly, by removing the many resource costs related to agribusiness. If we water the garden from a home rainwater tank then the water cost is zero, and the additional costs of maintenance using organic and/or permaculture principles are minimal compared to the environmental costs of the agribusiness supply chain.

BIODIVERSITY AND ECOLOGY

*Natural capital – the ecosystem services that must be maintained
so that human development can be sustainable.*

ANON.

Although energy, water, nutrients and materials are all important parts of the garden ecosystem, it is perhaps the way that gardens interact with the surrounding environment of living organisms that first comes to mind when we think of garden ecology. We will be looking at gardens for wildlife in later chapters, but here we want to consider why, in general, it is important to protect habitats and biodiversity, and how cultivated landscapes can be part of this endeavour.

Global sustainability

We have always known that ultimately we depend on nature for our well-being and survival. The seemingly limitless supply of free natural resources was once referred to quaintly as 'nature's bounty'. To emphasise the critical role of nature in human commerce and society, the language of sustainability science now uses the much less colourful and more market-orientated expression *ecosystem services* to mean essentially the same thing. This economic analogy compares natural resources to capital assets that cannot be depleted without serious consequences. It also highlights the fact that we treat many natural assets as if they had no value. Clearing forests, developing coastlines, adding CO_2 to the atmosphere and many other environmental impacts generally benefit only a small sector of society while overall natural capital is eroded. In recent years we have seen the increasing use of market mechanisms (carbon and water credits, taxes, financial incentives, etc.) to tackle the problem of nature being treated as an economic externality.

Underpinning environmental sustainability is the assumption that, for our own good, we must monitor and improve our custodianship of the other life-forms on planet Earth.

The Millennium Ecosystem Assessment divides nature's numerous ecosystem services into: provisioning services, regulating services, cultural services and supporting services (see Figure 8.1), and shows how they interact to affect our general sense of well-being. The figure demonstrates clearly why protecting nature must become a routine part of our daily

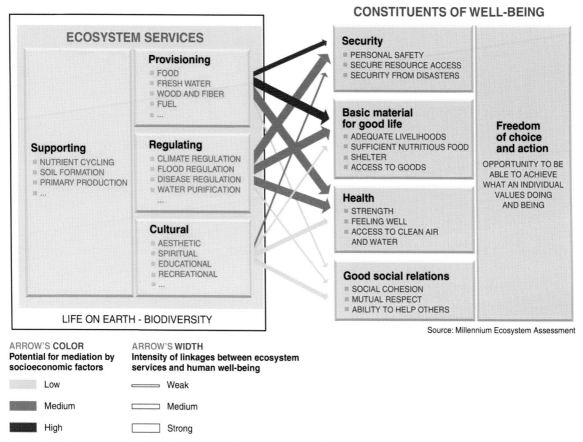

Figure 8.1 Ecosystem services and their links to human well-being.[1]

lives. Environmental sustainability must include biodiversity conservation, protecting the variety of organisms and their habitats at all levels of biological organisation from the biosphere down to the genes of individual organisms. Figure 8.1 also shows how ecosystem services provide us with vital 'constituents of well-being' that we easily take for granted – our health, resources, social and cultural relations, and sense of security.

International concern about environmental issues has generated a wide range of government and non-government organisations (NGOs). NGOs include organisations such as the World Wide Fund for Nature (which plays a major role in drawing environmental issues to the attention of the general public), and information centres such as the Worldwatch Institute. The United Nations in particular has led the way through the United Nations Environmental Program (UNEP). Some organisations, like the Intergovernmental Panel on Climate Change (IPCC), formed by UNEP and the World Meteorological Organization, specialise in a particular topic; while others, like Friends of the Earth and Greenpeace, are more concerned with direct political action.

There are now many international environmental treaties, agreements, conventions, protocols and the like. The following are some of the more important international agreements.

The Convention on Biological Diversity (CBD) is an important international document directed at protecting biodiversity. It was formulated under the charter of the United Nations and seeks to conserve the diversity of life on Earth and to ensure that this diversity continues to maintain the life-support systems of the biosphere. It aims to:

- conserve the world's biological diversity,
- promote the sustainable use of the components of biological diversity,
- provide for the equitable sharing of benefits from the use of biodiversity, including providing access to genetic resources and the transfer of relevant technologies.

The *United Nations Conference on the Human Environment* was the first major global environment meeting. It was held in Stockholm in 1972 and focused on air and water pollution, but it was not until 1992 that the CBD was adopted in Rio de Janeiro. It has subsequently been ratified by 180 countries, including Australia. The CBD is a binding international regulation for the signatory countries. The ultimate authority of the CBD is the Conference of the Parties (COP), consisting of all governments and regional economic integration organisations that have ratified the treaty. COP reviews progress, identifies new priorities and sets work plans for members. The first global periodic report on biodiversity, *Global Biodiversity Outlook*, was published by the CBD Secretariat in 2001 with a second report in 2006. Both reports attempted to assess the status of global biodiversity and progress in implementing the CBD.

UNEP produces periodic assessments of the interaction between environment and society through a series of global environment updates. The most recent is *Global Environment Outlook (GEO-4)*, released in 2007.

Agenda 21 is a United Nations-initiated program of action for sustainable development to be taken globally, nationally and locally by organisations of the United Nations system, including governments and other major groups in every area in which humans impact on the environment. It was initially adopted at the 1992 Earth Summit where the Commission on Sustainable Development (CSD) was established to monitor effective follow-up.

The World Conservation Monitoring Centre is the world biodiversity information and assessment centre of UNEP. Based at Cambridge in the UK, this organisation provides information for policy and action to conserve the living world.

However, it is the Millennium Ecosystem Assessment of 2005 and the Millennium Development Goals, already discussed, that appear to carry the greatest international commitment.

The initial chapter discussed how environmental management can be implemented at many levels of human and biological organisation. We have already referred to two general management strategies:

- tackling the direct impacts
- addressing the underlying pressures that lead to the impacts.

> ## INFO BOX 8.1: DETRIMENTAL HUMAN IMPACTS ON THE GLOBAL ENVIRONMENT
>
> ### Atmosphere (air)
> *Atmospheric changes*
>
> - ozone depletion
> - global warming due to a build-up of greenhouse gases such as carbon dioxide mainly through the use of fossil fuels and changes in land use
> - acid rain
> - chemical air pollution (smog, photochemical pollution).
>
> ### Geosphere (land)
> *Land degradation*
>
> - salinisation (sodification)
> - desertification
> - erosion
> - clearing for agriculture, urban development, etc.
>
> ### Hydrosphere (water)
> *Degraded water bodies*
>
> - eutrophication (excess nutrients)
> - contamination with industrial wastes
> - depletion of freshwater reserves by large-scale freshwater diversion from natural habitats to agricultural or urban areas
> - temperature pollution from industry.
>
> ### Biosphere (organisms)
> *Loss of biodiversity*
>
> - habitat loss through forestry or by clearing for agriculture, pasture and urban development
> - increased rate of extinction
> - introduction of non-indigenous invasive plant and animal species.

Direct impacts

In one sense, protecting nature is simply about improved management of land, water bodies and the atmosphere. Info Box 8.1 is a list of direct impacts of concern at a global level.

Indirect impacts

'Start of the pipe' demand management of human consumption is now a primary concern of sustainability science because, as we have seen, it is the indirect impacts of human consumption that make up by far the largest part of our resource use. But managing human consumption is, in many ways, more complicated than cleaning up an oil spill or altering a watering regime. This aspect of sustainability has been discussed earlier in the book.

Australia

The CBD was ratified by Australia in 1993 and its obligations are met through the *National Strategy for the Conservation of Australia's Biological Diversity*. Australia also has international obligations under the United Nation's Agenda 21 and the Organisation for Economic Co-operation and Development (OECD) to undertake environmental performance reviews. The 1992 National

INFO BOX 8.2: DETRIMENTAL HUMAN IMPACTS ON THE AUSTRALIAN ENVIRONMENT

Atmosphere (air)

Atmospheric changes

- global warming due to a build-up of greenhouse gases such as carbon dioxide mainly through the use of fossil fuels and changes in land use.

Geosphere (land)

Land degradation

- deteriorating soil quality with acidification
- nutrient depletion
- sodification and salination of land
- water and wind erosion
- contamination of the land with the residues of agricultural chemicals.

Hydrosphere (water)

Degraded water bodies

- negative changes in river processes with loss of environmental water flows
- negative influence on waterbodies of nutrients, salts and pollutants
- contamination of groundwater with nutrients, salt and pollutants
- sodification and salination of rivers.

Biosphere (organisms)

Loss of biodiversity

- loss of native species
- decline of remnant vegetation, especially waterside plants
- loss and fragmentation of habitat, especially on productive soils
- large-scale decline of paddock trees; decline in native pasture and rangeland
- negative impact on natural and cultivated systems due to invasive alien species.

Strategy for Ecologically Sustainable Development (NSESD) goes part way to meeting these obligations and has the following objectives:

- to enhance individual and community welfare by following a path of economic development that safeguards the welfare of future generations,
- to provide equity within and between generations,
- to protect biodiversity and maintain ecological processes and life-support systems.

Work at the national level is performed through the Intergovernmental Agreement on the Environment and the Council of Australian Governments (COAG). Since 1992 strategies and policies to implement the NSESD include: *National Heritage Trust; National Water Reform; National Strategy for the Conservation of Australia's Biological Diversity; National Greenhouse Strategy; National Oceans Policy; Regional Forest Agreements* and the like. The ABS has produced 14 headline sustainability indicators (six are environmental) as part of its *Measures of*

Australia's Progress report of 2006 as well as summaries of environmental issues and trends. State of the Environment reporting provides a periodic summary of the condition of the Australian environment and this form of reporting is also done by each of the states.

In 2001 an *Australian National Land and Water Audit* singled out the main problems for direct environmental management in Australia (Info Box 8.2).

Australia is home to more than a million different species of organisms, many of which are endemic and not found anywhere else. With its wide variety and unique species and ecosystems, Australia is privileged to be home to some of the most exciting, diverse and unique organisms the planet has to offer and it is listed as one of the world's 17 most mega-diverse ecosystems.

Monitoring Australia's biodiversity is still in its infancy and the use of biodiversity indicators has only just been established. In the Commonwealth *Environment Protection and Biodiversity Act 1999* measures of biodiversity include listings of threatened flora and fauna and threatened communities as well as assessments of critical habitats and threatening processes (see Tables 8.1 and 8.2).

The number of threatened ecological communities rose from 21 in 2000 to 36 in 2006, although this does not necessarily reflect conservation status. Five habitats critical to the survival of a species or ecological community were identified in the *Register of Critical Habitat* of 2006.[2]

Another way of assessing biodiversity is by listing key threatening processes where the process threatens or may threaten the survival, abundance or evolutionary development of a native species or ecological community. These may be listed when there is the potential for a native species or ecological community to become more endangered. This is the first step to addressing the impact. The number of listed threatening processes has increased from 6 in 2000 to 17 in 2006.

Local

At the local and community level there are some useful general principles of landscape maintenance.

When reserved land is isolated, it becomes extremely vulnerable to changes such as

Table 8.1 List of threatened fauna 2006

Threat	Frogs	Birds	Mammals	Fishes	Reptiles	Other animals
Extinct	4	23	27	–	–	–
Extinct in the wild	–	–	–	1	–	–
Critically endangered		5	2	2	1	4
Endangered	15	38	34	16	11	7
Vulnerable	12	64	53	20	38	6
Conservation dependent	–	–	1	–	–	–
Total – 384	31	130	117	39	50	17

Source: Department of Environment and Heritage.[3]

Table 8.2 Threatened fauna: change over the period 2000–2006*

Threat	2000	2001	2002	2003	2004	2005	2006
Extinct	53	54	54	54	54	54	54
Extinct in the wild	0	0	0	0	0	1	1
Critically endangered	0	3	11	12	14	14	14
Endangered	98	107	109	110	112	121	121
Vulnerable	172	183	184	187	189	192	193
Conservation dependent	0	1	1	1	1	1	1
Total	323	348	359	364	379	383	384

Source: Department of Environment and Heritage.[3]
* may reflect factors other than conservation status (e.g. taxonomic revision).

drought and climate change. Management of cultivated land, whether urban space or land used for forestry and agriculture, must be able to maintain biodiversity levels and ecological flexibility: it is not sufficient to rely solely on a few public reserves. Changes must also occur on private land so that ecosystems remain functional. A suite of strategies has been suggested to deal with this situation, with two main ones suggested below.[4]

1. *Managing the spatial distribution of vegetation* by maintaining large, structurally diverse patches of native vegetation that are connected through corridors, or stepping stones, of similar vegetation. This is more beneficial for biodiversity when the connecting vegetation is present in different structural parts of the landscape (slopes, gullies, plains), and when it includes the different environmental conditions of an area as well as being similar in structure to the native vegetation. Buffer zones can be created around sensitive areas.

2. *Managing ecological processes* including: protecting keystone species (those that affect the system more than others) and those that together ensure the ecosystem functions efficiently; managing disturbance regimes such as grazing and fire so that they have minimal detrimental effect; and controlling invasive species and site-specific threats as well as protecting individual species of special biological significance.

Adjusting to climate change

A key element of sustainable horticulture is planning for the effects of climate change. A technical report on the impacts of climate change on gardens in the UK provides a comprehensive coverage of issues for consideration, most of which are applicable to Australia.[5]

Practical ways to encourage native biodiversity in gardens will be discussed in the next four chapters.

INFO BOX 8.3: CLIMATE CHANGE

Climate

The more densely populated regions of Australia can expect increased average and summer temperatures with an associated increase in evapotranspiration; also, less rainfall that is more unevenly distributed across the year and with an increasing number of extreme weather events such as droughts, bushfires, high temperature days, waterlogging, high-intensity storms with floods and high winds. Managed garden plants will be less vulnerable to climate change than plants in the wild and in terms of immediate concern short-term extreme weather events pose a greater threat than long-term changes in temperature.

Effects

- Earlier dates of bud burst, flowering and leaf fall, altered periods of dormancy, growth season, emergence of butterflies and other phenological (phenology is the study of the times of occurrence of recurring natural events) changes in garden organism behaviour (pollination).
- Changes in the range and virulence of pests, diseases and weeds; generally an increase in severity, range and numbers. Warm-climate potential garden escapes, for example, will be favoured by climate change.
- Higher temperatures accelerating biological processing of organic matter with an increased release of CO_2 and nitrogen from the soil.
- Moisture content of unirrigated soils will progressively decrease over time.
- Increased CO_2 will generally increase plant growth but this response may be modified or nullified by other factors such as increased temperature. However, studies indicate that forests, especially in the Northern Hemisphere, are growing at an increased rate as a result of the elevated CO_2 levels resulting from climate change. In Europe coniferous and hardwood volumes have increased by up to 50% because of rising CO_2 levels, an extended growing season and increased nitrogen deposition with timber yields in the UK 20–40% higher in the 21st century.
- Rise in sea levels coupled with storm surge and high winds carrying salt spray may have adverse effects on coastal plant life and gardens.

Management

- Trees are the most vulnerable to the major weather events of climate change like wind, drought and fire, so care is needed in planning and replacement, selecting provenances best adapted to the conditions predicted by climate change studies (especially fruit and vegetable cultivars).
- Changing water/irrigation needs (infrastructure, rain gardens, stormwater etc.) as rainfall over gardening and agricultural regions decreases and evaporation increases. Fruit and vegetables will require more water.
- Adapting to mains water conservation and introducing new irrigation practices and equipment.
- Greater consideration of plants adapted to higher temperatures and lower and less evenly distributed rainfall and consideration of current plants that will be vulnerable in some localities.
- Gardens may assist natural systems to adjust to climate change by increasing biological

connectivity through green corridors and acting as green refuges.
- Opportunity for greater community involvement with revegetation and environmental rejuvenation programs, maintaining local provenance seed banks and the like.
- Heritage gardens, botanic gardens and major historical public parks may require special and more expensive management as many have collections that are adapted to a cooler climate. Here there is an opportunity to educate the community and set an example by adopting expanded sustainability and conservation goals. For example, in 2004 the Royal Botanic Gardens Kew started working towards the international standard of Environmental Management ISO14001 which included not only the gardens but also their offices, research labs, visitor catering and shops.
- Management of soil and tree carbon.
- More careful monitoring of water use to take account of desired landscape quality requirements of grass, landscape types, trees and sports turf.
- Engagement with programs like the *National Climate Change Adaptation Programme*.

A seaside garden by landscape designer Georgina Martyn, designed for family pleasure, low maintenance, low water and low energy. (Photo: Simon Griffiths)

DESIGNING LOW IMPACT GARDENS

Australian plants are the basis of our natural ecosystem and a garden containing local plants helps keep the ecological landscape intact.[1]

KEY POINTS

Design is critical to developing sustainable gardens and landscapes.

In well-designed gardens, the environmental impact of water, energy and materials use is low, natural biodiversity is encouraged and food harvested.

Designs take account of local and site conditions.

Ecological design encourages self-sustaining gardens.

New design approaches and technology are used where suitable.

Sustainable design will become standard practice for new gardens and parks. A garden's design determines its environmental impact both during its creation and throughout its life, so the designer has a vital role in setting the course of a garden towards a sustainable future.

Carefully planned gardens are usually the most successful ones, and the creators of these gardens generally have a clear vision for their garden from the beginning. Preparing a plan that considers visual appeal, sense of place, ease of maintenance and environmental impact will pay dividends later.

Careful design can reduce a garden's environmental impact *at the time of construction* through the careful choice of materials and methods, and will reduce its *ongoing environmental demands* by lowering the resources required for maintenance.

Thoughtful design will not only reduce the environmental demands of the garden itself, but can also have a positive influence on the associated buildings. For example, a design that integrates the house and garden can moderate the immediate environment of the house, reduce the need for appliances, create household energy savings, and have other benefits as well.

Elements that improve the sustainability of gardens (such as rainwater tanks, greywater systems, small areas of lawn or none at all, plant selection, etc.) are cheaper and easier to include at the design stage when planning a new garden. However, although less efficient and often more expensive, changes to existing gardens can be incorporated at any stage (like redesigning sections with smaller lawns, or replacing plants with ones that are drought-tolerant, disease-free or food-providing, by installing water tanks or planting shady trees).

Being more sustainable may mean minimising resource use but this need not stifle creative design. For each style, clever choice and placement of landscape materials and plants can lead to better environmental outcomes.

Sustainable design of gardens, cities and buildings is about selecting and combining materials with low environmental cost throughout their life cycle, and using them to achieve beautiful, functional spaces that we find comfortable and stimulating to be in. It is an exciting, rapidly developing new area, inspiring research and expanding knowledge. Monitoring the environmental performance of these new designs, methods and materials will be critical over coming decades.

A ready reference Summary Guide to sustainable garden design can be found in Chapter 13.

Fundamentals of sustainable garden design

Earlier chapters looked at the pressures on the environment that stem from our consumption of water, energy, materials and food, which collectively threatens biodiversity. Using these categories, we can now look at garden consumption and ways to minimise its environmental impact through creative garden design:

Water	• Minimise reliance on mains water (e.g. harvested and recycled water; soil water storage; no or low, efficient irrigation).
Energy	• Minimise need for use of fossil fuels.
	• Maximise energy efficiency of appliances, tools and machinery (e.g. pumps and lighting).
	• Reduce embodied energy in landscape materials, tools, machinery and chemicals.
Materials	• Reduce the use of materials.
	• Choose local over imported products if appropriate.
	• Choose materials that are produced sustainably.
	• Use materials that can be either re-used or recycled.
Food	• Reduce reliance on non-local foods; reduce food miles; avoid foods with known high life cycle impacts.
	• Maximise local food production to reduce agricultural environmental demands.
Biodiversity	• Promote a rich biodiversity within the garden that encourages the natural ecology and minimises negative impacts.

Consideration of these points throughout the design process will lead to the most sustainable outcomes for the new garden.

The design process

Beautiful gardens are the result of plans that are based on sound design principles that guide the construction of the garden.

Garden design principles and general design considerations

During design, we think of the garden as an abstract collection of three-dimensional shapes. It is how we assemble shapes of different sizes that will define boundaries and spaces. Creating a garden is similar to an artist creating a three-dimensional sculpture that you can walk around, or an architect designing spaces in a building, but with the shapes and sizes of the plants defining the spaces rather than plaster walls. It helps then to think in the abstract, so that the garden is seen as a series of shapes rather than actual plants or garden structures. It is with these abstract shapes that the garden's structure can be developed. Once shapes and their sizes are defined for the proposed garden, the non-living, hard landscape components, and soft landscape components, which include living plants, substitute for the shapes.

The creative process is an important one for developing a successful garden, and it is governed by principles applied to other art forms. Appealing gardens result from a careful consideration of **unity, repetition, simplicity, rhythm, sequence, focal points, transition, contrast, framing, balance,** and **proportion** or **scale** (see Info Box 9.1). These design principles are universal. Different designers will apply them in quite different ways, leading to unique but equally successful garden designs, even for the same site.

Just as a sculptor needs to know the properties of the metal, wood, stone or plastic for his artwork, the garden designer understands the site's location, soil, water flows, climate, an appropriate palette of plants, the qualities of hard landscape materials (for paving, decks, fencing, walls, pools, pergolas and other garden structures) and furnishings (such as seats, tables, pots and lighting). This knowledge helps in the creation of a flourishing, aesthetically pleasing and environmentally sound garden. Unlike a sculptor or architect, the garden designer relies heavily on the garden's relationship with the local environmental conditions for the garden to thrive.
Every site is different. The design needs to respond to its site and to changing conditions, like periodic droughts, seasonal winds, the threat of flood.

Preliminary questions to guide design development

There are several preliminary questions to be asked at the start of the design process to ensure that the garden serves the needs of its users. Every choice made will have environmental consequences, many of which are considered in later sections of this chapter. Whether the garden is designed by the user or by a professional, the process of answering these questions is still required:

1. What style of garden would you like? There are many styles including: contemporary Australian native, formal,

INFO BOX 9.1: GARDEN DESIGN PRINCIPLES

Unity

The visual cohesiveness of a garden made of individual elements comes from:

- Uniformity of form, colour and texture, e.g. massed plantings of same species or cultivar.
- A consistent character or style within the garden.

It needs to be balanced with variation to avoid monotony; diversity within a unified theme.

Repetition, simplicity and rhythm

Repetition:

- Comes through repeated colour/colour range, or use of similar leaf textures, plant sizes or shapes.
- Simplifies the design and avoids the sense of chaos caused by too much variety and too many details.
- If measured, can be used to create a rhythm in the design, to cause one's eye to 'dance' through the garden at regular intervals, possibly to a focal point or a borrowed landscape beyond.

Sequence and focal points

Garden designs that:

- Facilitate the movement of the eye along defined, often subtle trails, allow the viewer to experience the garden in a defined sequence, leading to a point of interest or focal point such as sculpture, a plant of different foliage colour, habit (e.g. weeping) or texture (large leaves against small).

Visual sequences, like the flow of paths, avenues of trees, or a glimpse of what is to come, can encourage the visitor to move through the garden.

Transition and contrast

Interest and a visual flow through a garden can:

- Be created with gradual changes in plant size (small to large), leaf texture (coarse to fine) or colour (whites to pinks to reds, mauves and blues).
- Play useful tricks on the mind such as reinforcing an impression of depth when fine textured plants (small leaves) are planted behind medium-textured, then large-textured plants.

Contrasting plants (e.g. silver foliage on dark green), can be used through the garden to:

- Promote a sense of rhythm.
- Create patterns.
- Define shapes as in parterre gardens.
- Create large living abstract canvases as did the 20th century Brazilian landscape architect, Roberto Burle Marx.

Framing

As artists 'frame' their subject on a canvas and photographers crop their prints, gardens too can be framed using hedges, walls and trees. These design elements define spaces and focus our attention, allowing us to concentrate on a composition within the garden, or as we move through it, a series of compositions, each carefully framed, and each worthy of our consideration.

Balance

Balance is achieved visually in a design when the elements on either side of a central axis have equal ability to attract the eye.

- Symmetrical or formal balance relies on the components being mirror images on either side.

- Asymmetrical or informal balance is achieved when the eye is equally attracted to objects of different shapes, sizes and textures, arranged in a seemingly random fashion on either side of an axis.

Proportion

The sizes of open areas, trees, massed plantings, buildings and garden structures like pergolas need to relate to one another.

- If one element is much taller, smaller, broader or narrower than the others it will appear disproportionate.

- A small ornamental pool in an inner city courtyard will seem out of place in the broad sweeps of a rural property.
- Tall, spreading *Eucalyptus* trees are fine in country landscapes or urban parks, but will dwarf a Victorian worker's cottage.
- In the home garden, relating to the human scale is a good way of checking if proportions are appropriate.

Italianate and cottage. If you are unsure, visit a range of gardens, such as those in the Australia's Open Garden Scheme. Borrow or buy garden design books. The style of the garden will be influenced by the 'sense of place' and by building style.

2. How would you like to use the garden?

   Is gardening one of your main leisure activities? Do you enjoy doing other activities outside? Is the garden mainly a setting for the house? Do you want to harvest food from it? Is the garden to be bird-attracting? Will a swimming pool or tennis court be required? Are pets to be accommodated?

3. What activity areas do you need/want in your garden?

   Will you be eating and entertaining outdoors, do you need children's play areas, will you be growing fruit trees and vegetables, what service areas do you need, do you need areas for storage (cars, boats, tools, etc)? Is disability access necessary?

4. How much time do you have to maintain the garden?

   Be honest: every weekend? One day every fortnight? Every month? Only occasionally?

5. How much money would you like to invest in your garden?

There are many books available that can help strengthen your understanding of garden design, and the materials, including plants, used in creating a garden.

Practical steps in the design process

Answering these preliminary questions will give a clear understanding of the qualities and components the garden will have, providing you with a 'wish list'. The next step is to understand the site.

Steps in garden planning incorporating a sustainable approach

1. Analyse the site.
   - If still to be built, identify the best location for buildings.
   - Note how the garden relates to the house:
     – What windows and doors open on to the garden?
     – Is the garden easily accessible from indoor living areas?
     – Are there indoor rooms requiring privacy from the garden?
   - Note how the garden relates to surrounding landscapes.
   - Identify the 'sense of place'. This will influence the style of the garden.
     – Is the area suburban Federation, rural agricultural fields or dominated by natural vegetation for example?
     – Are there views of landscape or vegetation that can be 'borrowed' for the garden?
   - Collect base data.
     – Note circulation patterns of people (adults and children), pets and cars through the site.
     – Identify soil type and pH.
     – Note levels, slopes and drainage lines.
     – Note services including water, electricity, gas, easements, boundaries.
     – Identify shading patterns.
     – Identify existing buildings, plants and trees and where vegetation, especially trees, will need protection.
     – Identify existing activity areas (eating, playing, composting etc.)

   Consider sustainability:
   - Identify microclimates (hottest parts, coolest parts, wettest parts, driest parts, prevailing winds, shaded areas).
   - Note slopes and drainage lines.
   - Note appropriate areas to capture roof run-off (location of downpipes) and to install water tanks.

2. Draw site plan.
   - Undertake a measured site survey (spaces, levels, buildings, structures, services such as water and drainage, paving, existing plants that are to be retained, e.g. trees).
   - Draw scaled drawing to be used as base drawing for design development.

   Consider sustainability:
   - Create tracing paper overlays over the site plan of existing conditions that identifies the microclimates, slopes and drainage lines, roof drainage and downpipes etc.

3. Develop design.

   Using tracing paper over the base drawing, using the overlays of existing conditions too, roughly outline areas of:
   - Hard landscape for walkways, sitting and eating areas, garden buildings, and
   - Soft landscape for garden beds and lawns.
   - Use pencil and be comfortable with erasing, replacing and playing with various shapes and combinations until the design feels right.

   Consider sustainability:
   - Identify where major plantings are to go, for example deciduous trees for

summer shade, and shrubs for screening and wind protection for the house.
- Where possible, place paving areas and garden structures on the driest areas.
- Design garden areas requiring more water into the moist parts of the site, and requiring less water into the drier, more sun exposed parts of the site.
- Plan the garden so that microclimates will be created by enclosure with walls, fences and plantings of more exposure-tolerant plants.
- Design the garden for low maintenance and with less need for powered tools or equipment like mowers. Ask questions such as:
  - Is a large, formal hedge needing regular clipping really necessary for example? If so, can it be clipped manually? Can an informal hedge be as effective?
  - Is a water feature requiring a pump the best option? Will a still water feature effectively create a calm, cooling atmosphere?

4. Create a planting plan.
   - Choose a palette of plants that suit your design and aesthetic requirements.
   - Place plants according to their size, form, colour, texture and function using the garden design principles.

Consider sustainability:
- Choose plants suited to the site that preferably require no, or at least low, irrigation and are relatively pest-free.
- Zone plants according to irrigation requirements so irrigation frequency is reduced to the minimum.
- Choose the right plant for the place in the garden and allow enough space (including height, for example, under power lines) for the plant to grow to its ultimate size, avoiding the need for frequent maintenance.

5. Prepare construction drawings.
   - Prepare drawings to guide the grading of the land and the construction of retaining walls, drainage, fences, boundary walls, pergolas and other garden structures. These can include details for irrigation systems and garden lighting.

6. Stage design implementation.
   - Identify the stages of implementing the garden design and plan the appropriate sequence for garden construction so earlier stages are not damaged or undone by later ones (see following chapter on landscape construction).
   - Assess site for opportunities to zone areas that will help the staging process by isolating sensitive areas from construction work at appropriate times or throughout the whole construction period.

Consider sustainability:
- The implementation process should be planned to minimise the use of resources (this is dealt with in more detail later in this and the chapter on landscape construction).

Factors to consider throughout the design process:

- Reduce the use of resources.
- Use recycled materials including recycling materials from the existing to the new garden.
- Source local materials.
- Use products that cause minimal damage to natural environments when extracted or manufactured.
- Include plants that act as carbon sinks, encourage indigenous biodiversity and contribute to the household food supply.
- Minimise the potential of environmental weeds.
- Protect existing large trees.
- Minimise environmental damage to areas surrounding the garden and maximise environmental advantages.
- Reduce maintenance needs, particularly those requiring powered equipment.

Environmental design considerations

No matter how good a garden looks, its value is diminished when it is established and maintained by heavy use of limited resources. With creative design, it is possible to overcome the constraints set by restricted resource use. In this section ideas are introduced that encourage consideration of the environment as an essential part of garden design. In the following sections we will discuss design in relation to water, materials, energy, food, biodiversity and plant selection in more detail. The final section of this chapter covers preparation for design implementation.

Considering ecology

Encouraging self-sustaining gardens through ecological design

Design based on ecological principles results in ecologically healthy sites that are resilient and can cope with some stress. They usually have a diversity of plant species which encourages the presence of other organisms, the total biodiversity providing a natural means of keeping pests in check with less need for chemical pest control.

Healthy sites provide ecosystem services or garden services such as:

- Climate moderation
  - Warmer in winter (e.g. wind protection) and cooler in summer, so that house heating and cooling respectively is reduced or eliminated. Vegetation providing summer shade reduces direct and reflected radiation onto buildings, and can provide a more humid cooling atmosphere.
- Water purification
  - By filtering water slowly through sites before run-off occurs, there is less pollution of waterways such as creeks, rivers, bays and oceans.
- Air filtering
  - Reducing airborne pollutants in urban locations.
- Food
  - Which may not only be grown free of pesticides, but if practised widely by a community, can reduce pollution by lessening the need for transporting food.

Designing buildings and gardens as ecological systems offers the benefit of lower resource use and greater sustainability.

Choice of site

If you have the opportunity to choose the site to be landscaped, then this is the time to weigh up the potential environmental consequences. Is the proposed site the most appropriate for your needs? Will your proposal result in short- or long-term environmental costs to you or your community?

Consider:

- The effect on a community's sustainability when a particular site is chosen.
  - Whether your site is on prime agricultural land, which might be best protected for local food production, and to reduce food miles.
  - If intact indigenous vegetation is going to be destroyed on your site. Private land with natural habitats is an evolutionary legacy that is easily destroyed, and impossible to recreate. Any natural bushland needs care to retain or improve its vigour. If house and garden plans cannot accommodate the retention of indigenous vegetation and the rich web of life it supports, alternative cleared sites may serve the purpose equally as well.
  - Whether the land is home to biologically significant natural habitat, or communities of rare or threatened plants or animals. Endangered plants may be legally protected on private land.
  - Whether alterations to your site (buildings, hard landscape, vegetation change, etc.) can impact on the established ecology of the plants and animals in the area and try to anticipate what these effects might be (changes to subsurface water flows, reduced nesting sites, disruption to movement patterns of animals, etc.).
- Restoring or 'greening' a previously damaged site can contribute to a community's sustainability. Industrial or commercially degraded land, old factory sites, neglected public land can all be reborn as 'green' sites and give an area a sense of renewal. Even areas once contaminated with toxic wastes, such as petrol stations, can be treated and rejuvenated.
- Can your goals be achieved on this site with minimal environmental impact? Is the site amenable to a sustainable lifestyle?

Within the site itself, a rearrangement of its components might improve environmental outcomes (e.g. relocation of buildings and hard landscaping to enhance passive solar heating; building over the most degraded soils; making space for an underground water tank; capturing the natural run-off from rainfall more effectively, etc.).

Remember, it is at this early stage that lasting environmental benefits can be locked in.

Protecting soil quality

The importance of soils is often overlooked. In the broad landscape, soils contribute to environmental services including the cycling of nutrients, the absorption of rainfall, the storage of water, the slowing of run-off, and the absorption of chemical pollutants including excessive nutrients. Well-structured soils support the healthy growth of plants that provide food, shelter, medicinal compounds and, importantly, the oxygen required for all life. The organic carbon maintained in soils reduces the release of carbon-containing, climate-changing gases. Landscape designers, horticulturists and gardeners can maintain and enhance these soil qualities as they design and work in the landscape.

Soils are the key to healthy gardens, and contribute to the garden services that improve the environment we live in, so protecting existing soils can save the time and resources needed to reinvigorate degraded soil at a later date. The soil's structure, along with its populations of micro-organisms, is easily damaged by inappropriate cultivation, vehicle and foot traffic, and the dumping of waste building products like concrete.

At the design phase, assess soil quality on the site and use this to guide how the site is developed. Assigning, where possible, the soils in best condition on a site as garden zones, and those areas with the poorest soils for buildings or hard landscaping, will reduce the additional resources needed to improve soils for vigorous plant growth.

Soil fertility has a strong influence on the kinds of plants you grow. Indigenous species are adapted to the nutrient levels of the area where they grow so the plants look healthy and green. Soil fertility does not necessarily need to be adjusted: increased soil fertility can be detrimental to some species. High phosphorus levels are toxic for many Proteaceae like *Banksia*. Increasing fertility may also increase weed growth at the expense of the plants being grown.

Artificially increasing soil fertility is best kept at modest levels. Grouping together those plants with similar nutrient requirements is an effective way of reducing the need for broad scale soil amelioration, restricting it to small areas that need it, and protecting the native soils and vegetation.

Efficient use of space

A well-designed garden can accommodate a range of activities and allow for future needs. A small area can, for instance, provide a setting for different activities at different times (games area in the morning, an eating area for dinner). Another space could also allow for changing family needs over time: from children's sand pit, to basketball ring, fire pit for teenage parties, and a terrace for family gatherings. (See Info Box 9.2.)

Resources and infrastructure

Alternatives to conventional resource-hungry infrastructure can be explored. Composting toilets and greywater treatment systems are proven alternatives dealing reliably with wastes and recycling scarce water while at the same time using fewer resources than a conventional connection to the sewage system. This deals effectively with a problem at a local level, on-site, rather than regionally. A similar approach can be used for solar power, mini hydroelectric generators and wind power. Solar photovoltaic panels can be connected to the electrical mains grid in some areas and the owner paid for the excess electrical energy they produce. These systems give the owner the flexibility of using electrical equipment without continually adding to greenhouse

> **INFO BOX 9.2: GARDEN ACTIVITIES**
>
> - Gardening
> - Growing food
> - Playing (equipment, games)
> - Eating
> - Drying washing
> - Sitting
> - Reading
> - Entertaining
> - Observing wildlife
> - Storing equipment, tools, pots, potting media etc
> - Drying camping gear
> - Woodworking
> - Car, boat, caravan parking
> - Car maintenance.

gases. Even without installing a photovoltaic system that provides for all electrical needs, perhaps smaller appliances like a solar light at the bottom of the garden will provide adequate light, and will avoid the high embodied energy of electric cable and PVC conduit. In rural areas, on-site generation may be a feasible alternative to installing long private lines connected to the grid and the community infrastructure needed to service them. Alternatives like this may involve some inconvenience but are generally cheaper and allow for greater self-management.

Resources can be saved by careful attention to scale. The width of a driveway can be restricted to 3.5 m which still allows for the opening of doors of large vehicles; 3 m is sufficient for sections where vehicles do not park.

A drive can be made of two narrow strips of hard surface with lawn or other groundcover growing up to its edges. This allows for greater water infiltration, is less noticeable and uses fewer materials. A very keen gardener went even further by using only two narrow asphalt strips, just wide enough for the tyres, with lawn growing up to the edges. The asphalt strips almost disappeared visually, they had minimal impact on water infiltration into the soil and the resources required for their construction was about 15% of a full 3.5 m drive. By having a drive 3 m wide, rather than 3.5 m, over a cubic metre of concrete is saved for every 15 m of drive.

Many similar economies of scale can be made with a little lateral thinking.

Hard landscaping designed for healthy plant growth

Providing good growing conditions for your plants makes good economic sense and avoids the unnecessary wastage of resources needed to propagate and cultivate replacements. The design of hard landscaping can affect the quality of plant growth, and may need to cater for the special requirements of particular kinds of plants depending on their habit (tree, shrub, groundcover, herbaceous perennial) and the habitat they are adapted to (soil fertility and structure, availability of water, soil pH, etc.). Some plants may need special hard landscaping installed, as do plants needing well-drained soils when planted at heavy clay soil sites; they are best grown in raised beds.

Street trees

Reduce the area of hard paving to the minimum required for the particular landscape. The greater the area of exposed soil, the greater its aeration, the better the water infiltration, and the healthier the root environment. This applies particularly to trees which need a minimum volume of soil to support healthy growth, and do not grow well if cramped into a small hole surrounded by impermeable paving, or squeezed into inadequately sized planter boxes. The majority of tree roots are found in the top 30–50 cm of soil, so a depth of 1 m of soil should be sufficient. The distance tree roots spread when growing in unconfined spaces is not so clear-cut. As a rough rule, most large roots are within the dripline of the tree, but the fine roots important for water and nutrient uptake can spread far beyond. Sustainable tree maintenance is enhanced by avoiding planting into small volumes of soil and conditions that deoxygenate the soil. Permeable paving can often help when it is important to maximise the paving area (discussed later in this chapter).

Apart from choosing species and cultivars carefully, street tree growth can be enhanced by planting along soil-filled trenches. These increase the volume of soil available for tree roots, but the paving over the roots needs reinforcing as soil good for root growth doesn't have the same structural properties as the compacted base that is usually used, unless a *structural soil* mix is used. Structural soil mixes contain angular, even-sized, crushed rock fragments that lock together when compacted, leaving spaces not filled with fines. Structural soil mixes may not be suitable for drier climates due to the lower water holding capacity (see section on healthier urban forests in the next chapter).

Engineering with plants

Hard engineered solutions are often used to deal with slopes and the edges of water bodies. This may result in steep masonry walls of concrete or brick and concrete channels for waterways. A more environmentally friendly and less expensive approach is to use plantings as a long-term solution for those areas vulnerable to erosion: species can be used that require little or no maintenance, show a great resistance to damage and have the capacity to mend when a breach occurs.

Similarly, a living wall or living roof can reduce the heating and cooling requirements of buildings. Walls of vegetation discourage graffiti, change appearance, and are generally more attractive than masonry walls: they deaden noise, reduce glare and refresh the atmosphere.

In the USA, environmentally friendly and sustainable practices along these lines are referred to as bioengineering.

Stabilising slopes and controlling erosion

Plant roots stabilise soil, resisting water erosion down slopes and along streams. They reinforce the soil like the wire mesh used in concrete, binding soil particles together and giving it added strength. A combination of trees, with their more extensive root systems, shrubs and groundcovers, will usually provide sufficient root density in these situations. A mass of fine roots is stronger than a few large roots in much the same way as the mass of small fibres in a rope gives greater strength

Figure 9.1 Different materials used for soil stabilisation: Ecocell (top row), TensarMat (second row), Jutemaster (third row) and Enviromat (bottom row). (Photos: Geofabrics Australasia)

than a single strand of the same diameter. The longer a planting has been established the greater its stability.

In North America, live willow branches are used for slope stabilisation. Slopes are terraced and the lengths of willow are placed on to the levelled soil, with their base facing into the slope, and perpendicular to the contour. The terraces are refilled with only the tops of the willow branches protruding. This helps stabilise the slope immediately, and continues to improve with the rapid growth of roots and shoots from the willow. Appropriate species to meet local needs may be available that take advantage of this principle in Australia.

In the early phase of establishment additional measures may be needed. Mulching will reduce the force of raindrops on loose, exposed soil particles, and slow surface water flow on low-grade slopes. Organic mulches are relatively cheap and will decompose, improving soil structure. Products used on steeper gradients include mats made with biodegradable materials such as poplar wood, wool or synthetic materials. These can be placed over slopes to give immediate soil protection from rain erosion and surface flows, as well as providing a tough substrate for root establishment. Other products are made of sheets of many open-ended cells that trap and retain the soil while still allowing water and grass roots to pass through. The environmental sensitivity of the site may determine whether biodegradable mats are the better option, or if the long-term stabilisation offered by synthetic materials is more suitable. It is also worth considering whether locally sourced materials could do the job instead of materials manufactured and transported from elsewhere (Figure 9.1).

Erosion along drainage channels can be minimised by creating a series of ponds or wetlands to slow the flow during rainstorms. This also promotes water infiltration, while filtering out sediments and nutrient (see water sensitive urban design section in the next chapter).

As with any living landscape, its success will be enhanced with monitoring and maintenance, ensuring weak points are promptly repaired.

Green walls for buildings and level changes

Interior building temperatures can be modified by having additional outer, green living skins. In summer, climbers trained over wire mesh frames or self-clinging climbers growing directly on walls, reflect and absorb radiant heat, keeping the wall cooler. In winter they can protect walls from cold winds, although deciduous vines are useful when winter sun can warm walls and the building's interior. Screening vegetation that is close, but not attached to the wall can shade while still retaining a level of air flow (Figure 9.2).

Green Screen®, a company based in California (ArchBar distributors in Australia), produces wire mesh panels and accessories for their attachment to masonry building walls specifically for this purpose. These, and accompanying freestanding modules, have been used for large commercial projects and smaller scale houses (Figure 9.3).

Another type of green wall uses a plant growing medium that is contained by geotextile-lined, high strength plastic modules that were originally designed for underground water storage (see Figure 9.4). These green wall modules have been installed by the National Parks Board in Singapore and were surviving well after two years. The modules do require substantial galvanised metal frames anchored to walls for support which does add to their cost, and this may have slowed their acceptance in Australia. A lighter version is currently being developed by the manufacturer.

Fifth Creek Studio, a landscape architectural business based in the Adelaide Hills, is trialling South Australian native plants, using VersiTank® 250 Green Wall and Fytowall™ systems. Species being trialled include:

Figure 9.2 Vegetation used to screen the CH2 building in Melbourne.

Figure 9.3 Wire mesh framing supports climbing plants on walls and free-standing fences, shading walls and keeping them cool. (Photos: Green Screen®)

Ficinia nodosa (syn. *Isolepis nodosa*, knobby club rush), *Themeda triandra* (kangaroo grass), *Dianella revoluta* (flax lily), *Atriplex semibaccata* (berry saltbush) and *Enchylaena tomentosa* (ruby saltbush). If these local species are successful they will bring a special regional character to these green walls and act as a role model for other regions (see Figure 9.4).

Green retaining walls for stabilising abrupt changes of landscape level can be created in a number of ways. Crib walls, staggered block walls, gabions (small rocks contained in wire baskets), and strong honeycomb materials stabilise the soil and also provide pockets for plant growth. Whatever system is used, it is important to match plant selection to water supply (which may mean using appropriate irrigation methods), that an appropriate soil mix is used, that pockets are not left empty permitting weeds to colonise easily, and that

Figure 9.4 VersiTank® 250 Green Wall planting trial by Fifth Creek Studio in the Adelaide Hills using South Australian native plants. (Photo: Fifth Creek Studio)

it is adequately monitored and maintained after establishment.

Green roofs for climate control

Green roofs are extremely effective for climate control of buildings. In Europe, sod roofs have been a part of housing construction for centuries, and in recent years, because of their excellent insulating properties, modern, lighter weight versions have been introduced in both Europe and the USA. They are known as eco-roofs or green roofs, and in Germany and Switzerland, if a roof exceeds a specified area, a green roof is required by law.

Green roofs can be flat or up to an angle of 30°. They require little additional strength to a conventional roof and with the correct choice of species, in temperate climates at least, require no irrigation or fertilising. The plant covering provides coolness inside during summer and warmth in winter, and it stops or slows water run-off after rainfall as it acts like a sponge. This may be important for areas prone to flooding.

In Australia, there are few sod roofs, but the dug-out houses of Coober Pedy demonstrate how the internal climate of a building can be moderated by the insulating properties of earth. Parliament House in Canberra has perhaps the best known green roof in Australia. There are others, including those at Charles Sturt University in Albury (Figure 9.5), Wharf 11 in Sydney, the Sydney Conservatorium of Music and the Victorian Marine Science Consortium in Queenscliff, Victoria (see Figure 9.6).

Figure 9.5 Green roof at the Thurgoona campus of Charles Sturt University, Albury. (Photo: Peter Vine)

Green roofs are of two types: *extensive* with shallow soils, and typically with low growing drought and sun-tolerant plants, and *intensive*, which are also referred to as roof gardens and can potentially be used as outdoor recreational areas. The latter often need stronger construction to support large planters, deeper soils and garden structures, as well as requiring irrigation for shrubs, and occasionally, trees. Extensive green roofs

Figure 9.6 Marine Science Consortium green roof in Queenscliff.

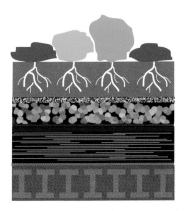

Figure 9.7 Diagramatic representation of a green roof.

provide a low carpet of living vegetation rather than a roof designed as a garden to walk, sit or entertain in.

The essential elements of green roofs are waterproof membranes, root barriers, drainage layers, filtering layer, lightweight growing media and the plants (Figure 9.7). It is essential that green roofs have root-resistant, waterproof membranes or if the membrane is not root-resistant, a root barrier must be installed to protect it. The waterproofing needs to extend up walls to at least 150 mm above the final soil level and should be tested prior to installing the green roof. Green roofs can be installed using modular, easily carried, pre-planted trays that lock together in a grid. The Landscape Research, Development and Construction Society in Germany has released widely accepted guidelines on the *Planning, Execution and Upkeep of Green-roof Sites* in both German and English.

Australia's climate and erratic rainfall may slow the widespread adoption of green roofs, but modifications of the concept may work. One of the authors allowed a deciduous Boston ivy (*Parthenocissus tricuspidata* – naturalised near Perth so beware) to climb over the wall and roof of his unattractive cement sheeting garage, cooling the internal summer temperatures in the process (Figure 9.8). The lightness of the creeper was no threat to the structure, and the garage was to be demolished so did not warrant additional expenditure on waterproofing membranes, although this would be the usual procedure. The roof leaked a little, but only with prolonged heavy rainfall. Over the years humus built up, and the creeper's roots grew between the overlapping cement sheets of the roof, allowing water to seep through. Although in this case waterproofing was not important it illustrates the need for a sturdy structure as a foundation for a green roof: it must be designed and constructed to resist penetration of plant roots and stems, and especially water – a major enemy of buildings.

Plant selection for green roofs is critical to avoid weeds that might spontaneously colonise and compete successfully with the desired plants. A careful analysis of the

Figure 9.8 Deciduous *Parthenocissus tricuspidata* (Boston ivy) insulates a garage roof.

Figure 9.9 This newly planted Adelaide roof garden is designed to encourage animal life in the city. (Photo: Fifth Creek Studio, South Australia)

indigenous flora of your area may identify appropriate plants. Plants more likely to contribute to a successful green roof will tolerate drought and low nutrient levels.

The growth substrate for green roofs needs to be lightweight, shallow and with lower fertility to discourage weed colonisation.

Roof gardens, as opposed to extensive green roofs, may serve as outdoor recreational areas as well as having insulating characteristics or the ability to slow water run-off. Adelaide's Fifth Creek Studio's landscape architect Graeme Hopkins has designed a rooftop landscape that not only provides an attractive open space for the building's residents, but offers a range of habitats to small birds and insects, and includes a bat box to encourage micro-bats. Fifth Creek Studio are proposing a series of stepping stone rooftop habitats to bring wildlife back into the city centre (Figure 9.9).

Major features

Swimming pools, tennis courts and outdoor kitchens are major features increasingly incorporated into domestic gardens. They all contribute to active and enjoyable outdoor lifestyles, but they do come with environmental costs. Major features such as these require many resources, each with its environmental impacts including embodied energy and greenhouse gas emissions. There may be important resource demands through the feature's life too, like the water for pools, and for surfaces of tennis courts that need regular watering.

Swimming pools

Private swimming pools encourage fitness, provide a central recreation area for families,

Figure 9.10 Swimming pools can have a large environmental cost.

are attractive settings for social occasions and are generally great fun, but they do have a large environmental cost (Figure 9.10).

Embodied energy in a standard domestic swimming pool (~ 3.5 x 10 m) could be 250 000 to 300 000 MJ or more, including 20–30 000 MJ for the reinforced concrete, and 200–250 000 MJ for the excavation. Add to that 50 m^2 of brick paving and another 150–200 000 MJ have been embodied in this new landscape feature. Then there are the continuous energy costs to maintain the pool. Pool pumps, being run over long periods, can use more energy than refrigerators over a year.

A pool of this size needs in the order of 50 000 L of water and will need regular topping up during the warmer months due to evaporative losses and splashing. Mean daily evaporation from a Class A Pan in southern Australia can range between 4 and 7 mm in summer. So assuming a similar evaporation rate from the surface of a 35 m^2 swimming pool this would mean a daily loss of between 140 and 245 L through evaporation alone.

Depending on the type of pool, chemicals, usually chlorine or bromine-based, are used to reduce the microbial content of the water to protect the health of the swimmers. Often the chemical used is sodium hypochlorite (better known as household bleach). Some manufacturing processes for sodium hypochlorite produce toxic wastes or by-products.

Saltwater pools produce chlorine on site by passing an electric current through a common salt (sodium chloride) solution. A new salt-based sanitising system replaces sodium chloride with a combination of magnesium chloride, potassium chloride and magnesium sulphate. Unlike sodium chloride, pool water sanitised with these salts is suitable for gardens, meaning the water from backwashing the pool filter can supplement irrigation water.

Backwashing sand pool filters can be a great consumer of water yet if it is not done periodically, the clogged filters can make the pump work harder, consuming more energy. It is worth selecting pool filters that use less water for backwashing. The larger pore sizes in glass-based filters require less water for backwashing and offer less flow resistance, making pumps run more efficiently.

Pool water can also be sanitised using ozone, which is created by a high voltage discharge or UV light splitting O_2 (oxygen) molecules to form single O atoms which collide with O_2 to form highly reactive O_3 (ozone). Residential electrical discharge systems using less energy than UV light systems are available. Being very toxic, any remaining ozone must be filtered out of the water before being returned to the pool. Chlorine is used in combination with the ozone in these systems.

In Europe, natural swimming pools have been developed since the 1980s, particularly in Austria, Germany and Switzerland. Their construction also requires resources and maintaining them relies on using pumps for extended periods. However, they can be cheaper to install, their running costs may be up to a third that of a conventional pool, they use less water, and to sustain water quality, instead of adding chemicals, the cleansing properties of plants and micro-organisms are used in a 'natural' water habitat. For smaller pools, there are two areas of roughly equal size, the swimming area and the planted aquatic ecosystem, with either a wall low enough for water to extend across both or, in some cases, with the two areas completely separated (two pot system). The biological zone is proportionally smaller for larger pools. A pump is required to circulate the water through filters and then through the planted cleansing or regeneration zone back into the swimming area.

Maintenance of a healthy aquatic ecosystem may not be as simple in these as in conventional pools, but some natural pools in Europe have been managed successfully for a number of years. One of the main maintenance objectives is the prevention of nutrient build-up within the system.

Another advantage of natural pools is the increase in biodiversity they support. There are few natural swimming pools in Australia, but the number is increasing as their benefits are recognised (see Figure 9.11).

Designing for sustainable water use

Optimising the site's natural water resource

With so many of us living in urban areas serviced by reticulated water, it is easy to forget that every garden has its own natural supply of water. This is delivered to the site as rainfall, underground water movements from adjacent land (especially on sloped land), and for larger more rural properties, sometimes via streams. All this is a part of the global water cycle. We need to understand how the water cycle is operating in our area if we are to tap into this resource effectively, and minimise our dependence on water diverted from elsewhere, sometimes hundreds of kilometres away. We need to ensure, through imaginative design, that the landscape elements of the site are working towards water conservation, not against it (Figure 9.12).

Understanding the water resources of a site is a critical element in the design process, and therefore a major contributory factor to the ultimate success of the garden. To start with, it

Figure 9.11 Two BioNova Australia natural swimming pools in urban and semi-rural locations. (Photos: BioNova Natural Pools Australia)

Figure 9.12 The columns supporting the upper theatre of Park Güell in Barcelona channel water to irrigation reservoirs below. The park was designed by Antonio Gaudi in the 19th century and opened to the public in the early 20th century. (Photo: Dave Cooksey)

will help if you understand the rainfall in your locality and region. The Bureau of Meteorology has long-term climatic data for recording stations across Australia. Data for your closest station can sometimes be supplemented with the knowledge gained over the years by long-term local residents, some of whom may have recorded the daily weather.

Altering the landscape can change above- and below-ground water flows. Thoughtful design can avoid the often destructive force of water and can result in more effective use of rain falling on a site. Water can be directed into the soil to recharge the soil water reservoir. It can be collected as surface water in features such as ponds and wetlands, or it can be directed as increased run-off. It is critical to consider these factors and to determine what the correct balance is for a particular site and design.

In most parts of Australia harvesting rainfall will be beneficial, particularly as climate

Figure 9.13 Water from roofs, paving and roads flows quickly away from our home gardens and residential areas. From top: downpipe and paving drain; stormwater from residential properties exits into street gutters where it flows into drains and is carried along concrete drains to the sea.

9 – DESIGNING LOW IMPACT GARDENS

change is expected to produce less rainfall for many regions. Conventionally, rainfall is channelled away as quickly as possible through stormwater drains with little chance for it to infiltrate and recharge the soil water. In times of plentiful rainfall and well-stocked dams, the effects are hardly noticeable. But this appears to be a luxury of the past (Figure 9.13).

Harvesting rainwater means depriving your local environment of run-off so this must always be taken into account, and if necessary releasing some water to maintain environmental flows.

It is possible to design gardens to avoid the use of mains water altogether. The mains water system, based on large reservoirs and probably distributing water over large distances, interferes with natural ecosystems. Valleys are flooded, and habitats below dam walls suffer perpetual drought, when the flows are limited and periodic flooding has been eliminated.

We must consider in turn the rain that falls on the various surfaces of our land: on the garden, on hard paving surfaces and on roofs. It can be directed into garden beds from hard surfaces if they are constructed sloping down towards the garden. Swales (shallow linear depressions) can help capture excess run-off and channel the water into an appropriate area of the garden. Rainfall from roofs can be collected and stored for use in times when natural rainfall does not meet the garden's needs or when we want to reduce the demand on mains water for other household functions (see below).

Figure 9.14 shows how the water can be used in a household. Harvested rainwater can be used for irrigation, the toilet, laundry, and possibly hot water system, but water can also be re-used, as greywater, to flush the toilet and with caution, for irrigation.

Harvesting stormwater from roofs

Table 9.1 shows the volume of water that can be collected from the various surfaces of a typical sized dwelling in each of Australia's major cities. As 1 mm water falling on 1 m^2 of surface area produces 1 L of water, tens of thousands of litres can potentially be harvested by capturing rain from the roof (see Info Box 9.3). This can meet many of the household water needs in most of our cities.

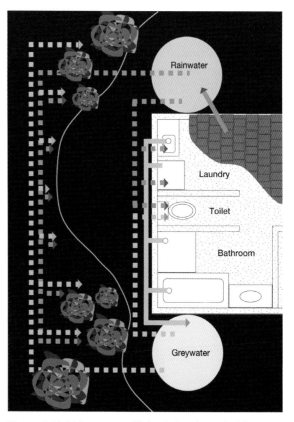

Figure 9.14 Using water efficiently in a household.

Table 9.1 Water that can be collected over one year for a typical urban house in Australia's major cities (note losses of up to 15% through spillage, evaporation or leakage may occur)

City (mean annual rainfall (mm))	Water collected on 150 m² roof (litres)	Water collected on 120 m² paving (litres)	Water collected by garden 380 m² (litres)	Total water for 650 m² house block (litres)
Adelaide (530)	79 500	63 600	201 400	344 500
Brisbane (1146)	171 900	137 520	435 480	744 900
Canberra (633)	94 950	75 960	240 540	411 450
Darwin (1535)	230 250	184 200	583 300	997 750
Hobart (569)	85 350	68 280	216 220	369 850
Melbourne (653)	97 950	78 360	248 140	424 450
Perth (869)	130 350	104 280	330 220	564 850
Sydney (1217)	182 550	146 040	462 460	791 050

Anticipating monthly rainfall from Bureau of Meteorology averages helps in developing a water budget. Knowing how much rain you can harvest, when you can harvest it, and how and when you need it will guide decisions about the area of garden to be irrigated and the tank capacity required.

Using the first month with excess rainfall over use (March for the clay loam and April for the sandy loam) as the starting point, Figures 9.15 and 9.16 predict for different tank sizes when all rainfall can be captured, when overflows occur, how much water can be lost through insufficient storage capacity, and when insufficient water remains for irrigation or other household uses.

Under this model, those households with water-efficient appliances, sandy loam soils, and which require harvested rainwater for both irrigation and household needs (toilet and laundry), will have an empty tank for a period of 2–3 months if water is stored in anything less than a 25 000 L tank, and this right at the time when the water is needed most for irrigation. Only the 25 000 L tank is likely to capture all rainfall and store sufficient water to maintain an irrigation schedule through the year, and inadequate tank capacity will lead to the loss of thousands of litres of potentially harvestable water during the wetter periods.

> **INFO BOX 9.3: DISCOVERING YOUR OWN WATER SOURCES**
>
> 1 mm of rainfall over 1 m² is equivalent to 1 litre of water
>
> So rainfall collected on your roof for one year:
>
> Rain harvested (l) = Average annual rainfall (mm) x Roof area (m²)
>
> Similarly, rain falling on the hard paving that can be directed into garden beds or underground tanks:
>
> Rain harvested (l) = Average annual rainfall (mm) x Paving area (m²)
>
> The total rainfall resource includes rainfall on garden areas:
>
> Rain harvested (l) = Average annual rainfall (mm) x Land area (m²)

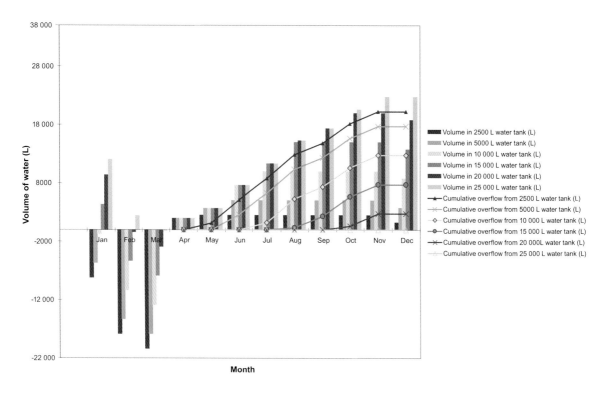

Figure 9.15 Water storage and cumulative overflow for various water tank sizes in Melbourne (sandy loam).

Clay loam soils result in a significantly different pattern of household water capture and use. Although at each watering approximately the same amount of water is applied per square metre (through fewer drippers, but over a longer period) the greater water holding capacity of the soil increases the time between each irrigation. A 10 000 L tank will therefore adequately meet both household (laundry and toilet) and garden irrigation needs.

Although the ideal tank size for the household with a sandy loam soil would be 25 000 L and for the one with a clay loam soil 10 000 L, it may not be possible to install such large tanks. Having any tank, however, will reduce the dependence on mains water which is good for the environment, and may help householders through periods of mains water restrictions. Tanks can also help reduce peak stormwater flows during and after rainstorms and therefore lessen the likelihood of flooding and erosion.

Although it could be argued that mains water restrictions should apply for both inside and outside uses, changes to restrictions are unlikely at least for the time being. Water restrictions impinge most on a householder's ability to irrigate a garden. For gardeners, using tank water for irrigation must be a high priority, and the irrigation needs for a garden during the driest times of the year will be a major determinant of tank size for any given climate.

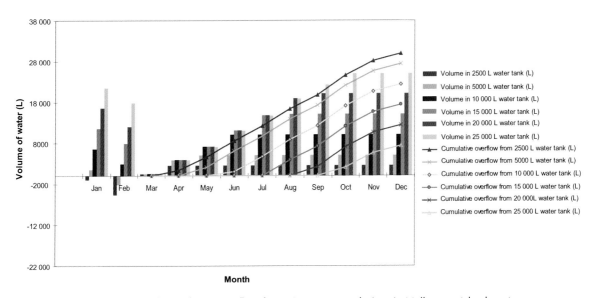

Figure 9.16 Water storage and cumulative overflow for various water tank sizes in Melbourne (clay loam).

Even with a less-than-ideal tank size, the householder can still optimise harvested water use. When rainfall exceeds the capacity of the tank, as much water as possible can be diverted indoors to lower the amount of water overflowing from the tank. Towards the end of the rainy season, indoor water use can be reduced, ensuring that the tank remains close to capacity. Perhaps water at those times will not be used either for the laundry or for the toilet, and at the times when irrigation needs are most critical, all tank water may be reserved solely for garden use.

It is possible to lower the irrigation frequency for sandy loam soils by increasing the organic content of the soil, using water-absorbing crystals and mulch. It is also possible through careful plant selection to design gardens with sandy loam soils that require less or no irrigation, gardens relying only on natural rainfall. Irrigation demands for clay loam soils can be reduced in the same way.

Appliances needing replacement offer an opportunity to significantly lower water demand. Front-loading washing machines can use as little as 40–50 L per wash which is about a third of that used by older top-loading machines. Dual-flush toilets have been around for a long time, but those manufactured in the last 10 years or so are designed to use just 3 or 6 L. Even more water can be saved by restricting toilet flushing for faeces and only occasionally for urine.

In terms of water storage, the important questions to consider when designing a garden are:

- Is irrigation required at all?
- Which plants will require irrigation?
- Can they be grouped into similar water requiring areas of the garden?
- What are the yearly water requirements of the irrigated part of the garden?
- When is the water required?

Table 9.2 Estimated change by 2030 to annual average rainfall, average seasonal rainfall and annual average potential evaporation for regions of Australia

Region	Estimated change to feature					
	Annual average rainfall	Summer average rainfall	Autumn average rainfall	Winter average rainfall	Spring average rainfall	Annual average potential evaporation
SW Australia	−5 to −11%	−3 to −7.5%	−3 to −7.5%	−5 to −11%	−5 to −11%	+1.9 to +4.3%
Southern SA	−3 to −7.5%	−3 to −7.5%	−1.5 to −3.5%	−5 to −11%	−5 to −11%	+1.6 to +3.7%
Victoria	−1.5 to −3.5%	0%	−1.5 to −3.5%	−1.5 to −3.5%	−5 to −11%	+2.2 to +5%
Tasmania	+1.5 to +3.5%	−3 to −7.5%	0%	+1.5 to +3.5%	−1.5 to −3.5%	+1.9 to +4.4%
NSW	0%	+1.5 to +3.5%	+1.5 to +3.5%	−3 to −7.5%	−3 to −7.5%	+2.4 to +5.6%
SE Qld	−1.5 to −3.5%	0%	−3 to −7.5%	−3 to −7.5%	−3 to −7.5%	+2.4 to +5.6%
NE Qld	−1.5 to −3.5%	+1.5 to +3.5%	−3 to −7.5%	NA	0%	+1.6 to +3.7%
NT (Top End)	0%	−1.5 to −3.5%	0%	NA	+1.5 to +3.5%	+1.6 to +3.7%

Source: CSIRO Marine and Atmospheric Research (2006).

- Will mains water be available to make up any tank water shortfall?

Calculations for tank size requirements should allow for long dry periods and for the expected changes to long-term averages under human-induced climate change. CSIRO scientists are continually refining predictions of the conditions we are likely to experience as we live through the 21st century, and although the actual figures may change, trends for a drying climate for a number of regions can be seen in Table 9.2. Southern mainland Australia and Queensland are expected to receive less annual rainfall. New South Wales and the Top End of the Northern Territory are predicted to receive the same annual rainfall. Tasmania is expected to receive more annual rainfall, although with less in spring and summer. All regions are likely to have increased evaporation.

Even with these changes occurring, it is worth noting that the water harvesting capacity for Australian households remains significant, and can make an important contribution to water management for a household and for a community.

Choices for storing water

The traditional corrugated steel water tanks that are so much a part of Australia's rural landscape are reliable, economic and will suit many situations, but they do take up valuable space, and they can corrode (Figure 9.17). In

Figure 9.17 Traditional galvanised steel rainwater tank.

recent years there has been an increasing choice in the sizes, shapes and colours of water tanks. They can now fit physically and unobtrusively into tight corners of the garden. Tanks are made from steel, plastic (polyethylene, food grade), reinforced concrete and fibreglass. No matter what the material, tanks come with an embodied energy cost.

In an LCA study done by the Centre of Sustainable Technology at the University of Newcastle, water tanks made from steel, high density polyethylene (HDPE) and concrete were compared. Steel tanks were found to have the least environmental impact, having the lowest embodied energy, lowest emissions of greenhouse gases and sulphurous oxides, and had low embodied water and emissions of nitrous oxides. HDPE had the highest embodied energy, and reasonably high emissions of greenhouse gases, but the lowest embodied water and emissions of nitrous oxides. It must be remembered though that HDPE like other plastics is made from non-renewable fossil fuels, and recycling the plastic at the end of its life may not be practical due to the long exposure to UV light. Concrete tanks had the highest of all categories assessed apart from embodied energy.

Some modern tanks have an architectural function as well, storing water and becoming 'feature' walls, perhaps free-standing, that help define garden spaces, and introduce new textures. They are available in a number of colours (Figure 9.18). Free-standing models need to be installed on concrete footings. Shapes include tall, narrow tanks that cover very little land area, those that fit under eaves

Figure 9.18 Waterwall® comes in 1200 and 2400 litre sizes. One model can be used as a freestanding wall in the garden. (Photo: Waterwall Solutions)

and so on. The narrowest models are less than 400 mm deep and can hold up to 1200 L. Although certain models are low capacity, they might have a modular design so additional units can be connected to increase the water storage capacity.

Underground tanks are more expensive to install and require heavy fossil-fuel powered excavation equipment to do so, but they can be placed under lawn areas. Some models are strong enough to be placed under driveways. They can be made from concrete, food grade plastics or fibreglass. The plastic models generally have a capacity of 3000–5000 litres (Figure 9.19), although larger ones are available, and often it is possible to have multiple units that can increase overall storage capacity (Figure 9.20). There are fibreglass models that store 5000 L or more.

Another load-bearing underground water storage system (for example, Rainstore®) is made up of modular, thin-walled, recycled

Figure 9.19 Underground water tank. (Photo: Tankmasta®)

polypropylene plastic cells (1 × 1 × 0.1 m) that stack to the volume required and that can be enclosed in a watertight membrane (or filter fabric if the system is to be used for retaining and filtering stormwater before continuing into urban waterways). The surrounding membrane upon which the cells are stacked goes over a prepared base of compact sandy gravel. The cells can be installed below lawns and are strong enough to go under driveways. Being modular, this system can be made to various sizes. It also has a greater effective storage volume for the size of excavation compared with other underground storage systems because of the rectangular shape and low volume of plastic per cell volume.

An additional or alternative source of water to feed underground tanks captures the rain falling on some permeable synthetic grass surfaces installed over an impervious membrane that collects and directs the water into the storage unit (see Figure 9.30). The system is potentially useful for sports fields using synthetic turf.

A recent development is the use of bladder systems akin to the bag in a wine cask, swelling in size when filling and collapsing as the water drains (Figure 9.21). The geomembrane that it is made from has high tensile strength and is highly resistant to tearing, puncturing, abrasion and microbial attack. The bladder is housed in a geotextile blanket supported by a steel frame. These systems can be installed on the ground, under the floor and between the stumps of the house or under an outside deck. They come in standard sizes between 2200 and 7000 L in different dimensions depending on the spacing restrictions between the stumps where the system is to be installed, but can also be made to specification. Additional bladders can be added to increase capacity. The system needs a minimum vertical clearance of 750 mm. Having the storage system under the house makes it easier to direct the rainwater from all or most of the downpipes around the house.

Figure 9.20 It is possible for some underground tanks to be joined, increasing overall capacity. (Photo: Action Tanks)

Figure 9.21 Bladder systems like this Rain Reviva one can collect water from all or most downpipes for storage under houses or decks. (Photo: New Water)

New housing built on a concrete slab can also have under-house water storage in the system known as waffle raft construction. Prior to pouring the concrete, interconnected modular tanks of approximately 600 L are installed in a regular grid among the spacers that create the waffle-like structure of the slab. Roof gutter storage systems are also available. These larger-than-normal gutters are cheaper when installed on new houses rather than existing ones.

No matter what water storage solution you choose, your plumber should be able to connect it to the irrigation system, laundry, toilet, and in some cases, the hot water system. It is also possible to install an automatic switching mechanism to revert to the mains as the source of water when the level in the tank becomes too low. Other systems will top up the tank from the mains, but water restrictions may apply to water from these mains connected tanks.

For maximum harvest we need to collect water from all downpipes. If this cannot be done then we need to connect those collecting rainfall from the largest roof areas. Underground or under-house storages allow collection from many points remembering that the stormwater piping can be installed to cross under the house. Positioning of the water storage will also take into account where the water will be used, and allow access to power if this is needed for pumps.

It may be useful to use U-shaped piping under paths if it enables better positioning for the tank (Figure 9.22). The U-pipe in this system (also known as a wet system because the pipe between the roof guttering and tank will always have water in it) ideally needs all entry and exit points to this pipe, as well as the tank itself, to be shielded with 1 mm mesh

Figure 9.22 Downpipes can extend under paths before rising into above-ground tanks.

to prevent mosquitoes breeding, although this is not always practicable. Similarly, mosquitoes need to be excluded from the tank water of dry systems, although because the pipe drains completely of water after the rain has stopped, the pipe does not need to be protected. In some areas, neglecting to control the breeding of mosquitoes may have more than nuisance-value. Diseases like Ross River Fever and Dengue Fever can be transmitted by mosquitoes in some parts of Australia. Excluding mosquitoes from the stored water will also exclude other animals, like birds, that might contaminate the water.

It must be remembered that having a water tank does require maintenance. However, by including a first-flush diverter and filters, and by installing leaf guards on the spouting or using guttering designed to exclude leaves and other debris, the inflow of organic materials and other contaminants from the roof can be minimised (Figure 9.23).

In most situations, the water pressure from gravity-fed tanks will be too low for effective watering and indoor use. Automatic pressure-activated pumps for irrigation systems, for example, will be needed, or if space or the site slope allow, the water can be pumped to a higher header tank. Using a header tank allows water to be transferred using a solar powered pump during the day even though the water may not be needed until later, but this minimises the need for using mains electricity. Specifying an energy-efficient pump is desirable if it is to be run from the mains.

Harvesting stormwater from the landscape

Soil is an important water reservoir for a garden. It can hold many times more water than even the most elaborate on-site tank system, and established plants are able to access the soil water (green water) during prolonged dry periods. The importance of soil water became obvious during the lengthy drought of the 1990s and early 2000s, supporting much of our natural and planted landscapes, although the drought was sufficiently severe to cause water stress and occasional death in established trees.

Adding to the soil water storage can be as easy as draining downpipes directly onto the garden beds as is done as standard practice in Canada (Figure 9.24). Local council regulations need to be checked before this is done in Australia, and it is also important not to have excess stormwater running off your property onto your neighbours. There are simple fittings that can be installed into downpipes for redirecting water from the roof into storage containers or into fitted hoses that direct extra water onto the drier parts of the garden (Figure 9.25).

Figure 9.23 The Smartflo guttering system designed to keep leaves and other debris out of the water storage. (Photo: Smartflo)

Figure 9.24 Downpipes in Canada usually empty directly onto lawns or garden beds. (Photo: Donna Samco)

Garden beds and lawns can be contoured to collect and direct the garden's surface water into parts of the garden where it can accumulate in swales, infiltration trenches, soaks, rain gardens, or ephemeral wetland features that support plants requiring a higher water availability, at the same time as providing an infiltration point for the water to soak into the soil.

Concrete, brick or stone paving is usually impervious to water. This encouraged the rapid removal of water from the site, leaving little opportunity for it to recharge the soil water. Nineteenth century street gutters were better, with the gaps between the stone cobbles allowing some water to percolate into the soil (Figure 9.26). Paving can be easily sloped towards garden beds or lawns, rather than towards stormwater drains, allowing the harvesting of a considerable quantity of water. A typical garden may have 150 m^2 of hard landscaped surfaces which, over the course of a year in a region with an average annual rainfall of 600 mm, can provide up to 90 000 L of water.

Increasing the permeability of paved areas, rather like the 19th century gutters, is another way of harnessing natural rainfall, and can be achieved in a number of ways. Although not as widely accepted in Australia as in Europe where it is popular, porous

Figure 9.25 Fittings installed into downpipes can redirect water into the garden rather than the stormwater drains.

Figure 9.26 Nineteenth century kerbing made from cobblestones allowed some water to infiltrate into the soil.

Figure 9.27 Decking allows rainfall to reach the soil underneath while protecting tree roots from the damaging effects of pedestrian traffic. (Photo: Royal Botanic Gardens Melbourne)

Figure 9.28 Pavers laid with spaces between let rainfall penetrate into the soil beneath, and can be interplanted.

paving is potentially very useful for enabling the infiltration of water into the soil, and lends itself to Water Sensitive Urban Design (WSUD) (see next chapter). Porous paving can be used for roads, roadsides, parking and pedestrian areas, and can be made from many materials including timber decking, concrete, asphalt, stone, chipped wood, and aggregates of even-sized particles bound together by an adhesive compound.

Instead of a hard-surfaced paved terrace, installing a wooden deck with its regular spacing allows rainfall to penetrate to the ground below (Figure 9.27). Garden paths can be made of masonry pavers such as concrete that are laid on a sand/gravel base. They can also be set into a mud mix (cement/sand), so their edges are not abutting, and with some of the space between pavers free from the mud base to allow water infiltration and crevice planting (Figure 9.28).

Driveways need not be paved across their full width. Using only paved strips for vehicle tyres will reduce the area covered by paving by about a third or more.

Gravel is an attractive alternative to the hard surfaces of concrete, brick or stone, and has a natural ability to drain. Stabilising the gravel may be important, particularly in well-trafficked areas, and can be achieved using products which retard the movement of the gravel particles by containing it in under-surface cells (e.g. Gravel Pave (www.invisiblestructures.com.au) uses recycled plastic rings bonded onto a geotextile filtering material).

Porous concrete is suitable when a stable, strong paving material is needed, and can be coloured before pouring or stained after setting to blend with the landscape. It is made by binding a graded even-sized aggregate, up to ~10 mm in size, with cement. The aggregate selected is important for forming the pores in the concrete that allow the water to pass through the paving. The ingredients are essentially the same as standard dense concrete although more cement is used in the mix, making porous concrete more expensive and with a higher embodied energy than standard concrete. Its differing characteristics also need experienced people to lay it, further adding to costs. The overall expenditure of a project using porous concrete may be cheaper though, if its use eliminates the need for other infrastructure to process stormwater collection and treatment. The void space in well-constructed porous cement is usually 11–21% and water infiltrates at well over 100 mm per hour.[2] Micro-organisms on the large surface area within porous concrete are able to reduce organic carbon in the stormwater. Nutrient load can also be lowered – a study in Florida showed levels of phosphorus down to 42% and nitrogen to 52% compared with water coming off dense concrete. If porous concrete becomes clogged with sand or organic material reducing the infiltration rate, it can be cleared by using high-pressure washing with water followed by sweeping. Vacuuming may also work to remove sand.

Porous asphalt is similar to porous concrete, with even-sized graded aggregates, but the binding agent is asphalt rather than cement. Asphalt is a flexible paving material refined from petroleum products to various levels of viscosity chosen to suit the local climate. Heat softens the asphalt when preparing the aggregate paving material and for installation. The amount of asphalt used for binding the aggregate is critical, with enough needed to coat all particles but not so much that the porosity of the pavement will be reduced. Recycled rubber can be incorporated into the asphalt.

Porous asphalt has been used since the middle of the 20th century, but not without problems. Earlier versions softened on hot summer days, and over time, the binding asphalt drained from the top of the aggregate, clogging the pores lower down. Current porous asphalt preparations have developed sufficiently to largely overcome this problem using additives of liquid polymers, or mineral or cellulose fibres which retard asphalt drain-down. Aggregate sizes have also increased to as much as 19 mm or more with corresponding increase in the pore sizes and better infiltration. Porous can cost more than dense asphalt to install, but overall project costs may be less if other stormwater infrastructure is not required. The porous asphalt pores have been shown to host microbial ecosystems that can metabolise stormwater chemicals like organic carbon and nitrogen. Clogged porous asphalt can be cleared using high-pressure water (if restrictions allow), sweeping or vacuuming. Water infiltration is an important environmental advantage of porous asphalt. It also improves road safety by having higher friction in wet conditions and has been shown to reduce tyre noise of traffic by 3 decibels or more.

Clear bonding agents are also used to bind aggregates of decorative gravel that can be laid *in situ* or pre-manufactured as pavers in similar sizes to standard landscape pavers and laid on a well-prepared, freely draining base. It has been used around street trees in the centre of Melbourne and around *Brachychiton rupestris* in the Children's Garden in the Royal Botanic Gardens Melbourne (Figure 9.29). A collaborative project between Monash University, CERES and Dymon Industries is

Figure 9.29 Porous paving made from bound aggregate placed around trees protects the tree roots from pedestrian traffic but still permits water infiltration to the root zone. These examples are in the Children's Garden of the Royal Botanic Gardens Melbourne (top) and Swanston Street Melbourne.

currently working to refine the design and installation specifications of porous paving including the development of modular designs that can be easily installed and adapted for the harvesting of stormwater for later use in toilets and for irrigation.

In turf areas there are systems like Grasspave® that enhance the percolation of rainwater through turf to subsurface soil levels, adding to the soil water reservoir, and reducing water run-off particularly during heavy rainfall.

The Eco Logical Lawn Water Capturing System (patent pending) allows rainwater falling on a synthetic lawn surface to infiltrate into the soil or pass to a sub-turf membrane which directs it to a subsurface water tank underneath or adjacent to the lawn for later garden use (Figure 9.30).

Chipped wood or bark mulches can also be used to create rustic porous paths. Being organic they will decompose and need to be topped up regularly. In freshly laid paths, the high carbon and low nitrogen content of these materials can cause nitrogen shortages for nearby plants as the micro-organisms compete for the nitrogen during decomposition. Maintenance of paths made from loose materials may be high because of the activities of people or animals causing scattering. Granular rubber made from shredded old car tyres has also been used for paths. It is claimed rubber does not leach toxic chemicals; however, the rubber's surface can hold toxic chemicals which need to be washed off prior to use. Recycled rubber used as mulch in a display bed at the Royal Botanic Gardens

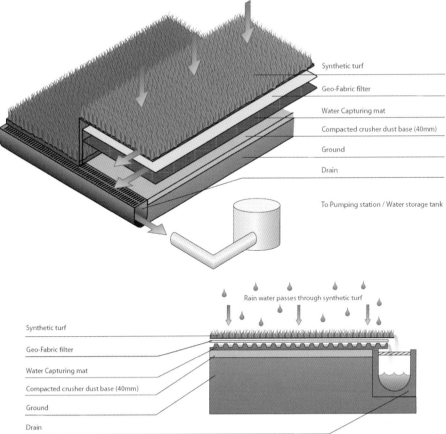

Figure 9.30 Rainfall is captured through the synthetic turf of the Eco Logical Lawn Water Capturing System and directed to underground storage. (Photo and illustrations: Pro-tech Corporation Pty Ltd)

9 – DESIGNING LOW IMPACT GARDENS

> ### INFO BOX 9.4: THE DO'S AND DON'TS OF UNTREATED GREYWATER IRRIGATION
>
> **Do:**
>
> - Check regulations for greywater use with your local council and state government authority to ensure greywater re-use is legal in your area before establishing a greywater system.
> - Where possible use the safest water sources – e.g. the rinse cycle from the washing machine and shower waste.
> - Use untreated greywater sparingly.
> - Use untreated greywater for subsurface irrigation only.
> - Filter the greywater to remove hair and lint that may block irrigation lines.
> - Choose your laundry detergents, soaps, shampoos and hair conditioners to minimise environmental or garden plant damage.
> - Use the minimum quantity of laundry detergents, soaps, shampoos and conditioners to achieve the outcome you need.
> - Design irrigation systems to distribute greywater according to plant needs.
> - Keep children and pets away from greywater.
> - Wash hands after contact with greywater.
> - Take account of soil type (slope, absorptive capacity).
> - Turn off greywater diverters when there is rain or the soil is sodden.
> - Carry out regular maintenance.
> - Monitor the effects of your greywater system, looking for greywater pooling or changes in plant growth or soil structure.
>
> **Don't:**
>
> - Include nappy washing water which contains faecal deposits.
> - Include water containing harsh sanitising chemicals like bleaches or disinfectants.
> - Use dishwashing water which may have high levels of fats or water from dishwashers containing highly alkaline dishwasher detergent.
> - Allow greywater to run from your property.
> - Store greywater for longer than 24 hours.

Cranbourne was found to leach chemicals that were toxic to plants.

Recycling water

Greywater from the bathroom and laundry (**not** blackwater from the toilet) can be used directly for subsurface irrigation of the garden in some areas of Australia, but local council and state government regulations should be checked. Greywater contains micro-organisms, mostly benign, that have been washed from our skins or from our clothes, but illness-causing micro-organisms may be included in the mix, and therefore human contact with the greywater should be minimised. Untreated greywater should definitely not be used for above-ground or overhead irrigation, and if stored prior to use, it can be stored for no more than 24 hours before the storage unit is completely evacuated (automatically for some systems), thus reducing the likelihood of a build-up of the microbial population. Greywater must not be allowed to drain away from your property into others. (See Info Box 9.4.)

There are various ways of diverting untreated greywater from the sewage system including: simple rubber funnels that can be temporarily inserted through the inspection point of drains, plastic hoses to extend the washing machine wastewater hose into the garden, PVC fittings with a manual switching valve that can be fitted permanently into drainage pipes to allow diversion when needed, as well as sophisticated systems that collect suitable greywater from the laundry and bathroom, and then automatically feed the untreated greywater through drip irrigation lines in the garden.

A cautious approach to the use of unprocessed greywater is strongly advised. Apart from the potential health issues caused by its microbial content, greywater contains the chemicals we add when showering with soaps, shampoos and hair conditioners, and when we are washing with laundry detergents. All these have the potential for polluting our soils and soil water, and for detrimentally affecting plant growth in our gardens. If untreated greywater is to be used for irrigating the garden, it should be done so sparingly, interspersed with freshwater irrigation so that the ability of the plants and soil organisms to absorb and metabolise the chemicals in the greywater is not exceeded. The quality of the greywater itself should be as high as possible, meaning minimising the amount of chemicals we use and ensuring that the chemicals we choose will be least harmful in the environment.

Unless great care is taken to minimise chemical use, our recommendation is to use untreated greywater sparingly in the garden. If everyone begins using it on a large scale, there is potential for groundwaters to become quite contaminated with the range of chemicals in soaps, detergents and shampoos used in the average household.

Some people recommend that kitchen wastewater is not used due to the fats they may contain being difficult for the soil micro-organisms to break down. However, collecting the water when rinsing fruits and vegetables provides a fine source for gardens.

The number of commercial cleaning chemicals available in our supermarkets and used in our homes increased rapidly during the 20th century. However, many of our cleaning tasks can be achieved using only a few well-chosen products that minimise negative impacts on our immediate and broader environments. There are a number of 'green cleaning' guidebooks available that can help us develop less reliance on using large numbers of chemicals, and of those that we use, choose the least harmful. The microfibre of some cleaning cloths now eliminates the need for any chemicals at all in many cleaning applications, needing only water to clean effectively. That directly improves the quality of our greywater, and the chemical load on our environment.

Laundry detergents are complex mixtures of chemicals, the ingredients varying from one brand to another. It is difficult to measure all the chemicals in commercially available laundry products for comparison but Lanfax Laboratories of Armidale, NSW has undertaken unique studies that are now referred to internationally.[3] Lanfax has looked at the levels of phosphorus, nitrogen, salts (electrical conductivity), sodium and the pH of various commercially available laundry detergents designed for both top-loading and front-loading washing machines. In recent data these parameters were measured for the total washing cycle, and some of the findings are summarised below, but it is worth regularly checking the Lanfax Laboratories website for more recent studies that account for changes in manufacturers' formulations.

In the study, most liquid detergents fell within the pH range of 5–9 which is considered by some to be acceptable, although various natural systems will be adapted to narrower pH ranges, and repetitive addition of greywater with a different pH could be detrimental. Powdered detergents are generally very alkaline, with over 85% of the front- and top-loading products tested with pH values of at least 10. Apart from the effect on the metabolism of individual organisms adapted to pH ranges well below this, the high pH could disrupt the normal functioning of natural ecosystems. Gardens could similarly be affected. Higher pH can increase the cation exchange capacity of soil particles, that is, total amount of positively charged ions such as Mg^{2+} and Ca^{2+} that can be held, but not beyond about pH 8.5, and a pH of 10 can reduce the availability of some nutrients such as nitrogen, iron, manganese and boron. It can also disrupt the growth of beneficial soil micro-organisms including the spread to and infection of seedlings by ectomycorrhizal fungi.[4]

In general, liquid laundry detergents had lower salinities than powdered detergents. When used in washing machines at the recommended rate, the average electrical conductivity (EC, an indicator of salt content)

> ### INFO BOX 9.5: SOIL pH AND PLANT GROWTH
>
> For almost all soil types there will be plant species that can grow in them. Some plants will be well adapted for growth in high pH alkaline soils and others in low pH acidic soils. However, for most species a soil pH range of 5.5 to 7 will result in healthy plant growth.

of liquid detergents was about one-tenth of the powdered ones, although lower salinity detergents are also available in powdered form. Of greater importance is the measurement of sodium in the detergent. If greywater contains high levels of sodium there can be a direct impact on plant growth and the potential to make the soil sodic. Irrigating with high-sodium waters is a major contributor to soil damage as the sodium ions replace other ions on the soil particle surface, causing the soil structure to deteriorate. Water infiltration is slow in sodic soils potentially leading to greater surface run-off and less effective rainfall harvesting into the soil, and when water of low salinity is applied to sodic soil particles they disperse away from each other resulting in poor soil structure.

High levels of nitrogen and phosphorus compounds entering natural waterways via greywater can cause eutrophication, and increased biological oxygen demand. The subsequent ecological disruption that it can cause was realised in the 1960s when Lake Erie, one of the Great Lakes of North America, was declared 'dead'. The increased flow of plant nutrients into the lake from sewage, which included high-phosphate detergents, caused an explosion of the algal population and a lowering of oxygen availability, diminishing native aquatic populations, some species of which were wiped out. Subsequent international agreements between Canada and the USA have controlled the effluent entering the lake and the problem has been reduced, but it is a reminder of what can happen on a large scale, and in water bodies that are connected to our gardens.

High phosphorus and nitrogen levels can also negatively affect growth of some plant species. Although most Australian taxa will respond with increased healthy growth when additional phosphorus is added to Australia's often phosphorus-deficient soils, many species in the family Proteaceae, for example, and some species in the genera *Acacia, Baeckea, Bauera, Beaufortia, Boronia, Bossiaea, Brachysema, Chorizema, Daviesia, Eutaxia, Hypocalymma, Jacksonia, Lechenaultia* and *Pultenaea,* will find elevated levels of phosphorus toxic.[5] Nitrogen in the form of ammonium can cause toxicity for species in the genera *Acacia, Alyssum, Banksia, Dianthus* (carnation), *Impatiens, Lactuca* (lettuce), *Pilea, Protea, Salvia, Tagetes* (marigold), *Thryptomene* and *Verbena*.[4]

Ultrasonic washing machines requiring no detergent for lightly soiled clothes are now available in Japan. These leave the wash water completely free of detergents, unless the clothes are heavily soiled.

Similarly to laundry detergents, soaps, shampoos and hair conditioners are also variable complex mixtures of chemicals, but there has not yet been similar studies to Lanfax Laboratories' laundry detergent study for these bathroom products, so it is difficult for a consumer to make choices that will help to determine which product will have the least environmental impact when processed by the sewage system of a community. Soaps are usually largely composed of sodium salts of fatty acids (triglycerides) formed through the saponification reaction resulting from the addition of sodium hydroxide to fats or oils. During the cleaning process when we are using soap, the Na^+ ions dissociate. As already noted, sodium ions can lead to sodicity and a resulting decline in the quality of soil structure, so untreated shower greywater should be applied cautiously and sparingly.

The School of Applied Sciences at RMIT University is currently undertaking a preliminary study with the Alternative Technology Association that will help improve our understanding of greywater interactions with soils and the long-term effects of using untreated greywater under different conditions. Another project at the University is looking at the sodium ion adsorption properties of peat, their work potentially contributing to systems lowering sodium in greywater that is used for irrigating.

A way to avoid these issues is to process the greywater before use so the water applied to gardens is of a sufficiently high quality. There are now a number of greywater processing systems that work by an initial filtering of solids, and then deactivation of the remaining microbial population to bring the water to class A standard (Figure 9.31). These systems may be in the order of AU$8–10 000. There is also a new system of processing that will soon be available in a domestic scale model that can process blackwater without the initial

Figure 9.31 The Oasis GT600® and the Aqua Reviva process greywater at least close to Class A standard. (Photos: Nubian Water Systems and New Water)

filtering, and that uses pressure and shearing forces to rupture the micro-organisms.

A problem with greywater treatment systems is that the equipment contains a large amount of embodied energy, they require energy to run, and some have filters to replace. It is also worth enquiring about the salt levels of the treated water to ensure that the water quality is adequate for irrigating plants. Even with these treatment systems, a cautious approach to household chemical use is important.

> **INFO BOX 9.6: CONSIDERATIONS FOR GREYWATER TREATMENT SYSTEMS**
>
> - Ensure the system complies with state authority and local council requirements.
> - Obtain local council permits if required.
> - Treatment systems can be complicated, costly, and have high levels of embodied energy.
> - Make sure the installation is simple, effective and produces a water quality suitable for plants.
> - Carry out regular maintenance.

The main steps of the domestic greywater treatment systems are: prefiltering to remove lint, hair and other larger material that may block the system; collection in a pre-treatment reservoir; treatment in a unit that uses physical, chemical, adsorption and microbial processes; disinfection with UV light or bromine; and storage of this Class A quality water for later use (not for drinking). Systems may require backwashing or periodic changes of cartridges. The systems take up little space under the eves and may use under-house storage.

Greywater systems can also use the filtering capacity of natural wetland systems if there is sufficient space in the garden. In Clovelly House in beachside Sydney, Kennedy Associates Architects designed a Green Wall that acts like a vertical wetland, an appropriate solution for smaller gardens (Figure 9.32).

The Mobbs House in Sydney treats all its wastewater, including blackwater, using a

Biolytix® filter system, a created ecosystem contained in a tank where worms, insects and micro-organisms in a soil-like matrix of humus and coco-peat, filter and process the wastewater as it trickles through layers of filter beds.[6] The passive system is energy efficient, requiring little power to operate. The water at the base of the tank is collected and disinfected with UV light, and can be used for the garden and toilet.

When considering greywater treatment systems where the water will be used on gardens, it is important to ask the supplier of the system about the chemical composition of the treated water. Although it is important to reduce the microbial population to minimal levels for human health reasons, it is equally important that the remaining chemicals in the treated water are at levels that will not be toxic to the plants to be watered.

Optimising irrigation

Ideally a garden should be designed to avoid any supplementary watering over and above the natural rainfall or what can be harvested from it. This can be done by appropriate plant selection, reducing wind using shelterbelts or walls, increasing shade with garden structures or spreading trees, matching the plants to the varying microenvironments of the site and so on. There are many reasons why we may need to water but whatever the constraints, we can minimise that need.

Selection of plants with low water demand is a good start. For example, indigenous plants are adapted to survive in a particular region and can be retained or reintroduced. To these plants can be added those that are not native but are adapted to similar climatic conditions and will not require irrigating, although caution is needed to avoid introducing potential environmental weeds.

The designer can group plants with higher water requirements into zones of similar water

Fig 9.32 Clovelly House in beachside Sydney has a green wall greywater treatment system. (Photos: Bart Maiorana via Kennedy Associates Architects)

> ### INFO BOX 9.7: CLASS A WATER
>
> Approved greywater treatment systems process wastewater to Class A standard which will have extremely low levels of potential pathogens. Wastewater can be processed on-site or, in some new subdivisions, can be collected from a community, processed centrally and returned to individual residences via a third 'purple' pipe system (in addition to fresh drinking water and sewage pipes).
>
> Class A water is required to meet a defined level of water quality with specified micro-organisms present in lower concentrations than a defined threshold. There are no restrictions on the type of irrigation system Class A water can be used in, and it can therefore be used in overhead watering for all food crops including those consumed raw and grown close to the ground.

needs. The plants can be placed in a low point of the garden where the natural drainage accumulates, or on the shadier southern side of walls, or, if supplementary irrigation is needed, assembled into small areas so the irrigation system can be kept to a minimum and water will not be wasted on plants with little need of additional water.

The irrigation efficiency of different systems is an important factor to consider. In general, drip irrigation systems have the lowest water demand (subsurface irrigation being the most efficient drip) followed by microspray which can lose a significant amount of water through evaporation of the water from the soil and plant surfaces, and due to wind drift. Drip systems can be more problematic on very sandy soils as the soil wetting pattern has little of the lateral spread that is found on heavier soils with water tending to percolate straight down. However, many of these problems can be overcome by decreasing the spacing between drippers, increasing the application rate, and ameliorating the soil with organic matter to increase its water holding capacity. There should be no need to consider other irrigation systems for most domestic gardens, but public landscapes may require more robust systems.

The recent prolonged drought introduced us to what climate change predicts for much of Australia. Increasingly severe water restrictions were imposed, with even the standard non-drought conditions of water use becoming more restrictive, and outdoor water use not possible at all in many areas. Even where water could still be used outdoors, it was restricted to hand-held hoses or drip irrigation systems at very limited times through the week. For many areas of Australia, it is likely that gardeners will be increasingly cycling through different levels of water restriction as rainfall remains low, and if an irrigation system is required for maintaining the health of a particular landscape, the most effective way of delivering water is via a drip irrigation system. Indeed, at the times when irrigation is most needed, drip will be the only permissible system if mains water can be used at all. Correctly designed drip systems are the most effective method of delivering water to plants using the often limited water harvested and stored in tanks on site.

No matter how well designed the garden and irrigation system, its effectiveness is also influenced by how it is used. Monitoring the weather (including rainfall, temperatures and evaporation), the soil moisture and plant condition will help guide when irrigation is needed, and there are no automatic systems that can totally replace our involvement.

The hardware of irrigation systems is predominantly made of plastics which come with high embodied energy. PVC and other plastics have embodied energies of over 100 000MJ/m^3 which means the energy in a 6 metre long 40 mm diameter PVC pipe is enough to power a 50 W light globe for half an hour. It is important therefore to design the irrigation system using the minimum of materials. The manufacture of PVC, an organochloride, also produces toxic compounds and the solvents used for its installation can also pose health problems.

Xeriscape

The word *Xeriscape* (from the Greek *xeros* meaning dry landscape) was introduced by the Denver Water Department, Colorado, in 1981 to promote the ideas behind water conserving landscape design. The use of the term and the practice of Xeriscape design have become more widespread in the drier parts of the United States, and it has also been adopted to some extent in Australia. In the early 1990s, the Royal Botanic Gardens Melbourne in collaboration with Melbourne Water installed the Water Conservation Garden modelled on Xeriscape principles (see Figure 9.33). The seven principles are:

- Planning and design for low water use.
- Improving soil to increase water infiltration and water-holding capacity.

Figure 9.33 Water conservation garden at the Royal Botanic Gardens Melbourne.

- Irrigating efficiently.
- Selecting appropriate plants and grouping together plants of similar water needs.
- Using mulches to reduce evapo-transpiration and weed competition.
- Creating practical turf areas in size and turf type.
- Maintaining appropriately.

Water features

Water features within landscapes provide a calming and cooling ambience, especially in warm, dry climates. They have been used effectively since antiquity in places like the Moorish gardens of southern Spain. There is definitely a place for the use of water in Australian gardens too, developed in a way appropriate for our needs, our garden spaces and water availability. In some regions, if planned well, a water feature may be included in the garden's water budget and may help to modify the microenvironment. However, it is worth noting that all water features will need regular replenishment of water, so their size and style need to be considered carefully. Even small, still, reflecting pools will have significant rates of evaporation on hot days, even in shady corners. Fountains and moving water features are subject to greater levels of evaporation as the spraying of the water actually increases the water's surface area exposed to the air. Any feature requiring a pump also adds to our energy use.

We can treat water features as being ephemeral, capturing rainfall in pools or along water-courses in times of excess. We can divert rainfall from paving or the roof through a water feature on its way to our soil water storages. The feature may be a periodically flooded wetland that cycles through wet and dry times, like many Australian natural creek beds.

Energy and landscapes

Landscapes reducing energy use in buildings

Houses surrounded by vegetation, particularly gardens with layers of groundcovers, shrubs and trees, sit in their own microclimate, and in warm or hot climates, indoor temperatures can be made significantly cooler, particularly with well selected and placed shade trees. Deciduous trees offer summer shade and winter sun. Having more vegetation than paving can reduce reflected and radiant heat. Plantings shielding buildings from the full force of prevailing cold or hot winds can moderate indoor temperatures.

These are examples of garden services which, like environmental services, provide our communities with benefits that may not otherwise exist or would need an engineering solution to achieve the same result. The examples show how garden services can eliminate or reduce the need for air conditioning and heating.

The effects can be seen across a city community too. The public parks and gardens, street trees, and the private gardens of the dominant detached housing, ensure large areas of Australian cities are also urban forests. Although we appreciate living in our verdant cities, we may not be fully aware of all the benefits they provide. Imagine walking on a hot day down the street in your area, with all the vegetation gone.

Shade is the biggest benefit trees offer. The Western Center for Urban Forest Research and Education in California, using methods they developed, found trees reduced the use of energy for annual cooling by 10.9%, peak cooling by 6.1% and annual heating by 0.7%.[7] They also found that trees shading the western side of a building gave the largest annual and peak energy savings, and the next largest savings were from trees shading the south-western (equivalent to our north-west) and eastern sides of the house.

Evapotranspiration probably also contributes to the cooling effect of a residential landscape if the planting is dense enough and is multilayered under trees, the garden acting rather like a giant evaporative cooler. The system is reliant on adequate soil water for evapotranspiration to occur. It is interesting that mains water restrictions may affect the operation of this cooling system, but do not apply to electrically powered evaporative coolers that can use 50–60 L/hour (that is 1200–1440 L/day) to cool a 140 m^2 house at 35% relative humidity.[8] What is not clear is which system may be more efficient at cooling per litre of water used. This is an area that requires additional research. It is worth noting though that evaporative coolers may also contribute to greenhouse gas production if green electrical energy is not used.

Landscapes complementing energy efficient buildings

Buildings account for a large proportion of Western society's energy use and greenhouse gas emissions. For an average Australian household, excluding embodied energy of products, energy use in the home causes the greatest amount of greenhouse gas production, followed by the household's transport needs and then waste.[9] Low energy homes are not new but are now being encouraged through legislation and an improving market value; they incorporate technological solutions, passive design principles, or both.

It is essential that a building's accompanying landscape or garden enhances its energy efficient properties, and certainly does not work against them. Designing their landscape requires an awareness and understanding of the energy efficient qualities of a building.

Passive solar houses

Passive solar houses, for example, are designed with overhanging eaves to keep hot summer sunshine out but to capture winter sunshine through north-facing windows. The winter sun directly warms internal masonry floors and walls during the day, and these re-radiate the warmth inside the house during the evening, maintaining a comfortable temperature without using additional energy sources. Bricks and stone, with their ability to hold a lot of heat energy (they have high thermal mass), are ideal for this purpose. This system relies on exposing the walls and floors to the maximum amount of the sun's radiant energy as possible. Vegetation shading the windows during the cooler months of the year will reduce the solar energy gain and therefore the warmth of the house, increasing the need for supplementary heating.

The angle of the sun at the winter solstice will dictate the maximum vegetation height for various distances from the building. Figure 9.34 shows the various maximum vegetation heights at different distances from a building for Australian cities at different latitudes that will ensure that the sun will still radiate through north facing windows on the winter solstice (the shortest day of the year).

Complementing careful placement of plants on the north side of houses can be banks of vegetation that protect houses from cold prevailing winds, reducing their penetration into the house. In southern parts of Australia, this means planting on the south-west, south and south-east sides of houses.

Solar hot water and photovoltaic panels

Preventing the shading of solar hot water panels and solar photovoltaic cells is also important if they are to work at maximum efficiency. The number of months that a solar

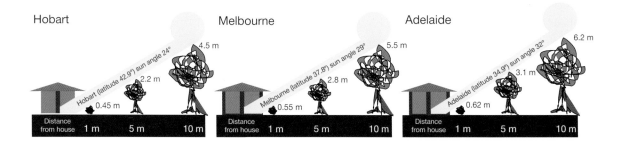
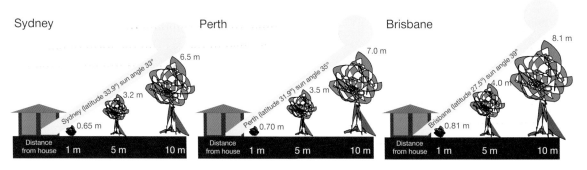

Fig 9.34 The noon sun angle at the winter solstice determines planting distance/height for houses passively solar-heated – examples for Australian cities.

hot water system can operate without using a gas or electrical backup will be extended if the collecting panels are never subject to shade. Likewise, photovoltaic panels will generate more electricity if never shaded. Panels with monocrystalline or polycrystalline cells may have a considerable decrease in output, even if only one cell of a panel is shaded, it acting like a switch turning off the circuit for a number of cells.[10] Shading can also cause hotspots and thus damage to certain solar panels. Amorphous photovoltaic cells may be able to function with some shading.

On-balance decisions

As always, competing needs may need assessment according to the particular circumstances of the site. So, for example, although deciduous trees go a long way to accommodate the need for summer shade and winter sun, even bare branches will block some of the weak winter sun from reaching the internal thermal storage floors and walls of the building, or the external solar hot water or photovoltaic panels. Is summer shade more important than winter sun? Are high temperatures more bearable than cold? Can some shade be tolerated without causing damage to the equipment? If deciduous trees are to be planted across the line of winter sun, species with less and finer branching may be the better choice.

Trees themselves dry the soil, can create root problems for drains and building foundations, disrupt paths and other structures, and their leaves block gutters and drains. In an average-sized garden the welcome summer shade can also cut out light needed for important plant groups such as vegetables and natives.

Balancing these and other competing questions is very much tied to the individual needs of the occupants, and the specific site conditions.

Clothes drying
Including appropriate service areas in a garden can also reduce a household's energy consumption. An area for outdoor clothes drying eliminates the need for electric clothes dryers (and their embodied energy), and for most parts of Australia there is little need for these energy-consuming appliances.

Reducing embodied and maintenance energy
Embodied energy
Every material we bring in to a garden contains embodied energy as seen in the chapter on energy. Gardens can contain a surprisingly large amount of this in their paving, fencing, retaining walls, irrigation, drainage, decks, terraces, pergolas, sheds, pools, fountains, furniture, pots and other features. Designing is the time to ask whether a structure is really needed, if the size can be reduced or an alternative material can be used. A garden's embodied energy can be significantly lowered at this stage.

Comparing the embodied energy for common landscape and building materials will guide choices (see Table A3 in the Appendix) along with a consideration of a material's expected life span and the energy input needed to maintain it. A lower initial embodied energy may be outweighed by the need to replace the material periodically or its need for frequent upkeep, each of these adding to the overall energy costs. For example, 30 m^2 of hardwood decking, nailed onto hardwood joists and supported by concrete stumps has an embodied energy in the order of 1700 MJ, and oiling the deck adds around 290 MJ. After six years of annual oiling, the same amount of energy is embodied in the deck in the form of oil, as all the other materials. This is still far less than concrete paving of the same area which has an embodied energy of over 13 000 MJ. Gravelled paved areas like the Lilydale toppings used locally in Melbourne can have relatively low embodied energy and can be easy to maintain.

Natural stone usually can be extracted without a high energy input, and as long as it does not have to be transported over long distances, will not have a high embodied energy. Using local stone reinforces the landscape's sense of place, and increases the viability of local quarries.

It is worth noting that some products may have a high embodied energy as do adhesives, but the quantities used may be very small.

Often different information sources will record varying embodied energy values for the same materials, and that is why data from Australia, New Zealand and the USA have been included in Table A3 in the Appendix. The variation reflects the impact of sourcing raw materials and processing them in different parts of the world, each region being influenced by the proximity of resources and markets, and the efficiency of their energy production and use. It is also an indication that the methodologies for calculating embodied energies are still being refined.

The materials in Table A3 have been ranked in order of increasing embodied energy within their landscape categories, making it easier to compare the embodied energy of different materials. Metals, plastics and paints have some of the highest embodied energies and are best used moderately. For example, designing gardens to minimise the extent of irrigation systems, which are mostly plastic, can reduce a garden's embodied energy significantly.

Paving is often a major landscape feature contributing a sizeable proportion to the total embodied energy of a garden. Asphalt has the highest embodied energy. Clay bricks have less, but added to the energy in the bricks is the embodied energy of the materials they are laid on: the mortar mix of cement and sand, or the sand combined with the energy needed to level them with a compactor. Concrete pavers or *in situ* poured concrete have a lower embodied energy per kilogram, but the concrete pavers also need to be laid on a base and poured concrete may need metal reinforcing.

New processes will change the relative environmental impacts of materials. Steel manufacturing is responding to climate change by developing low-temperature compact blast furnaces in Japan, Ultra-low CO_2 Steel (ULCOS) in Europe, and HIsmelt for continuous ironmaking in Australia.[11] Rather than using coal or coke, CSIRO Minerals is part of a team investigating the use of charcoal made from the native hardwoods being planted to reduce soil salinity. The CO_2 released during steel processing can be reabsorbed by the replacement trees, avoiding the release of carbon from fossil fuels. It is also expected the furnaces can be operated at lower temperatures and increased production, if charcoal is used.

Re-using or reprocessing materials saves large amounts of energy although it is estimated that 30% of salvaged building materials are not re-usable. Re-using materials without reprocessing saves the most energy (over 90% of the embodied energy is saved for a variety of materials), but even reprocessing is worthwhile for many materials. Aluminium and steel reprocessing respectively saves 79% and 75% of the energy that would have been required to produce the same quantity from raw materials. Glass saves 22% of the energy.

Life Cycle Assessment (LCA) (Figure 9.35) of the materials and furnishings used in gardens offers a clear way of finding out the energy impacts of different designs. Figure 9.36 summarises the major landscape stages that can be analysed through LCA.

The science of LCA is relatively new and the methodologies are still being refined. Many economic sectors have few or no LCA studies published. The landscape industry is currently poorly served, and although some landscape materials have had LCAs, the materials have often been analysed as components of buildings rather than in the very different landscape environment where they can behave quite differently.

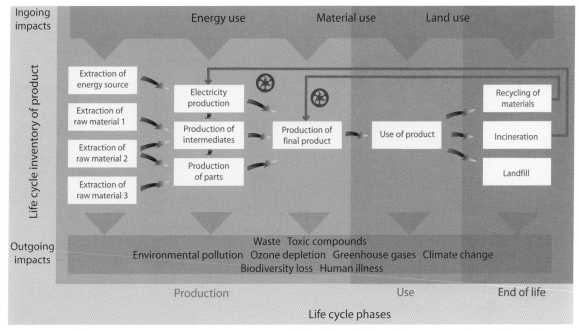

Figure 9.35 Schematic representation of a product system's life cycle with data collection points for product and waste flows (orange-red-yellow arrows from left to right). Also illustrating the impact assessment of the resource consumption coming in at the top, emissions going out at the bottom, and opportunities to recycle.[12]

LCA will be a useful tool for choosing the most appropriate materials to lower environmental impact. In 2004, the life cycle of a teak garden chair available in a European retail outlet was undertaken as a pilot study for the European Commission, which has a policy to reduce environmental impacts of products through their life. The main stages of producing the chair, the transport distances and important environmental effects are summarised in Figure 9.37. The manufacture of the chair is largely non-mechanised, so transport

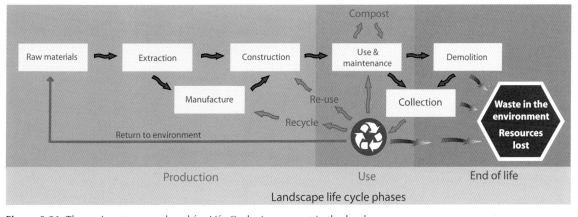

Figure 9.36 The major stages analysed for Life Cycle Assessment in the landscape.

9 – DESIGNING LOW IMPACT GARDENS

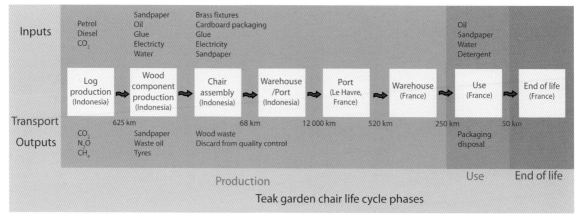

Figure 9.37 The life cycle components for a teak garden chair defined for a European Commission study.[13]

contributes most to non-renewable energy use, and greenhouse gas production. It was suggested embodied energy from transport could be reduced per chair by modifying the design and by reducing the packaging needed for shipping so there are more chairs per container or pallet. Using local suppliers will also reduce transport energy. Energy can also be reduced by encouraging the users of the chair to have a high standard of maintenance to prolong the product's life and reduce the environmental cost of more frequent replacement. Another issue raised by the study was the importance of sourcing timber from licensed suppliers with good forest stewardship, although this can increase costs.

The teak garden chair report is useful, but as with many analyses, not all the appropriate information was available for the study, and the results of the project are more useful for the retailer than the purchaser. There is a need for many more LCAs to be completed for landscape-related materials and for the information to be freely available for landscape architects, designers, horticulturists and the general gardening public.

A simpler but less rigorous method already widely used is Life Cycle Costing (LCC). Its strength is in comparing the costs of different alternatives, and although often based on dollar costs, energy costs can be reflected too. Instead of comparing just the up-front capital cost, LCC includes five major costs over the equipment or project's lifetime: capital, maintenance, fuel, replacement and salvage. So the cost of the project then becomes:

LCC = Capital + Maintenance + Fuel + Replacement − Salvage

Capital – initial cost including design, engineering, construction.

Maintenance – all annual operating costs.

Fuel – petrol, electricity, gas, etc.

Replacement – including repairs and major components that are not annual.

Salvage – positive if there is resale or recycling value; negative for disposal costs.

INFO BOX 9.8: USING LIFE CYCLE COSTING (LCC)

Purchasing a new vehicle, to be driven 15 000 km/year for 5 years

Vehicle 1		Vehicle 2	
Capital	$39 000	Capital	$25 000
Maintenance	$300	Maintenance	$5000
Fuel	$4000	Fuel	$9750
Replacement	$2000	Replacement	$2000
Subtotal	*$48 500*	*Subtotal*	*$41 750*
Salvage	$20 000	Salvage	$12 000
Total	$28 000	Total	$29 750

LCC can be done as either an energy or monetary cost, but not both. It is a way of integrating embodied energy in the calculation of the 'cost' of a project – a way of combining both monetary and environmental 'cost'.

Energy for maintaining the landscape

In addition to embodied energy, energy is required to maintain and use gardens, for powered tools, lighting, pumps for tank water, pool pumps and the like.

Optimal garden designs are those that can be maintained easily without powered equipment. The temptation to use powered hedge shears on large, clipped formal hedges is high whereas a small hedge may be manageable with manual shears. Less vigorous species will need trimming less frequently, as will an informal hedge. Reducing lawn areas also reduces the energy needed to run powered mowers, and can increase the feasibility of using manual push mowers for home gardens. Choices when designing can have quite different energy outcomes through the life of the garden.

Some garden lighting is usually required for safety, and to extend outside activities into the evening. Solar garden lights with inbuilt solar panels are increasingly being used although the light output is often decorative rather than functional (see Figure 9.38).

LEDs (Light Emitting Diodes) are very efficient at converting electrical energy to light energy with very little heat being

Figure 9.38 Solar garden light.

Figure 9.39 The variety of LED garden lighting. (Photos: Collingwood Lighting)

produced. When a voltage is applied, LEDs produce light as electrons pass from an electron rich to an electron poor semiconductor. Different wavelengths can be created by changing the diode's semiconductor chemistry. The range of lights available is increasing and the price – they are currently quite expensive – is coming down. LEDs are already used in traffic and car lights, torches and bike lights, and interior lighting is developing. Outdoor LED lights can be powered by 240 or 12 volts. LED light fittings come in different colours, can be programmed to change colour, and come in various formats including LED strips for indirect back lighting (Figure 9.39). Some LEDs are coupled with solar panels and incorporated into waterproof paving-sized 'bricks' that can be placed within pavers (Figure 9.40). They come in standard 200 × 200 × 60 mm and 100 × 200 × 60 mm sizes. LEDs are much more efficient than incandescent globes, needing about half the energy to produce the same light, but they are still less efficient than fluorescent. White LEDs rely on blue light hitting a phosphor that emits the white light combination of green, red and yellow wavelengths. This process produces some heat. Another potential way of producing white light more efficiently is to combine red, blue and green LEDs to form the white light without needing the less efficient phosphor step, but efficient green LEDs are yet to be developed.[14] A characteristic of the blue phosphor LED is its starkness, and developing LEDs with warmer tones may increase market acceptance. LED technology is still rapidly evolving and will eventually solve these problems. Keith Higgins at Granville College of TAFE, NSW has developed a new LED that produces more light and uses less energy than standard LEDs, and has made powerful outdoor lights.

Fluorescent outdoor lights are the most energy efficient, needing only about 20% of the energy of incandescent lighting for the same light level, and come in many formats besides the traditional straight tube (Figure 9.41). There are many attractive outdoor fittings that suit compact fluorescents and some styles normally associated with halogen lamps have been designed for compact fluorescents. Although manufacturers of compact

Figure 9.40 Hotbeam SolarBrick™: a self-contained waterproof light with solar panels, LEDs and an ultracapacitor for electrical storage. The manufacturers claim they will last more than 10 years without maintenance. (Photos: Hotbeam)

Figure 9.41 Fluorescent tube light fittings come in many formats.

fluorescents have reduced the amount of mercury in them, they still contain small amounts, equivalent to about a pinhead. Disposal of large quantities into landfill can potentially pollute groundwater and suitable alternatives are needed as compact fluorescent use keeps increasing. Manufacturers are working on the development of mercury-free compact fluorescents.

Halogen lamps are not as energy efficient as either LEDs or fluorescent lamps; however, replacement bulbs using IRC (infrared coating) technology reduce the power usage for a given light output by refocusing heat back onto the filament and making it operate more efficiently. A 35-W IRC halogen bulb produces the equivalent light to a standard 50-W halogen.

The higher reflectance of light-coloured paving and wall surfaces means it is possible to reduce outdoor lighting requirements, translating into lower wattage or fewer fittings. Concrete is said to reduce the lighting level needed by 30% or more over asphalt.[2]

Installing 240-V light fittings may require installation of the electrical cable underground involving the use of PVC conduit and perhaps a trenching machine, both adding to the embodied energy.

Pumps are used in garden water features and for pumping water from rainwater tanks into the laundry, toilet or irrigation system. Water tank pumps powered by solar or wind energy are available but even carefully selecting an efficient, durable pump will reduce ongoing energy use. When choosing pumps, the motor's power (in watts) should be low, and efficiency high. Some pumps are advertised to be 90% efficient, although most would be lower.

Swimming pools are found in 10% of Australian households and on average produce 1.3 tonnes of carbon dioxide each year although it could be a lot more if the pump is left on all day throughout the year.[15] A typical saltwater pool will use 2200 kW of electricity annually, making it one of the largest energy consumers in the household.[16] Pool pumps contribute most to a pool's energy use as they circulate pool water through a filter and, in saltwater pools, an electrolysis cell (now in about half of all pools and increasing). Commonly a pump motor of 1 kW is able to pass 50 000 litres through the filter in 3–4 hours, which in peak season is recommended to be done twice a day (2–3 hours per day in winter). Pump energy use can be reduced in two ways: with a more efficient pump design, and by using more efficient motors. Unfortunately manufacturers do not generally provide information for easy comparison of these features, and so it is difficult to identify which are the better pumps. The difference in energy use between the least and most efficient pumps can be as much as running a refrigerator. Energy can be saved by using smaller pools with less water to be filtered and heated, by using low pressure filters, large diameter pipes and sensors for monitoring feedback.

The sanitising equipment for saltwater or ozone pools use about 17% of the energy when compared with filter pumps. The electrolytic cells of saltwater pools use less energy than ozone generators.

It is worth considering solar photovoltaic or wind energy systems to offset the energy needed for running the pool's electrical equipment or, if this is not considered feasible, to subscribe to an accredited green energy plan from a power provider.

Heating the water increases energy use even with solar heating, which requires either an additional pump that can cope with the head of pressure required to circulate the water through the solar collector, or for the main pump to be run longer than it would otherwise need to be to capture the solar gain. If a separate pump is used for solar heating, it is important to select one with an efficient motor. There are Australian standards for solar heating systems for swimming pools and the materials they are made from: rubber or polyvinyl chloride. Some solar systems boost the water temperature further with gas heating systems that are rather like large instantaneous hot water systems. Australian Standards require them to be at least 70% thermally efficient, although pool heaters in the United States are required to be 78% efficient and some rate at 88.5%, so there could be considerable energy savings made through careful choice of gas heaters. Heat pump pool heaters are very efficient for heating pools year-round, but may be no better than gas heating with respect to greenhouse gas production.[16] Less energy is needed to maintain pool temperature by regularly covering the water with a thermal blanket when not in use, particularly at night. This is now compulsory in some states in Australia. Lismore City Council in New South Wales captures the energy from their public pool pumps to heat the shower water.

Spas can also use significant amounts of energy for pumps and heating. The 1000–2000 litres usually needs to be heated to the mid-thirties Celsius rapidly. Solar is unable to meet this requirement, so spas are therefore reliant on gas and/or electricity. Heat loss can be reduced if the spa is in a sheltered position, it is adequately insulated, and a well-fitting, highly insulating thermal cover is used when the spa is not in use. Outdoor spa pumps are generally more powerful, and therefore energy consuming, than pool pumps in order to supply the pressure required at the multiple inlet nozzles, but they are not run at high pressure for prolonged periods. If water is kept in a spa, the water needs low-level circulation through the filter.

There is a proposal to introduce a Pool Energy and Water Rating Scheme that will indicate the energy and water consumption of pools and spas by considering the overall design, the efficiencies of equipment and their integration.

Choice of materials

Apart from the embodied energy in landscaping products an understanding of the environmental impacts that have been caused by their extraction, transport, manufacture, installation and ultimately their disposal will help us to choose which products may serve our design purpose with least effect. Major environmental effects can occur at any stage through a product's life. Ideally we should know the embodied water and energy, and the environmental and human toxicity of manufactured goods and their by-products of manufacture. These are still largely undocumented for the many landscaping products that may be used in a garden, listed in Table A4 in the Appendix.

Plants contain embodied energy from heated propagation beds and houses, and a significant amount of embodied water through irrigation. Intuitively, the more advanced the plant when purchased, the more embodied water it contains although scientific studies are yet to confirm this. Water is used in many manufacturing processes, including for the manufacture of plastic plant pots.

There is generally little information available about the sustainability of extraction and/or processing of the products we purchase. Timber is one of the best researched areas and some timber products are identified as coming from sustainably managed forests. As consumers, it is worth reminding our suppliers that we prefer our purchases to have minimal environmental impact by asking them about their sources. SGA-registered nurseries identify environmentally better products (see Info Box 9.9).

> **INFO BOX 9.9:
> CHOOSING MATERIALS**
>
> For every material used, ask:
>
> - What is its embodied energy content?
> - What is its embodied water content?
> - What are the environmental effects of its manufacture? Are toxic chemicals a by-product? Are large quantities of energy required?
> - From how far does it have to be transported? Is there a suitable source closer to the landscape/garden site?
> - Does the extraction or harvesting of the material threaten natural ecosystems?
> - What other materials and processes are required to install the product, and what are their environmental impacts?
> - What is the material's expected life? (longer life means fewer replacements over a period)
> - Is it easily maintained?
> - Is it easily repaired?
> - If necessary, can it be recycled or disposed of with minimal environmental impact?

Timber

The Good Wood Advisory Centre (GWAC) is part of a worldwide network of groups working to halt the destruction of rainforests, old growth forests and other native forests of high conservation or social value. Formed in 1989 as a nonprofit, incorporated association, the group's aim is to sustain forest values by changing the ways people think about wood. With this aim in mind, the group researches and distributes information about the social and environmental contexts of wood, its production, processing, application and recycling.

Much of the impetus to form the GWAC rose out of the efforts of people protesting about the importing of rainforest timbers during the late 1980s and early 1990s. As a result of these protests, the connection between Australia's timber consumption patterns and global rainforest destruction quickly became evident to many people, prompting a demand for information on sustainably produced timbers, specifically:

- Where does the wood we find in shops come from?
- What are the environmental and social impacts of the woods we use?
- How do we minimise the waste of wood?

The mainstream industry was doing little to answer questions such as these, so in 1991, the GWAC published its first edition of the *Good Wood Guide*. The book adopts the view that there are good social and environmental reasons to purchase timber which is grown, cut, or milled locally by an environmentally aware supplier. It reasons that purchasing timber from a local supply enables consumers to achieve a closer connection with their timber, the people who cut it and the forest from which it comes. When the source of timber is close, and the number of links in the supply chain are few, it is easier for consumers to directly assess the environmental impacts and take responsibility for the use of the timber in question. Consumers are in a better position to influence the practices of the supplier. Less energy and resources are used in transport, packaging, shipping and other processes, all of which unnecessarily increase total energy costs.

The *Good Wood Guide* fosters a responsible attitude toward timber and timber products by reduction of use, responsible use, recycling and reforesting. Anyone can join the GWAC, which links timber consumers, architects, builders, timber workers, craftspeople, furniture makers, timber merchants, sawmillers, foresters, farmers and teachers.

Certification schemes for timber that promote sustainable forestry management are still evolving in Australia, but cover about 5.7 million hectares of forest.[17] The Australian Forestry Standard (managed by the Australian Forest Certification Scheme (AFCS), accredited by Standards Australia and with international links) is one scheme. Another is associated with the Forest Stewardship Council (FSC), an international certification scheme currently delivering certification in Australia via the bodies SmartWood™, a program of the Rainforest Alliance™, and Woodmark, a program of the Soil Association. Both certification bodies apply the international FSC principles and criteria adapted to Australian conditions. The Forest Stewardship Council Australia (FSC Australia) is in the process of developing a stakeholder-endorsed national standard that will eventually replace these interim adapted standards, with a draft for comment expected in 2008–09.[18]

Australia imports timber from some southeast Asian countries that are known to have poor forest stewardship. The decking timber merbau (*Intsia retusa*) is a commonly available timber that may come from these forests.

Radially sawn Australian hardwood timber reduces wastage during the milling process, producing more high value timber per log, and has led to new timber profiles that have convex upper faces that easily shed water, concave lower faces that have less contact with the joist, retain less moisture and are less prone to rotting, and tapered radial sides that are less likely to trap debris and retain moisture, all leading to longer life.

Knowing the longevity of various timbers can help improve selection for gardens and can reduce the frequency of replacement. The CSIRO has published in-ground durability ratings for native and plantation timbers.[19] *Callitris glaucophyllus* (white cypress pine) and *Allocasuarina luehmannii* (bull oak) are quite durable, the former expected to last greater than 21 years in southern Australia and more than 14 years in the north. Some *Eucalyptus* species have more in-ground durability than others, *Eucalyptus camaldulensis* (river red gum) being durable and *Eucalyptus regnans* (mountain ash) not, although it is very suitable as flooring. *Corymbia maculata* (spotted gum) is a durable timber for in-ground use.

Treated pine is more durable because it has been impregnated with copper, chromium and arsenic. Copper controls fungal growth, arsenic controls copper-tolerant fungi, chromium fixes the chemicals in the wood, and the chemical copper chromate controls termites. Health concerns of treated pine have lead to partial or complete bans in a number of countries including: Indonesia, Japan, Germany, Sweden and the USA. From 2006, Australian Pesticides and Veterinary Medicines Authority regulations have restricted the use of pine treated with copper, chromium and arsenic. It is no longer to be used when frequent human contact is possible, as with garden furniture, hand rails, decking, play equipment and picnic tables.[20]

Arsenic is not carcinogenic, but it can facilitate action of other chemicals that are. The fixing process, taking days or weeks depending on the temperature, ensures most of the arsenic does not leave the wood through the life of the product. Since 2006, manufacturers in Australia need to ensure that the chemicals are sufficiently fixed before distribution. However, some leaching still occurs and higher levels of arsenic can be found in the soil 10–20 cm from installed treated pine. Studies have shown that root crops like carrots can have increased arsenic levels if grown in this soil, but fruits like grapes, tomatoes and cucumber will not.[20] Although some recommendations suggest a satisfactory distance from treated pine is 100 mm for the growth of root crops, until further studies confirm this, gardeners may be better to use other materials near food crops. Treated pine should not be used as a garden mulch, as the heavy metal content is likely to exceed the Australian Standard.

Pine treated with other less harmful chemicals is available, although not as widely and it is more expensive. Alkaline copper quat (ACQ) protects against fungi and insects. Light Organic Solvent-borne Preservative (LOSP) uses compounds like zinc, copper or tin for fungal protection and synthetic pyrethroids for protection against insects. Tanalith E is a copper azole compound impregnated into softwoods under vacuum.

Plastic

PVC

Polyvinylchloride or PVC is found in many landscape products including water and drainage pipes, downpipe fittings, garden hoses, electrical conduit, lawn edging and lattices. It is a cheap, easily used material, but some concerns are being raised about its safety. Vinyl chloride, used in the manufacture of PVC products and to glue PVC pipe fittings is implicated in liver, blood and brain cancers in people working in PVC manufacturing plants.[21] PVC manufacture also produces carcinogenic dioxins, and some products are stabilised by adding lead, tin or cadmium, all neurotoxins for humans. PVC pipe can contain 0.1–1.8% lead by weight. PVC is made flexible by adding phthalates. The commonly used diethylhexyl phthalate is rated by the US EPA as a carcinogen, and shown in studies to be released during manufacture and disposal of PVC in landfills or by incineration. Little PVC is recycled at present, but there are significant opportunities to recover PVC from demolition sites and as offcuts on construction sites. Reducing the use of PVC is being actively considered in Europe and the US.

Alternatives to PVC with less environmental and health impacts published by Environment Canada in 2002 include high density polyethylene (HDPE), and cross-linked polyethylene. HDPE does not require chlorine in its manufacture, has less additives and can be more easily recycled. It is possible to reduce PVC use. The Sydney Olympic Stadium contains no PVC in the seating, cabling, floor coverings, wall finishes or plumbing.[21]

There are alternatives to PVC of course, and it is worth investigating them.

Recycled plastic

Plastic being a long-lived material, resistant to decay, lends itself to outdoor landscape use, but it does have high embodied energies and its raw materials come from non-renewable oil. Recycling (or more accurately, downcycling) plastic reduces the plastic going to landfill, and is used for garden and park furniture, sign posts, bollards, wheel stops for car parks, speed humps, fence posts, decking, retaining wall panels, stair treads and speed humps (Figures 9.42 and 9.43).

Figure 9.43 Recycled plastic used for signage.

Cement

The use and manufacture of cement produces 5–10% of global CO_2 emissions through the firing of the limestone (chemical liberation of CO_2 from calcium carbonate and fossil fuel energy used), and the energy use for mechanical equipment.[22] Between 1990 and 2004 production modifications have reduced CO_2 emissions by 21%. Another innovation being developed in Hobart, Eco-Cement™ containing magnesium oxide, reabsorbs the CO_2 released from the cement as it sets, and continues to absorb CO_2 from the atmosphere for up to a year.

Lighter weight aerated concrete panels (ACC) are increasingly used in building and can be used in the landscape for walls or garden buildings, although being porous they will benefit from rendering with an ACC compatible mix. Its greenhouse gas and embodied energy impacts can be 20–25% of standard concrete.[23] HySSIL (High-strength, structural, insulative, lightweight) has been developed by CSIRO and industry, and is cheaper and stronger than the similar aerated autoclaved concrete. Being lighter than dense aggregate concrete, HySSIL reduces energy needed for transportation of pre-moulded concrete and for construction, and has good thermal insulative properties. It is also easily recycled.

Figure 9.42 A recycled plastic speed hump.

It has been estimated that 40% of landfill is made up of building waste, including concrete. Recycling concrete can be as simple as using broken concrete slabs as the 'stone' for retaining or garden walls as the landscape architects did at Steele Indian School Park in south-western USA, the rough surfaces of exposed aggregate giving the walls an interesting texture.[24] Concrete is also milled to produce aggregate that is useful for paving base courses, but the concrete remaining on the old pieces of aggregate can weaken newly poured concrete and its use has been limited to non-structural situations. A guide to the use of recycled concrete is available from Standards Australia.[25]

Organic mulches and composts

Lower environmental impact mulches and composts are made from resources that are potentially renewable. Arborists can be a great source of chipped tree prunings, although the variety of plants chipped will give variable qualities to the mulch. Some local councils allow their residents free access to chippings from street tree work.

Commercially available organic mulches vary in their environmental impact, but it is worth keeping in mind that many come from forest trees, not all of which will be from sustainably managed plantations. Old growth forests are definitely not appropriate sources for garden mulches.

Spent mushroom compost made from straw provides good organic matter to improve soil structure. It can have a high pH initially but studies have found that acids produced by microbial breakdown restore it to a more neutral pH. It can also have a high salt content, but this can be allowed to leach out before use.

Pea straw and sugar cane mulch are good if you can source them locally; they can break down quickly though.

Inorganic mulches

Most of the washed river pebble gardens of the sixties and seventies have disappeared, many of the pebbles probably ending up in landfill – a resource taking many thousands of years to create is essentially lost. This is an excellent example of the waste resulting from changing fashions, and our society's quick-fix solution for getting rid of them in the haste to move to the next big garden style. There is another way. Some years ago, one of the authors and his wife lifted, sifted, washed and bagged the pebbles he inherited in his garden – all 230 bags worth; a slow but rewarding process to know that the pebbles gained a new life in other gardens, were appreciated by the new custodians, and saved at least a little bit of a creek bed from being destroyed.

Smooth pebbles currently feature in landscapes again, although the style may be different in the 21st century. Australian pebbles are now quarried rather than dredged from active river beds, with some being a by-product of the sand extraction industry. Are these a limited resource? Probably. Just how many ancient, pebbled river beds are buried in the Australian landscape is unknown, so treating them as a rare commodity is recommended. Many pebbles are imported from Asia and the Pacific, adding high transport energy to the environmental impact. Although some of the imported pebbles are rolled and smoothed stone chips from sculpting and stone masonry, others are taken from active river beds, and it is not always simple to confirm the origins.

Glass pebbles made from recycled Australian clear glass may serve as a suitable substitute for some landscape needs. Coloured polyurethane coatings give many colour choices although they also add to the embodied energy of the product. Recycled glass has also been used as an exposed aggregate in pavers, giving them a different reflective quality.

Brush fencing

Brush fencing made from *Melaleuca uncinata* was first used in South Australia in the 19th century, but is commonly used in other Australian states (Figure 9.44). It is harvested from Crown Land in NSW or freehold land in South Australia, Victoria and NSW. The brush is cut just above the ground using a machete or chainsaws with 50–70% of a clump taken to promote better regrowth. Stem diameter is 8–15 mm and the height 1400–1800 mm. Some is mechanically harvested now. The industry is actively encouraging the establishment of plantations to relieve pressure on remnant bushland and to increase reliability of supply. Some growers in South Australia and Victoria have worked towards improving *Melaleuca uncinata*, selecting for narrow-forked branching, lack of flowering, frost tolerance and disease resistance. Brush

Figure 9.44 Australian brush fencing.

fencing in South Australia has been known to remain in good condition for 50 years or more. Being locally sourced, from a regenerating shrub, and with increasing use of plantations, brush fencing has great potential for having low environmental impact.

Some brush fencing is imported from Asia which adds embodied energy through the longer transporting distance. One imported product harvested from a species in the important Australian Myrtaceae family (*Eucalyptus*, *Callistemon*, *Melaleuca*) has been found with flowers and fruit, raising concerns about its potential as a weed source (Figure 9.45).

Landscapes for food

Our foods are increasingly grown at ever-greater distances from their markets and transported over hundreds or thousands of kilometres, even though most people in Australia live in areas where the conditions for food production are very good, with

Figure 9.45 Imported brush fencing material from Asia has been known to have flowers and fruit – a potential new weed source?

adequate water, sunlight and temperature ranges. It is true that much of Australia has a net primary production in the lower range of 0–0.3 kg carbon/m^2/year, but the coastal band east and south of the Great Dividing Range, and Tasmania have net primary productivities between 1.5–2.2 kg carbon/m^2/year.[26] The global variation is between 0–3 kg carbon/m^2/year. There is no biological reason why we should not grow and source most of our food locally, and there are very good reasons to avoid the greenhouse gas production that transportation and other food production-related activities cause. Table A5 in the Appendix documents how food production contributed to about 25% of energy consumption in the USA in 1980.

Lower food miles are good for our environment, and you can't get lower food miles than if you grow your own food. If many suburban gardens included some food plants as earlier generations did, there would be the collective benefit of lowering the demand for food grown elsewhere. Many homes were built on former orchards or market gardens anyway.

A standard suburban garden can provide a worthwhile proportion of the food energy and nutrients we need. Most familiar food plants are able to be grown in Australia, although for each region we can be selective for those food plants that are suited to local conditions, needing minimal input of water and fertiliser, and little need for chemical pest control.

Importantly, the aesthetic qualities of the garden do not need to be compromised by including food plants. They can be used in feature plantings (banana, paw paw), for screening (macadamia, avocado, loquat, mango), and to provide summer shade and winter sun (grape, Chinese gooseberry (kiwifruit), fig, medlar, mulberry, quince). Fruit trees can have spring blossom (peach, plum, apricot, apple, pear, almond), autumn colour (persimmon, blueberry) and infuse the garden with fragrance (lemon, grapefruit, mandarin, lime). Their flowers can be attractive (passionfruit, feijoa), and their fruit decorative (cherry, pomegranate, olive). Vegetables do not have to be planted in rows, but interspersed with other plants. They can introduce colour (rainbow chard), texture (garlic, zucchini), and form (taro) or can be used as edging plants (parsley).

Figure 9.46 The Kitchen Garden at the Royal Botanic Gardens Melbourne.

What productivity can we expect? In a standard suburban garden, it can be difficult to supply all food needs, but every fruit, nut or vegetable harvested is one less that has to be produced and transported from elsewhere, and it is surprising what quantities can be grown for relatively little effort.

In one Melbourne garden, a single Marsh's Seedless grapefruit produces at least 100 and sometimes over 300 fruit a year, with little or no additional watering and little feeding. That covers breakfast for two people for a quarter to three-quarters of the year and provides each with over 200 kJ per serve. An Avocado 'Fuerte' produces 50–100 fruit a year averaging one to two per week for the kitchen, and the harvesting can be spread over a year as the fruit can be held on the tree until needed. An avocado has around 627 kJ per fruit. A feijoa, again with little irrigation, usually provides hundreds of fruit that can be eaten fresh or cooked. The same garden produces more including limes, mandarins, lemons, blueberries and vegetables. Excess produce from crops can be bottled and stored in cool ambient conditions requiring no energy after the initial cooking. Freezing is better for some crops but preferably the electrical energy is from renewable sources.

The interest in Australian species for food plants, often referred to as bushfoods, has grown in recent years. These developing new crops are adapted to Australian conditions, needing less additional resources than many traditional foods for good growth, and can be easily grown at home in many parts of Australia. The nutritional quality of many Australian species is high. Macadamias (*Macadamia integrifolia*), well known both here and overseas, have high nutritional content including protein, carbohydrates, twice the percentage of monounsaturated fats than almonds (the healthy fats found in olive oil) and more. Quandong (*Santalum acuminatum*) is still being developed as a crop, the fruit rich in vitamin C and the seed high in protein and oils.

Among the many other potential food species are muntries (*Kunzea pomifera*; fruit), lemon myrtle (*Backhousia citriodora*; leaf for flavouring), aniseed myrtle (*Anetholea anisata* syn. *Backhousia anisata*; leaf for flavouring), bush tomatoes (*Solanum* spp.; fruit), Midgen berry (*Austromyrtus dulcis*; fruit), wattleseed (*Acacia* spp.; seed for flour), Warrigal greens (*Tetragonia tetragonioides*; leaf for vegetable); mountain pepper (*Tasmannia lanceolata*; leaf for flavouring). The sap of cider gum (*Eucalyptus gunnii*) can be used like maple syrup.

In 2007, CSIRO published the results of research that showed a number of fruits of

Australian species have high levels of antioxidants, with the Kakadu plum (*Terminalia ferdinandiana*) having the highest level of ascorbic acid.[27]

There can be great pleasure harvesting from your own garden. Children learn, without realising, that the ultimate source of food is not the supermarket. You can also control the fertilisers and pest control methods used and limit chemical use.

Designing for biodiversity

Our individual gardens are a part of the greater landscape although we mostly don't think beyond our own fences. The plants we grow join with those of our neighbours, the street trees, the public parks and remaining indigenous vegetation to form a complex web providing food, protection and nesting sites for animals. Micro-organisms, many beneficial, are also a part of the system. It may be easier to visualise this broad urban habitat by imagining our suburbs from a bird's eye view (Figure 9.47). It behaves rather like natural habitats do, but usually with different species combinations.

Our garden design and plant selection will determine what other organisms may live in it or use its resources from time to time. Suburban areas can be quite rich with native animal species, birds probably being the most obvious, but there are many more: mammals, reptiles, amphibians, insects and other invertebrates like worms, millipedes and centipedes.

Studies in Sheffield indicate the biodiversity is similar in inner city compared with suburban

Figure 9.47 The interconnectedness of Australian suburban gardens.

gardens, and the mix of native compared with exotic species is not important.[28] Floristic diversity increases biodiversity, and trees and shrubs expand the garden's volume upwards, providing more habitat, more shade, more leaf litter, and therefore more diversity of habitat niches.

There have been limited equivalent studies for Australian gardens (mostly linking bird distributions to urban vegetation patterns), but it might be expected that biodiversity can be encouraged in similar ways: high floristic diversity, inclusion of trees and shrubs, maintenance rather than removal of leaf litter. We need not be restricted to indigenous plants, although we know they support the local fauna. Wattlebirds harvest nectar from many non-indigenous *Grevillea* species, for example, and also from non-Australian species like South Africa's red hot pokers (*Kniphofia*), and the grey-headed flying fox is known to feed on the many exotic plant species introduced into the urban and surrounding horticultural areas during the last 200 years or so, and are well established

200 SUSTAINABLE GARDENS

in our cities. Possums are regular feeders on our ornamental and fruit plants.

The characteristics that increase biodiversity in a garden – the trees and shrubs, leaf litter, the dry, damp, sunny, shady, enclosed and open areas – give robust, complex, garden ecosystems. There will be greater numbers of birds, butterflies, moths and other insects, mammals, amphibians and reptiles contributing to the vigour of the garden ecosystem, and benefiting the gardener, through the provision of garden services (like avoiding the need for chemical pesticides, and having the right pollinators present to increase fruit yields). Even increasing biodiversity in small gardens contributes to the neighbourhood ecosystem.

Existing trees are valuable for biodiversity. They also add value to a house just as gardens as a whole do – in the US a mature garden can increase the property value by up to 75%. Think carefully about removing trees as they take many years to reach maturity, to give the shade required for summer, the amenity year-round and the habitat niches required to support a rich diversity of life from insects to birds to mammals and micro-organisms. Trees alone can support an abundant web of hundreds of species of plants, animals and micro-organisms. Trees also lock up carbon.

Dead trees, stumps and shrubs not only decay, returning nutrients to remaining garden plants and animals, but they also offer habitat niches for ephiphytic plants (orchids, mosses), nesting animals (birds, possums), or ground animals (geckos, lizards). Although dead plants may be aesthetically unacceptable in smaller gardens, they can often be accommodated in larger gardens where they are screened by living plants. Nesting boxes to encourage animals to remain in your garden can be installed if enough suitable natural sites are not present. There are different designs for encouraging different species.

Indigenous vegetation may still exist on many building sites, perhaps connecting with surrounding public land. Indigenous fauna is supported by maintaining as much indigenous vegetation as possible, particularly where it links to similar adjoining stands. Building envelopes – areas protected from building works – can be designated at the planning stage, so all proposed built structures (house, paths, drives, other hard landscaping and the space required for the builder's equipment and supplies) are excluded from the most important vegetated areas of the site.

Indigenous vegetation can be re-introduced into the landscape to encourage greater visitation by a diversity of animals. Local indigenous nurseries, which are quite numerous in some parts of Australia now, can supply not only appropriate species, but also plants with the correct provenance for your area (that is, plants that are from the same geographical area and within the range of the genetic variation found there), and information about their growing requirements.

A still under-researched area is the effect of artificial light of urban areas on ecological processes. However, artificial light, either directly or through the sky glow over urban regions, has been demonstrated in various species to disrupt physiological processes, cause disorientation or misorientation, alter reproductive behaviours, change communication patterns, and modify predation and competition between animal species.[29] The well-known attraction of moths to light has been shown in Germany to kill large numbers of insects, removing species critical for pollination and as members of food webs. Landscape designers need to be mindful of the potentially negative influences of artificial night lighting, minimising it and using light sources with wavelengths that impact less.

Plant selection

Plant selection remains one of the most powerful tools available to plant users to contribute to the development of landscapes that are manageable both technically and economically, as sustainable as possible and yet rich and meaningful to site users.[30]

Plant selection is an important factor in creating sustainable landscapes. Sustainable plantings have low energy, water and other resource inputs (during production, establishment and management), are locally appropriate and contribute to wider ecological processes.[31] These plantings must suit the site, be dynamic, diverse and aesthetically pleasing.[32]

The reality is, very few plantings are designed to have these qualities. Commonly they have little diversity and require significant resources to maintain them without which they fail. Hitchmough[30] outlines a systematic process for plant selection using clear objectives for evaluating plants and considering as equally important the aesthetic, biological and functional requirements of the planting.

Successful plant growth requires a careful analysis of the site so plants with appropriate biological tolerances can be selected. Sites with particular characteristics that could limit plant performance need to be matched with taxa having the right biological tolerances relating to climatic factors (macro- to microclimate), specific soil stresses (water logging, compaction, salinity, etc.) and environmental stresses (atmospheric pollution, salt-laden winds, etc).

Plant selection needs to serve the users of the landscape and those maintaining it. As well as reflecting community needs, landscape character, local planning schemes, social, cultural and heritage values, the functional requirements of plants include shelter, screening, erosion control and weed suppression. They should have suitable growth rates, longevity, and the ability to be pruned and rejuvenated, and be resistant to pests and diseases. Waste from the chosen plants must be easily managed and preferably an organic resource to be returned to the garden. Plants need to match available site maintenance and have low resource input (water, energy and chemicals).

Plants should be non-weedy, contributing to and certainly not detracting from ecological processes and biodiversity needs of the area. Plants selected for their pollen, nectar, flower or fruit can support (directly or indirectly) birds, mammals, reptiles, amphibians, insects and more. Plants may also provide protection or nesting sites for animals.

The plants selected would also complement the energy efficient attributes of buildings.

Good plant selection relies on quality information; however, little of the landscape design literature is geared towards evaluating plant performance. More information is needed, particularly that which collates and reviews plant performance across a range of situations and climatic zones. Some of the best sources for plant selection have comparative tables of plants against climatic data, such as temperature and/or rainfall, as can be seen in the *Plant User Handbook*.[33]

The plantings we choose can and should be multipurpose: attractive, moderators of the environmental conditions of the garden and adjacent buildings, frameworks for biodiversity, low users of resources and even providers of food.

Ecological gardening

Natural landscapes entail complex interrelationships between many organisms, each contributing to the health of the whole ecosystem. The complexity of these ecosystems provides a strong buffering capacity against change, such as temporary variations of rainfall and temperature or the competition of invading plants or other organisms that may cause disease or the physical damage of grazing animals. Different ecosystems resulting from long periods of evolution can have reasonably stable species compositions, but they are not fixed, and the form they continue to exist in is governed by their internal interactions and the external influences around them.

Having some level of understanding how these processes work enables us to realise that often the land we are to create a garden on has had its natural ecosystems severely disrupted, if not destroyed. There may be no buffering capacity left, and it is up to us to re-establish a complex, interacting and buffering system of species that will withstand the external buffeting with minimal input from us.

A garden is a living entity composed of the plant species we choose to have in it, the other organisms already present and able to tolerate the altered conditions of the new garden, and those organisms that become a part of it after suitable niches establish as the garden grows.

If we look at gardens in this way, rather than as a collection of individual non-interacting plants, our plant choices and the way we assemble them are more likely to lead to healthy, low input gardens.

Plant adaptation

Recognising diversity in plants is easy. They grow to different sizes, and have various habits, leaf shapes, amounts of hair, flower colour, flowering or fruiting seasons and so on. Plant diversity represents the many ways plants have evolved, their form and biology enabling them to survive and reproduce in particular natural habitats and in association with certain other organisms like their pollinators or associated mycchorizal fungi or as a shrub amongst tall trees. So when we bring them into cultivation, each plant species comes with its own suite of tolerances. We recognise this when we read plant labels in nurseries that display simple symbols for preferred planting locations in full sun, partial shade or shade, or when a plant is marked as water saving.

Adaptation of a plant can mean it comfortably grows in a region with a defined climate pattern for example, but adaptation may also lead to very specific niches where a plant can grow within a region. Perhaps, within a mosaic of soil types, growth of a species is limited to islands of well-drained soils within predominantly poorly drained clay. Knowing the conditions to which the plant is adapted really helps in plant selection and placement within a garden. Equally, knowing the conditions within your garden can guide the selection of suitably adapted plants. Many horticultural books don't provide any more than the briefest of notes about preferred conditions. Look at the garden plants in your area, particularly in neglected gardens or parks. Do they thrive in drought or just survive it? Do they self-seed? How do they cope with competition from other plants? Can they grow well in shade? On hot days, do they wilt readily? How do they perform after a frost? Some species have very narrow tolerances to changes in environmental conditions, while other species' tolerance is broad. Natural distributions are not always good indicators.

Indigenous plants

The opportunity still remains for many indigenous plants to act as a visual link to the broader landscape. The indigenous vegetation can be enhanced or re-established, or blended with non-indigenous natives or exotic species to give a garden its style.

Growing plants that are indigenous to your area has a number of advantages. It encourages biodiversity by providing food, shelter and habitat to the local animals. It maintains the local gene pool. It provides a sense of place by working with the local landscape rather than setting itself against it. Generally, plants indigenous to your area require less maintenance, water, chemicals and nutrients than exotic plants.

Indigenous plants established in gardens, along roadsides, creeks, reserves, and other public land, provide a web of habitat that might ultimately re-connect with wilderness and parks, providing continuity of habitat for native animals and providing some relief from the massive clearing operations that have taken place since settlement.

Natives versus exotics

The question as to whether we should use native or exotic plants in our parks and gardens arises from time to time. Should we acknowledge the damage a city has done to the natural environment and demonstrate awareness of its Australian context by replacing parks of elms, planes and oaks with Australian indigenous trees? Are we proud to display and retain plantings that demonstrate our European heritage or should we establish our own national identity by planting Australian trees? Are exotic trees too verdant and water-hungry? Are native trees irregular, unpredictable, inclined to grow too fast, drop limbs, and be generally ill-adapted to city conditions?

These are difficult and complex questions because they involve conflicting community values, especially those relating to nature, heritage, culture, creative taste and national identity. Questions about relative plant performance can be settled scientifically, but there are no 'correct' answers to questions of value. The debate surfaces regularly in relation to the character of parks, street trees and gardens in our cities and the development of policy about their future management.

It might well be argued that caring for our environment and its finite resources is a different matter from preferring wattles to rhododendrons, or plane trees to eucalypts. The views and values presented in this book certainly emphasise the overriding importance of natural over artificial landscapes and point out that whatever we grow in our gardens will have some impact, no matter how small, on the wider environment. Even so, does this view justify the removal of existing historical European-style exotic trees from the centre of our major cities and their replacement with indigenous trees? No doubt this debate will continue to act as a litmus test of community values for some time.

Most gardens nowadays contain a mix of native and exotic species. We are fortunate in having an exciting and unique Australian flora that is adaptable to horticulture. But growing some exotic plants is not necessarily a threat to the natural world, or a rejection of our natural heritage or nationality – for some people there is still the simple lure of plant diversity. Exotic plants can be weedy, but so too can displaced native plants.

In our own gardens we like to make our own decisions. The home garden is one of the few places where we can freely express our creativity, and that is fine as long as we are mindful that some of our choices may have unintended environmental consequences. The message here is to be ever vigilant.

Environmental and economic weeds

Keep your exotic plants, and especially those from Mediterranean climates, locked securely within your enclosure, along with the cat, and have all of them neutered.[34]

Human dispersal of non-indigenous, sometimes invasive, plants and animals around the globe is a direct result of the progressive increase in international trade, travel and tourism. The rapid spread of diseases like HIV, bird flu and SARS illustrate clearly the potential dangers of long-distance, high volume modern transport systems. Dutch Elm disease, which decimated European and North American trees, has spread as far as New Zealand. In their natural habitats most plant pathogens are relatively innocuous, but transferred to another ecosystem they may reach epidemic proportions.

After land clearing and climate change, invasive species present the greatest threat to biodiversity world-wide. Exotic invasions also reduce ecosystem function and agricultural production as well as having negative impacts on human health. Current estimates suggest that the cost to Australia's primary industries in lost production and control now exceeds AU$4 billion per annum.

Naturalised plants and environmental weeds

Every country will have a number of naturalised plants. These are plants that originated elsewhere but which have become established and are reproducing themselves as though native to the area. They do not necessarily impact adversely on the environment. Those naturalised plants that do have a significant negative impact on the environment are generally referred to as invasive environmental weeds and, historically, these have attracted less attention than agricultural weeds which have an obvious economic impact through their interference with the growth of agricultural crops.

About 10% of the Australian flora is naturalised. In New Zealand this figure is 43%.

In Australia, quarantine restrictions, increased public awareness and additional government finance for invasive species management are still making little impact on the problem. So, in general, the problem of invasive species is getting worse. The percentage of naturalised plants is increasing, across the country and in each state. In addition, under international law, bans on potentially invasive plants can be overturned if it is considered that quarantine restrictions are being used solely as trade barriers.

So what is the answer? One solution is to ban much of our international plant-trade or to apply large-scale destruction of risky stock, but this would be a draconian, unpopular and unworkable approach. More realistic is the improvement of public understanding of the risks involved combined with more stringent screening and monitoring at all levels.

What is a weed?

'Weed' is a term used for a plant growing where it is not wanted, generally with detrimental consequences. This is sometimes referred to as growing 'out of place'. What is 'out of place' will depend on the context, so we have environmental weeds, garden weeds, agricultural weeds, etc.

Noxious weeds are plants that are legally banned under noxious (or declared) weed legislation as plants that cause, or have the potential to cause, environmental or economic damage. They derive from all weed categories (environmental, garden, agricultural, etc.) and are subject to government management. Environmental weeds can be defined formally as non-indigenous plant species that have invaded non-agricultural areas of natural vegetation and are presumed to impact negatively on native species diversity or ecosystem function. They can reduce biodiversity – competing successfully with the native plants and animals to produce a degraded habitat and sometimes they can cause genetic pollution by breeding with the indigenous plants. Their spread globally has coincided with the huge explosion in human population and mobility that has taken place over the last 200 years. Unless these weeds are managed, native plant communities will progressively disappear. Accidental introduction may be by transport with livestock and as contaminants of grain and agricultural produce, in transported soil, on machinery, vehicles etc. Deliberate introduction, not uncommon in the past, may be for agriculture and horticulture, for fodder, erosion control, etc.

Environmental weeds and gardens

As horticulturists of any persuasion, we have a compelling environmental responsibility not to import, cultivate or promote known or potential weeds

Geoff Carr

Restricting the sale and cultivation of potentially weedy plants may seem drastic but it is done to protect our native ecosystems. In less than 200 years, humans have been responsible for the introduction of over one-quarter of the plants listed as growing 'wild' in Victoria. Through the combined effect of agriculture, forestry and horticulture, more than 65% of the area of Victoria carries wholly or predominantly exotic plant species.

The majority of environmental weeds in Australia are plants that were deliberately introduced for agriculture, forestry or horticulture, escaping from cultivation into the natural environment.

Examples of well-intentioned but environmentally devastating plant introductions in Australia include: the purported spread of blackberry seed, *Rubus fruticosus* sp. agg., in the wild by Ferdinand von Mueller; the escape, possibly from Darwin Botanic Gardens in the 1830s, of *Mimosa pigra*, giant sensitive tree; and the escape from private gardens of *Echium plantagineum*, Paterson's curse, Salvation Jane. Unfortunately many of our most damaging environmental weeds are (or were) garden favourites, e.g. hawthorn, periwinkle, *Erica*, *Watsonia*, *Gladiolus*, English broom, gorse, St John's Wort and many more.

Environmental weeds present us with difficult horticultural decisions and challenges. For example, it seems horticulturally sensible to grow plants that do well in the Australian climate so it is tempting to encourage Mediterranean climate plants from South Africa, the Mediterranean, the Canary Islands and California and more temperate plants from the UK, South Western and Middle Europe as well as parts of North America. Confronting severe droughts, we promote plants that conserve water such as cacti and succulents. Unfortunately plants that are adapted to Australian conditions will probably also be successful in the natural environment, so we need to be extremely careful about the species we choose. Many garden styles and fashions inadvertently promote potentially weedy plants. Grasses and grass gardens are a good example. The family *Poaceae* is a major source of many rapidly reproducing potentially weedy species. About one-third of the grasses growing in Australia are non-native species. Enticing photographs of beautiful European gardens can lead us in a dangerous direction unless we closely monitor the plants we select.

The natural environment, once destroyed, is difficult or impossible to restore. A reduction in the palette of ornamental plants does not seem a huge price to pay for more intact wetlands, grasslands, heathlands, woodlands and forests.

Strategies for combating weeds

There are a number of strategies underway to reduce the problem weeds found throughout the country. In 1999, the Australian National Weeds Strategy Executive Committee announced a list of Australia's worst weeds. Out of the list of 20 weeds, 14 are garden escapees.

A further list of 52 'garden thugs' was produced in 2001 by the nursery industry. See Table A6 in the Appendix.

Groups including Weed Warriors and Weed Spotters exist to raise awareness and tackle outbreaks of weeds, and local councils and botanic gardens are invaluable providers of much information about how to identify and destroy potential problem weeds.[35]

Plant selection summary

It would be very useful to produce recommendation lists of environmentally appropriate plants. However, these lists would vary from region to region and place to place. Individual gardens have their own microclimates that may be unlike the typical regional climate. Even plants of the same species have different propensities to naturalise, being genetically variable in terms of their climatic tolerances, this often relating to their provenance. Plants that show no sign of naturalising in one area can become major environmental pests in another, such as Australia's east coast *Melaleuca quinquenervia* in Florida, USA.

For each region, plants need to be evaluated for their environmental suitability.

Environmentally friendly plants:

- Require little, if any, additional water.
- Are largely free from pests and diseases that need treatment.
- Require little energy expenditure to maintain
- Encourage local biodiversity.
- Will not naturalise.

In general, annuals and other bedding plants, roses, and lawn grasses can be resource hungry.

The Australian Urban Street Tree Evaluation Program (AUSTEP) is a useful, continuously updated local street tree evaluation project.[36] It helps people choose the right tree for the right location. The evaluations, undertaken periodically in various areas of south-eastern Australia, are done by professionals in horticulture, arboriculture, landscape architecture, urban design and management, together with local government staff.

Energy expenditure related to plants occurs in two ways:

1. The direct use of energy by power tools and power equipment such as mowers, brush cutters (whipper snippers) and hedging shears powered usually by petroleum fuels or electricity. Selecting plants that minimise the need for power tools will help reduce our direct energy use. Lawns can be reduced or eliminated to avoid the use of motor mowers and if lawns are still a part of the design, they can be of a size easily maintained by push mowers. Power tools are often used to maintain hedges, but selecting smaller, slower growing species or cultivars can reduce the trimming frequency or make hand powered tools more feasible.
2. The energy that has been expended to produce the plants that we wish to include in the garden. This includes the embodied energy in the potting media,

in the pots, in the plants invested during growth by the wholesale nursery, and the transportation of the plants between the wholesale nursery, retail nursery and the place of planting.

We cannot directly lower embodied energy of the plants we purchase, but purchasing carefully may reduce plant losses and the need for replacement. We can also choose plants that have been propagated in local wholesale nurseries rather than those transported long distances. Indigenous plant nurseries are reliably local.

Warm season grasses such as buffalo, kikuyu and couch require little additional water and grow well during the warmer seasons in mild climates, and there are a number of cultivars available now. They do require constant vigilance in maintaining their edges with garden beds, however, and these maintenance demands need to be balanced with the savings of water that may be made. Australian native grasses that are less water demanding have been developed for commercial horticultural release.

Naturalising species are generally non-native, like *Crocosmia × crocosmiiflora* (montbretia), or *Lantana camara* (lantana), but there are a number of examples where Australian species have become weedy outside their natural range: *Pittosporum undulatum* (sweet pittosporum), *Acacia baileyana* (cootamundra wattle), *Eucalyptus cladocalyx* (sugar gum), and *Melaleuca nesophila* (showy honey-myrtle). Horticulturists can be watchdogs, warning of potential new weeds that are yet to be recorded. For example, seedlings have been found under young plants of *Ceratopetalum gummiferum* (NSW Christmas bush) in Melbourne gardens indicating a potential problem.

Consider including at least some food plants easily grown in the area, and are especially enjoyed by the occupants to ensure the produce won't be left rotting on the ground. Vegetables are an obvious choice; however, they require regular maintenance. Many fruit trees need less looking after, and can be an important part of the garden's structure. Select high-yielding cultivars that are genuinely productive food sources.

Woody perennials are, in general, low maintenance, low input plants compared with annuals, which need seasonal changes of display, result in the discarding of more plastic punnets, and need additional water and fertiliser.

Planning for design implementation

Specifications for contractors

The more specifications documented with the designs, the more likely these will be adequately addressed by contractors in their bids for a project, and ultimately during construction. At the least, specify with plans and in writing as a part of the documentation the following:

- The areas and vegetation to be protected, and how they will be protected.
- How soils will be protected, including from compaction.
- The location for certain activities, including methods for handling chemicals or waste products, and the storage of building materials.

- The period during which this will occur, preferably starting before construction and remaining in place as late in the project as practicable.

This will alert contractors to what is expected, and allow them to adjust their estimates to reflect the true costs that may be incurred. Some contractors may not be interested in modifying their methods. They will either pitch their quote ridiculously high, or will drop out of the bidding. This isn't a problem. It is better to eliminate these people at an early stage than to try to deal with their reluctance to cooperate during the project. Financial bonuses or penalties can be used to reward or penalise adequate or inadequate performance in this regard.

The designer can also have a strong influence on the level of environmental impact of a landscape through the choice of materials specified for the project. This requires a level of understanding of the life cycles of products including the extraction of raw materials, the manufacturing process, transportation requirements, and the disposal at the end of a product's life so that those products causing least environmental harm are used. A lot of this information is not yet available, and designers cannot be expected to know all the answers. However, they should have some awareness of the issues and be able to reason that, for example, natural materials like stone may be the most appropriate for some features or recycled materials like plastic in others, depending on the appropriateness of appearance within the overall design, the function that the material is being specified for, and the budget available.

Choosing contractors

If a project requires additional skills, select people with a respect for the landscape and who will be willing to protect the natural assets of the site as much as possible. Designers and contractors who can work well in a team often result in the best outcomes for the client (see Chapter 11).

Useful organisations

The Australian Institute of Landscape Architects

Sustainability goals are woven throughout the 2007 charter of the Australian Institute of Landscape Architects (AILA), reflecting the importance the profession places on them. Principles in the charter refer to the protection of biological diversity and ecological processes, addressing climate change and the global effects of using the landscape, and improving the quality of life for current and future generations. A strategic action is to support and undertake research that

'advances the achievement of sustainable environments and processes that protect and enhance Australian landscapes'.

The AILA website includes environment pages (www.aila.org.au/environment) which have the AILA Environmental Policies (21 policies from water to weeds, rainforests to raw materials), Sustainable Canberra Garden, Retrofitting Gardens for Climate Change and links to useful websites like www.OpenEcoSource.org that promotes the rapid exchange of sustainability information.

The Green Building Council of Australia

The Green Building Council of Australia (GBCA) (www.gbcaus.org), founded in 2002, aims to develop a sustainable property industry for Australia and drive the adoption of green building practices through the Green Star environmental rating system for buildings. Although the GBCA concentrates on the green credentials of buildings, landscape is often included in their assessments (e.g. GBCA Education pilot), and their goals of setting standards of measurements for green buildings, promoting integrated whole-of-building design, identifying building life-cycle impacts, and raising the awareness of the benefits of green building can equally be applied to landscape. The GBCA discourages the degradation of flora and fauna and encourages its restoration. The use of recycled materials, and construction methods that facilitate the re-use of materials at the end of the project's life are favoured.

The *Journal of Green Building* covers topics such as using greywater, and the effect of pavement thickness on diurnal temperatures.

Good Environmental Choice Australia

Apart from water, energy and organic food, ecolabels may not indicate independent, reliable assessments of the product's environmental impact. Good Environmental Choice Australia (GECA) is establishing an independent environmental labelling system to distinguish products that meet 'environmental, quality and social performance' standards. Launched in 2001, it complies with international standards for environmental labelling (ISO 14024) and develops scientifically based life cycle benchmarks (Australian Environmental Labelling Standards) that need to be met by a product before the ecolabel can be awarded.

An initiative of GECA with the Australian Environmental Labelling Association is the Australian Green Procurement Network Website for professionals wanting to source low environmental impact products in Australia (www.greenprocurement.org.au). The scheme is still developing; however, there are product categories relevant to sustainable landscaping and gardening.

Alternative Technology Association

The Alternative Technology Association (ATA) is a not-for-profit organisation with branches throughout Australia and New Zealand promoting technology and practices that lower environmental impact. The Association aims to improve the sustainability of our community through the information, technology and products available.

The ATA also undertakes research projects into topics such as greywater, energy conservation and renewable energy. The Association produces two magazines, *ReNew: technology for a sustainable future* and *Sanctuary: sustainable living with style*, and booklets on renewable energy.

Botanic gardens

Botanic gardens around the country incorporate sustainability in many aspects of their work – from recycling office, kitchen and garden waste to reducing energy use and

Figure 9.48 Green waste recycling in botanic gardens.

using new technology for conserving water (Figure 9.48). Education programs for students and the public include sustainability principles. Here are a few botanic gardens' programs.

Royal Botanic Gardens Melbourne and Cranbourne

The 19th century Royal Botanic Gardens Melbourne (RBG Melbourne) reflect the European origins of the new settlers, its landscape of sweeping lawns and garden beds. The landscape relies on irrigation, but the quantity applied has been halved over a 10-year period, through staff education, technology, changing practices and converting lawns to low-water turf, the saving achieved even though the period coincided with an extended drought. The new landscape of the Royal Botanic Gardens Cranbourne (also a part of the RBG Melbourne) is a 21st century landscape celebrating the beauty and diversity of the Australian flora. It was designed from the outset with low irrigation needs and is an excellent example of how landscapes can be beautiful without high resource use (Figure 9.49).

Figure 9.49 The Royal Botanic Gardens Melbourne and the Royal Botanic Gardens Cranbourne. (Photos: RBG Melbourne: RBG Melbourne; RBG Cranbourne: Janusz Molinski)

Ongoing research programs at the Royal Botanic Gardens Melbourne in association with universities and other organisations, are aimed at improving irrigation efficiency, particularly by improving understanding of soil water interactions. The RBG Melbourne is currently exploring sustainable alternative irrigation water sources, including stormwater capture and treatment. Green waste has been recycled for re-use on the site for a number of years.

Figure 9.50 A demonstration sustainable landscape in Unley, South Australia. (Photo: David Jarman)

The two gardens are important educational centres for visiting students and the general public with sustainability an important theme. Programs like the Master Gardeners at RBG Cranbourne give the practical strategies needed to help home gardeners with issues like water conservation.

Adelaide Botanic Gardens

In South Australia, the Sustainable Landscapes project promotes landscape design, plant selection and horticultural practices that are sustainable and appropriate for local conditions. Demonstration landscapes in the Adelaide region show how pleasing landscapes can also have low environmental impact (Figure 9.50). The initiative is supported by the Adelaide Botanic Gardens, Land Management Corporation, the Innovation and Economic Opportunities Group, Adelaide and Mount Lofty Ranges Natural Resource Management Board, and the SA Water Corporation.

Royal Botanic Gardens Sydney

In 2003 the Royal Botanic Gardens Sydney launched Towards Sustainable Horticulture to encompass a number of sustainable horticultural initiatives for both the Gardens and home gardeners. The Gardens has also implemented Integrated Pest Management to reduce or eliminate chemicals; soils have been improved without contributing to toxic residues; green waste processing and re-use has been raised from 90% to 100%; and water quality leaving the site has been improved. At Mount Annan Botanic Garden (an RBG Sydney satellite garden) the 'What's the Big Idea Garden' demonstrates how home gardening can contribute to conservation by using recycled or sustainable materials, and how to grow colourful native plants using water-wise and other sustainable gardening methods.

The water wall at Veg Out Community Garden in St Kilda, a garden where growing food blends with sustainability and a creative cooperative spirit.

SUSTAINABILITY IN THE BROADER LANDSCAPE

> **KEY POINTS**
> Sustainability is enhanced by community facilities and good urban planning.
> Making use of community gardens and landscapes.
> Water Sensitive Urban Design.
> Environmentally sensitive urban design.

Urban and town planning can be used to reinforce sustainability by making it easier for the community to work, travel, shop, play, learn and eat in more sustainable ways as they go about their daily lives.

Community gardens

Gardening is an act of faith in the future.
DON LINKE, FOUNDING MEMBER VEG OUT

There is a long European tradition of community gardens. Since the first in 1977, they have become increasingly popular across Australia. Each is run slightly differently, but generally the annual fees entitle members to a plot of about 15 m² although size varies. They are usually used to grow food. Australian City Farms and Community Gardens Network can help find the closest community garden and provide advice on how to start one.

Progressively more schools have vegetable plots for students to learn about food gardening and to enjoy the produce. The programs are being initiated by keen teachers and/or parents, with increasing support from governments.

Street tree selection can include food-producing species, the fruit being available for harvest by the nearby residents. The idea is not well developed in Australia, but there are some historic examples and some local councils promote fruit species in their lists of acceptable species.

Redundant industrial sites for parks

Under-utilised industrial land in cities presents opportunities for a change in land

Figure 10.2 Birrarung Marr – parkland replaces industry in Melbourne.

Figure 10.1 Veg Out Community Gardens St Kilda, developed after local residents became aware of the Council's intention to turn the space into a carpark in 1998.

use. Spaces, significant for their size or location, can be transformed to green areas, increasing public open space for recreation, and providing climate-moderating vegetation. Gardens, parks, community farms and vegetable gardens are all possible.

A recent Australian example of reclaiming open space, the park Birrarung Marr, is built over land where once commuter-filled trains moved daily across the railway lines between the city and suburbs of Melbourne. It is now a well-used 21st century park complementing the formal public spaces of Federation Square (Figure 10.2).

Water Sensitive Urban Design

Hard city surfaces such as roads and carparks capture a lot of water, along with pollutants and nutrients (Figure 10.3). In older areas stormwater is channelled rapidly away (Figure 10.4).

Water Sensitive Urban Design (WSUD) recognises that the water of urban areas is a part of the natural water cycle. Sensitive planning and design can lead to higher quality (less polluted) water flowing into our rivers, lakes, harbours and bays, and to healthier urban environments where stormwater is retained long enough to support the healthy growth of urban vegetation. Government departments are directing more resources to minimise the pollution of our natural water bodies through run-off from urban areas, and are setting water quality targets. Melbourne

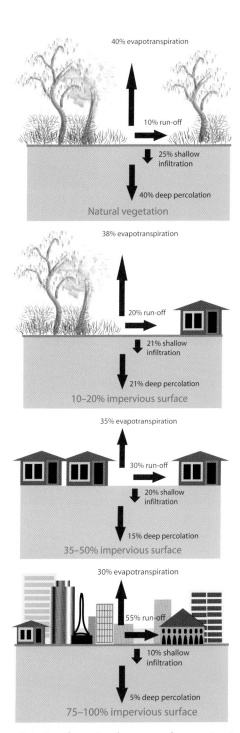

Figure 10.3 Transformation from natural vegetation (top) to cities (bottom) increases water run-off and decreases infiltration.[1]

Figure 10.4 Older parts of Australian cities were designed to remove stormwater rapidly through channels emptying into rivers and the sea.

Water, for example, aims to reduce suspended solids emptying into Port Phillip or Western Port Bays via stormwater by 80%, and nutrients like nitrogen and phosphorus by 45%.

WSUD principles aim to:

- Protect natural systems – including natural waterways and the ecosystems they support.
- Integrate stormwater treatment into the landscape – where natural creeks and artificial ponding are used in the landscape to process stormwater, and at

the same time act as recreational areas with increased visual amenity.
- Protect water quality – by removing pollutants close to the source through filtering and bio-retention, with higher quality water leaving urban areas and entering the broader environment.
- Reduce run-off peak flows – thereby reducing flooding events and the infrastructure to deal with them, using local retention and its more gradual release.
- Add value while minimising development costs – by improving the natural amenity of the urban area property values increase, this being achieved using methods that reduce the need for expensive infrastructure.[2]

WSUD can be adapted for an urban area or within individual properties. You can use rain gardens, porous paving, green roofs, water tanks, and:

- Swales – linear, often vegetated depressions between slopes that collect run-off and act as conduits for water transfer. They also filter litter and large particles as the water infiltrates, and they increase soil water (Figure 10.5).
- Infiltration trenches – these are excavated, lined with a geotextile material and filled with gravel. They are easily blocked and are best reserved for water with low sediment levels. They increase soil moisture, and can be covered with vegetation (Figure 10.6).
- Bio-retention systems – a number of water treatments collectively retarding the flow, and performing the primary and/or secondary cleaning of the water.

Figure 10.5 The Aurora Estate uses swales and infiltration to collect water from roadways. Top: The gutter design has been modified to direct water from the road into the grassy swale (arrow). Middle: The grassy swales allow infiltration through the soil and into the stormwater drainage. Bottom: Some swales are planted with plants such as *Lomandra* sp.

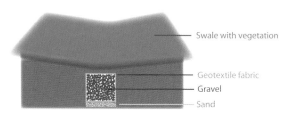

Figure 10.6 Infiltration trench.

Figure 10.7 Water run-off from a road at the VicUrban Lynbrook Estate is collected in a central bio-retention system dividing the road. The collected water is further purified through constructed wetlands that drain into a lake.

Figure 10.9 Gross Pollutant Traps remove litter for more efficient processing of stormwater.

Vegetation is an important contributor to the process. A bio-retention system could be grassed swales combined with infiltration trenches. Bio-retention is an integral component of stormwater processing in newer residential estates (Figure 10.7) and can be incorporated into existing infrastructure of older areas (Figure 10.8).

- Litter traps – a primary process for cleaning stormwater is to filter out litter greater than 5 mm across in Gross Pollutant Traps which can be cleared and the litter disposed. Further processing of the water is more effective after litter removal (Figure 10.9).
- Wetlands – these slow down the erosive forces of moving water and help purify the water by capturing sediments and, through the anaerobic micro-organisms they harbour, metabolise pollutants into less harmful compounds. They have: an inlet zone to slow the stormwater flow and for sedimentation of the larger suspended particles; a plant or

Figure 10.8 Bio-retention systems have been introduced into existing suburban areas, converting degraded drains for functional and attractive stormwater processing. At Wurundjeri Walk in Blackburn South, Melbourne, the drain channel was reformed and rocks installed to retard water flow. Geotextile material was laid and plants established along the bank to prevent erosion.

Figure 10.11 At the VicUrban development at Lynbrook, wetlands process the stormwater and form a part of the recreational environment for residents.

macrophyte zone with emergent and submerged aquatic plants that, along with the microbial flora on their stems and in the soil, metabolise the nutrients; and a deeper open water zone for sedimentation of finer particles and reduction in microbial populations. The water flowing from wetlands is of much higher quality than the water flowing into them and they are deliberately

Figure 10.10 Karkarook Park in suburban Heatherton, Melbourne, was a sand mine that was recently converted to a metropolitan park which features a wetland for processing the stormwater run-off from existing residential areas. Top to bottom: Inlet Zone including a Gross Pollutant Trap; Macrophyte Zone; Open Water Zone.

Figure 10.12 Docklands Park in central Melbourne combines wetland treatment of stormwater with a contemporary landscape, sculpture being integral to the wetland design.

constructed for residential areas now, to improve the quality of water flowing into natural waterways (Figures 10.10, 10.11 and 10.12). Although little used in Australia, subsurface wetlands offer an alternative to free surface-water wetlands. They are best able to treat the constant flow and high organic matter of wastewater but are less suited to the irregular flows of stormwater.[3] Their advantages include better public safety without open water, reduced mosquito breeding, greater capacity for treating water for the area of land used, and high ability to function in cold weather.

Rain gardens

The term rain garden can be applied to *'all the possible elements that can be used to capture, channel, divert and make the most of the natural rain and snow that falls on a property'*.[4] The basic concept for rain gardens is an attractive planted area, able to take surface run-off after rainfall, with plants able to tolerate periodically high soil-water levels. Rain gardens slow the flow of stormwater and begin the removal of nutrients and pollutants.

In Australia, rain gardens are increasingly being installed in private gardens and public areas. They can be as small as a metre square or larger than four hectares (Figure 10.13). In the new residential estate, Aurora, rain gardens have been installed in each house garden for the initial stormwater processing (Figures 10.14 and 10.15), in public areas and even to collect the excess water from a drinking fountain in the central square (Figure 10.16). In central Melbourne,

Figure 10.13 A square metre rain garden in the centre of Melbourne captures street water for the tree and improves the quality of water entering natural waterways.

Figure 10.14 Rain gardens in VicUrban's Aurora Estate collecting and beginning the processing of the stormwater run-off from the property.

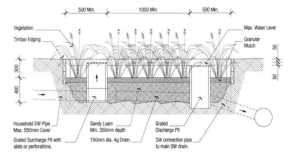

Figure 10.15 Design details for a rain garden indicating growing medium and overflow pipe. (Image: VicUrban)

Figure 10.16 A drinking fountain designed to flow directly into a rain garden in the central square of the Aurora Estate.

individual street trees are planted in rain gardens to capture the street water for use by the tree and to decrease pollutants. Larger rain gardens have also been incorporated into the streetscape, their function not immediately obvious (Figure 10.17).

The early 20th century Canna Bed at the Royal Botanic Gardens Melbourne, sited over a natural drainage line, was converted into a rain garden in 2007 to deal with periodic flooding and to improve the quality of the water flowing into the lakes (Figure 10.18).

Environmentally sensitive urban design

New subdivisions in Australia are incorporating many water and energy conserving features in their planning.

Some new housing estates collect wastewater, process it to Class A standard and return the water to the houses through a third purple pipe system for garden irrigation, toilet

Figure 10.17 Rain gardens at Victoria Harbour Melbourne. Top & Middle: *Cyperus papyrus* is planted in geometrical collecting basins. Bottom: An overflow drain is installed for occasions when run-off from rainfall is greater than the infiltration rate of the gravel-mulched rain garden planted with *Eucalyptus* and *Lomandra*.

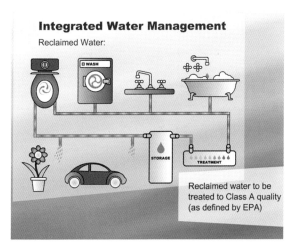

Figure 10.19 Aurora initiatives for water re-use. (Image: VicUrban)

Figure 10.18 The Royal Botanic Gardens Canna Bed (top) was reconstructed in 2007 (bottom) to become a rain garden.

Figure 10.20 Residential estates are incorporating the third purple pipe providing Class A standard water to premises for toilet flushing, irrigation and car washing.

flushing and car washing (see Figures 10.19 and 10.20). The system is completely separated from high quality mains water and distinguished by the purple colour, an internationally recognised standard for recycled water.

Installation of water tanks, solar hot water and/or photovoltaic panels for electricity production is mandatory for houses in some new subdivisions, and for all new houses in some states. Even in southern parts of Australia, solar hot water systems can produce hot water without gas or electrical backup for over 40 weeks every year. Renewable energy equipment can, of course, be designed into public buildings (Figure 10.21).

Even simple details like having pedestrian walkways through residential blocks can significantly shorten a journey to the shops

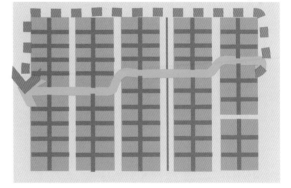

Figure 10.21 Solar panels incorporated into the public shelter at Aurora become an integral part of the contemporary design. An existing *Eucalyptus* was retained in this public square.

Figure 10.22 Permeability of residential blocks encourages walking rather than car travel.

Providing the conditions that support large populations of healthy trees can be achieved through:

- Maintaining a significant proportion of residential properties with large gardens that can accommodate trees.
- Designating open space to protect existing trees.
- Providing adequate space and soil volume for street trees.

and encourage people to walk rather than to hop in the car (Figure 10.22). Encouraging bicycle use can be as simple as providing suitable parking areas, and identification of safe bicycle routes (Figures 10.23 and 10.24). Introducing these into our current urban infrastructure may be possible for some suburbs, and straightforward when planning new subdivisions.

Healthy urban forests

The value of trees in the landscape, their role in reducing energy use for cooling, and their ability to capture and store CO_2 is known.

Figure 10.23 Bicycle parking facilities in commercial and public areas facilitate the use of alternatives to fossil fuel vehicles.

Figure 10.24 Safe bicycle routes encourage cycling in urban areas where road traffic can be intimidating.

Figure 10.25 Council House 2 uses green walls to shade north-facing balcony windows.

- Planning adequate public open space in all new developments and protecting existing parks and gardens.
- Encouraging commercial and industrial properties to include treed landscapes.

Residential properties

Encouraging increased population density in cities means that the existing infrastructure is used more efficiently, and it reduces the need for additional services for new subdivisions. A cost, though, is increased roof area and paving, and loss of vegetation, especially trees. The decreasing proportion of vegetation compared with hard surfaces will compound the heat island effect of our cities. Reassessing how we increase population density to retain a high proportion of urban forest throughout built-up areas could alleviate this. Planning controls in Barcelona, for example, allow for seven-storey buildings along transport routes, but lower density is retained between these.

Protecting existing trees

Identification, retention and protection of existing mature healthy trees in new developments recognises the great aesthetic and environmental value the vegetation can contribute to a subdivision that would otherwise take decades to achieve.

Street trees

Since the 19th century, nature strips have buffered houses from public spaces in Australian cities, providing good growing conditions for street trees (Figure 10.26). Retrospective tree planting in older, narrower city streets has often been in small squares cut through paving exposing poorly structured soils, not suitable for healthy growth of tree roots. An analysis of street trees in Montreal in the 1990s showed a life expectancy of only about five years after planting – costly both economically and environmentally.[5] Larger planting holes, the use of well-designed soils and increased aeration has improved survival. Street tree health in Australian cities is also

INFO BOX 10.1: COUNCIL HOUSE 2 (CH2)

This new 10-storey office building for 540 City of Melbourne staff, is rated as a 6 Star building, the highest level in Australia. It was designed to reduce gas use by 87% and electricity by 85% producing only 13% of the emissions that an equivalent sized building would normally use. Water is conserved by collecting rainwater, saving the water usually lost during the regular checking of the fire safety sprinkler system, and by mining the sewage both from CH2 and other city buildings via the nearby sewer. The 100 000 litres or so of blackwater predicted to be processed to Class A standard each day is used in the building for toilets, watering plants and the building's cooling system. Excess recycled water will contribute to the city's landscape irrigation needs and for its fountains. The balcony windows on the north side are protected by green walls (Figure 10.25).

Figure 10.27 Eucalypts planted in a continuous strip of soil reminiscent of suburban nature strips, will develop stronger root systems and be more vigorous in this central city location (New Quay, Melbourne). Gravel mulch planted with *Lomandra* has replaced the turf of a traditional nature strip.

better with provision of vastly improved growing conditions in planting holes or by reproducing the nature strip concept (Figure 10.27).

Structural soils are designed to avoid compaction yet maintain appropriate soil aeration and moisture levels, and thus better

Figure 10.26 Nature strips, a common feature of Australian cities since the 19th century, support healthy street tree growth.

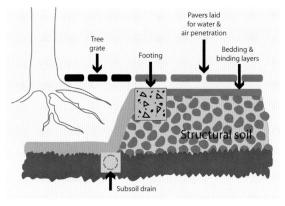

Figure 10.28 At the Homebush Olympic site in Sydney structural soils were used around the *Ficus microcarpa* var. *hillii* planting pits and under concrete pavers designed to leave holes for infiltration and oxygen exchange when abutting each other.

Figure 10.29 Paving materials like dense asphalt restrict oxygen exchange and water infiltration to tree roots. Tree roots can also break through the surface causing a safety hazard for pedestrians. Porous pavement was used in this landscape to overcome these problems. a) tree roots exposed after breaking through dense asphalt; b) original asphalt removed carefully, avoiding damage to tree roots; c) porous paving with recycled rubber aggregate replaces dense asphalt; d) junction of dense and porous asphalt; e) the replacement completed, it is difficult to tell the difference between the original and replacement paving.

soil conditions for trees in highly trafficked areas like the Homebush Olympic site in Sydney. Structural soils were used there for planting Fig trees in a central paved area. The soils contained even-sized basalt gravel that when compacted left spaces for sandy loam, air and water, good conditions for root growth (Figure 10.28).

Another way to protect tree roots, avoid soil compaction, and improve soil aeration and water infiltration is to use porous pavements (Figure 10.29).

Sustainable living beyond the home

The way towns and cities are designed can increase our capacity to live more sustainably. The provision of community gardens, pedestrian and bike access, efficient public transport and recycled water systems encourages more sustainable lifestyles. Communities based on environmentally sensitive design will be more sustainable. We have seen this already with the introduction of Water Sensitive Urban Design. The fabric of our cities will change as new approaches like this, based on a greater understanding of urban ecology, are applied.

Design is a critical element in the process of creating more sustainable gardens and public landscapes. Lowering their need for water, energy and materials, encouraging biodiversity and contributing to food supplies requires the consideration of many often competing factors. We have entered a new challenging and exciting period of horticulture where the blend of innovative technology and creative design will play an important role in securing our future.

CONSTRUCTING LANDSCAPES SUSTAINABLY

> **Key points**
>
> Construct landscapes using sustainable designs, in an appropriate sequence, for the long-term.
>
> Protect the natural ecology and site assets (soil, indigenous vegetation, mature trees).
>
> Reduce the impact of resource-use by using less, using low environmental impact materials (including re-used or recycled) and selecting quality materials that last.
>
> Prepare soil for healthy plant growth and source high quality plants.
>
> Install irrigation systems only where needed and that deliver water to plant roots efficiently.
>
> Minimise waste and recycle where possible.
>
> Monitor the performance of the new landscape.

Good design is an important precursor of attractive and more sustainable gardens. Construction can contribute to environmental problems, but it can also solve them. It is important to construct gardens in an appropriate sequence, protecting site assets, producing minimal waste and recycling where possible, and disposing of waste appropriately. Priority can be given during construction to using low environmental impact materials, installing low energy and low water equipment, and constructing for the long-term but with a view that materials can be easily recycled in the future.

Improving the sustainability of construction may involve changing currently accepted methods. This may mean using different equipment, altering the sequence of tasks, or sourcing alternative materials, sourcing supplies from companies that are managed more sustainably or are closer to the site. Sometimes we may already know the most sustainable approach, but other times we may need guidance or the information may not be available yet.

Choosing contractors who understand sustainability makes building a low impact garden easier to achieve. Such a contractor will be more aware of suitable construction methods, better materials and where to source them. Having a good working relationship with a like-minded contractor is more likely to lead to sustainable outcomes.

Landscape contractors specialising in sustainable gardens will gain a marketing edge, as more people prefer a green approach. The recent prolonged drought increased work for contractors who convert irrigation systems to drip, and who install equipment like rainwater tanks. But not all methods for overcoming the water shortage are sustainable. Just how many people can use bores that tap into groundwater that is only recharged by diminishing rainfall? It will be the challenge of contractors to be aware of the pros and cons of various landscape solutions and to advise their clients accordingly. As the market takes up greener solutions, the cost of lower-impact materials and equipment will decrease, the range and quality will increase, and they will become more competitive with traditional products.

We are only temporary custodians of the landscape. As it passes to younger generations, we should feel satisfied that we have at least maintained its quality or, better still, improved its health.

A guide to sustainable landscape construction

We cannot ignore the contribution that landscape industries make to our built human environment. The following points, adapted and expanded from the Santa Fe Green Building Council definition of green construction, are relevant for landscape contractors and home gardeners.[1]

During landscaping:

- Protect healthy sites, particularly those with indigenous vegetation, and/or mature trees that are growing well.
- Restore or improve damaged sites.
- Work with natural processes, not against them.
- Contribute to regional habitat conservation.
- Improve local quality of life and health.
- Reduce resource use.
- Reduce waste production, including pollution.
- Reduce transport and contribute to a diverse and stable local economy by using local resources and suppliers.
- Use resources in ways that increase their value.
- Select quality materials and equipment, and construct for the long-term.

Making better use of knowledge and technology

How do we keep in touch with developing ideas and products? There are a few organisations that can help, like Sustainable Gardening Australia (SGA), the Alternative Technology Association and the Green Building Council of Australia. They all offer ideas, training and/or information. Organisations like the Australian Landscape Industry Association and its state affiliates have an increasing awareness of sustainability. Also look out for local networks and industry groups. All of these can be found via simple web searches.

Green purchasing

When it comes to purchasing different pumps, soils composts, mulches, pavers, edging, timber or even plants, most suppliers do not give any indication about environmental impacts of one product over another. Choosing the right materials and equipment for construction would be easier if independent rating and ecolabelling systems based on life cycle assessments were better developed.

SGA has rating systems for some products, and its approved garden centres identify those that are better for the environment. Other nurseries may have no idea. The situation might change if there were greater demand for more detailed environmental information. Suppliers respond when they realise there is a need.

Good Environmental Choice Australia has approved a few landscape-related products and the Alternative Technology Association has articles in their magazine *ReNew* that may be relevant. Purchasing timber that meets the Australian Forestry Standard can avoid timber from non-sustainable Australian or overseas forests. Remember too, that the more we purchase value-added products like garden furniture made from local Australian timber, the more the timber will be prized, giving it a higher dollar value, so less timber will provide the same income as that derived from its use as woodchips or packing crates. The web-based Ecospecifier links the user to supplies of some products for landscaping including: organic and stone mulches, decorative pebbles, soils, irrigation, planters with subsurface irrigation, external lighting, benches, furniture, paving and very limited plant suppliers.[2]

In general, sourcing materials locally reduces transport energy, and using minimally processed natural products lowers embodied energy.

Ecolabelling

An 'ecolabel' indicates an overall environmentally preferred product or service. It is based on an independent third-party assessment of the environmental impact of that product or service in relation to other similar products and services. The International Organisation for Standardisation (ISO) ISO14024 standard provides verification of these third-party assessing organisations so they can be trusted to provide verifiable and accurate information that can be distinguished from various 'green' symbols or claims made by manufacturers or service providers themselves. Ultimately, reliable labelling encourages the demand and supply of products and services that cause less stress on the environment and has the potential to stimulate market-driven improved environmental performance. Other environmental management standards in the ISO14000 number series include management systems, performance evaluations, auditing, and life cycle assessment and labelling.

One example of ecolabelling is the Water Efficiency Labelling and Standards (WELS) scheme which applies to taps, shower-heads, dishwashers, washing machines and toilets. Another example is the Energy Star ranking of white goods. Small engines may have a rating scheme soon.

Considering the environment during landscape construction

Although many of the issues raised in this section may primarily be useful for landscape contractors, they will also be helpful for homeowners who may be constructing their own garden, employing contractors to construct components of the garden, or who want to have a greater understanding of construction processes when employing a landscape contractor who will be doing all the work.

Using well-planned sustainable designs

Chapter 9 described how to design a garden taking into account water, energy, materials, food and biodiversity. It introduced many ideas that will help even the non-designer understand how to make a garden more sustainable.

Clear plans documenting a sustainable design for the garden are the best way of guiding landscape contractors or home gardeners when constructing a garden. It is important that the plans are clearly interpreted to fully capture the designer's intent. Apart from having a general understanding of the concepts behind sustainability, particularly related to horticulture, the people constructing the garden may need to discuss the concepts and details more fully with the designer.

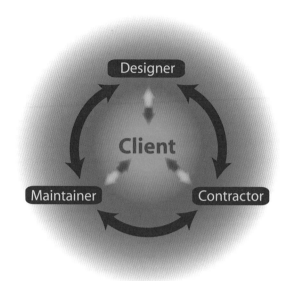

Figure 11.1 Better landscape outcomes result from good communication between the stakeholders.

Designers often appreciate ongoing involvement as the garden is being constructed. Designers may also help the construction process by providing more detailed specifications for construction methods, materials and their sources, the level of detail being dependent on the arrangement between the designer and the client. Very often the best landscapes result from a strong, dynamic relationship between the client, designer, contractor and ultimately the maintainer of the garden (Figure 11.1).

Matching designs and sites

Some sites are just not suitable for certain landscape styles or designs and it is better not to attempt to fit a square peg into a round hole. Choosing a different site or working within the constraints of the site are better options.

Even in small suburban gardens there can be significant variation in topsoil, subsoil, rockiness, soil moisture content, shadiness, sunshine hours, heat, wind, humidity and so on. A good garden design will have taken these environmental factors into account, and placement of hard landscape and planted areas will reflect the site conditions as well as the client's needs. An alert landscape contractor or knowledgeable gardener will pick up any obvious anomalies that may damage good soils or valuable existing plants, or that will establish vegetation where conditions don't suit. Sometimes it is only after beginning construction that limitations, like buried boulders, may be revealed, and modifications to the design may be needed.

Choosing a contractor

Contractors understanding the interaction between sustainability, landscape and horticulture are more likely to protect the natural assets of the site and minimise the impact of the garden's construction on the broader environment. At the very least, speaking to potential contractors will give some inkling of their approach. Ask how they would undertake specific tasks like sourcing materials, or their attitude to recycling, minimising waste or using toxic chemicals. If possible, visit landscapes they have already completed and talk to previous clients about their experiences with the contractor. Contacting the Australian Landscape Industry Association or its state affiliates may help find an appropriate contractor.[3]

In the future, designers or contractors may be able to supply their clients with documented environmental impacts of the proposed garden design complete with an LCA which includes possible alternative choices of materials that may lower environmental impact.

Sequencing works for efficient use of materials and equipment

The following sequence of tasks will help guide planning for efficient, low impact construction. The sequence will need to be refined to account for the variation between sites and plans. In general, keep a tidy work site, minimise waste, recycle where possible, use recycled materials when they can be sourced and new materials with low environmental impact.

1. Establish protection zones – fence off vegetation to be retained and soils to be protected from compaction and building waste.
2. Identify all services and infrastructure.
3. Demolish structures (recycle materials if possible).
4. Remove unwanted plants (retain healthy, mature vegetation that is suitable for the new garden).
5. Define works and clean-up areas.
6. Place boards over soil where foot traffic will be high. Temporary placement of gravel for vehicular traffic may also be needed.
7. Install work sheds/toilets if required.
8. Install waste and recycle bins.
9. Stockpile topsoil from construction zones in a fenced area for later use.
10. Undertake earthworks if required, e.g. for underground water tanks (remembering that using heavy equipment can add to

the embodied energy of the garden and CO_2 production).

11. Install tanks and associated reticulation into garden, and to house toilet and laundry.
12. Build structures like terraces, decks, pergolas and pools.
13. Dig trenches and install service lines like water, electricity and main irrigation lines.
14. Install hard landscaping like paving.
15. Decompact soil and reintroduce stockpiled soil.
16. Add amendments like compost.
17. Install irrigation system.
18. Place and plant plants according to plan – fertilise, if required, and water plants in.
19. Mulch garden beds.
20. Prepare soil for lawn areas if required.
21. Sow seed or install instant turf and water in.
22. Commission irrigation system and water features.
23. Maintain new plantings ensuring sufficient water is available until root systems establish.

Energy

Fortunately there are still many landscape jobs that are best done without the use of mechanised equipment. Human labour is much more energy efficient (see Table 11.1), and it is powered by renewable food crops rather than non-renewable fossil fuels. Many sites have restricted access, difficult slopes and small spaces to work in, and these favour hand labour. It is difficult to see how the creative fine-tuning of plant placement at the time of planting could be mechanised.

Table 11.1 Energy use of powered landscape equipment

Tool	Energy use (KJ/hour)
Human male	320
Stump grinder	95 000
Rotary hoe	85 000
Electrical circular saw	15 000
Chainsaw electrical small	15 000
Chainsaw petrol small	26 000
Chainsaw petrol large	37 000

Source: Thompson & Sorvig (2000).[1]

Fuel usage will depend on the type of equipment, its age, the conditions it is used in and the type of fuel it uses. Diesel fuel contains more energy per volume (39 000 kJ/L) than either petrol (35 000 kJ/L) or two stroke mix (petrol/oil 50/1; 35 000 kJ/L). Oversized electrical or petrol-driven equipment can consume excessive energy, compared with those that are the right size for the job.

Homeowners may already have green renewable electrical energy so sustainable construction involving electrical equipment is less of a problem. Contractors working on new sites may be able to source renewable energy from a local supply company during the construction period. In some cases, solar or wind power can be set up on site. One day, tool manufacturers may sell cordless tools able to be recharged from small portable solar panels temporarily set up by the contractor.

An aware contractor will respect sustainable energy features of the site and its buildings, and make sure they are not compromised during or after construction. The landscape should maintain or improve the energy efficiency of the associated buildings.

The environmental impact of energy use can be further lowered when sourcing materials, by purchasing from companies that manufacture using renewable energy, local resources and efficient transport vehicles.

Protecting topsoils

Topsoils are complex, living components of the landscape. They contain most of the plant roots mixed with a rich interactive population of millions of micro-organisms per gram, these being mostly bacteria and actinomycetes (organisms with characteristics intermediate between fungi and bacteria), but also fungi, algae and protozoa. Many micro-organisms, such as the mycorrhizal fungi associated with plant roots, provide benefits to the plants with which they share the soil. These include greater access to soil nutrients and water, and protection from disease, leading to faster, healthier plant growth. Regaining a healthy functioning soil after construction damage may not be possible, and even if it is, it will take many years and a lot of resources to achieve. It is better to protect existing soils as much as possible. It also makes more economic sense to avoid the extra work that would be needed to undo damage.

The living component of soils operates well when spaces between the soil particles allow sufficient aeration and water percolation. Soil compaction caused by pedestrian traffic, machinery and the dumping of building materials works against this and needs to be avoided by restricting access to sensitive areas by fencing off or placing boards along walkways and other heavily trafficked areas. Recycled plastic boards can be hired for this purpose (see Figure 11.2). Clearly marking restricted areas on plans given to potential contractors and making them available to all people working on the site reinforces the importance of protecting the fenced areas.

Figure 11.2 Soils can be protected from compaction by placing boards or solid plastic sheets on the surface.

The other major problem on construction sites is the highly increased rate of erosion, which is many times greater than normal rates on vegetated sites. Avoiding erosion is difficult. The very process of construction can reduce the surface protection and soil binding properties of vegetation, more compacted soils reduce the water infiltration rate and increase surface water, and drainage systems may be damaged or yet to be installed. Temporary drainage systems that collect water from builders' sheds or surface drainage should be considered. Sheets of hessian or similar material pegged across bare soil can protect from wind and water erosion, as can hay or straw. Mulches mixed with adhesives are good stabilisers. Keep vegetation destined to be removed for as long as possible during construction to protect soils. It is worth remembering to limit work when the soil is wet. This will prevent the soil structure being damaged, but it means avoiding undertaking construction projects in known wet seasons of the year, or at least during and immediately after a particularly wet period.

Protecting vegetation

Fencing around the root zones of any vegetation that is to be retained for the new garden will restrict access and avoid a number of problems that can affect plant health. These restricted areas should also be marked on plans to be given to potential contractors and to everyone who will be working on the site to ensure they know the importance of protecting them.

Plant roots can be damaged by discarded toxic chemicals percolating into the soil, and the anaerobic conditions caused by compaction or stockpiling of materials like soils and mulches over root zones which restricts oxygen reaching the roots. All these situations kill roots or retard root growth, subsequently reducing the vigour of the whole plant. This is especially critical for trees which can't be easily replaced.

It is good practice to contain chemicals on the worksite. This includes the sawdust from drilling or sawing treated timber, petroleum products, excess garden chemicals like herbicides, paints or the wash-up from painting gear, and excess concrete or the wash water from concreting equipment. Chemicals are best prepared over spill-trays to catch accidental spills. The contained chemical can then be disposed in a safe way. The Environment Protection Authority (EPA) or equivalent body in your state can provide information about disposing of chemicals safely.

Compaction of soils is caused by vehicles, pedestrian traffic, wheelbarrows, storage of heavy supplies such as timber, etc. Using just-in-time delivery avoids the need for stockpiling, but if it is used, it is better to stockpile in areas that will be permanently covered by paving, such as driveways or hard tennis courts. These are good areas for worksite parking too.

Protecting the natural water resource

Protecting waterways is as important as protecting soils. Poorly managed building sites can be sources of chemical pollutants that end up in creeks, rivers, lakes or the sea, damaging these important resources. The threat can be minimised by using toxic chemicals only when necessary, using the least toxic chemical for the purpose, preparing chemicals over trays so spills can be contained, preparing only the amount that is needed and disposing of any excess using approved methods.

The health of plants and animals reliant on natural creeks or rivers flowing through or near construction sites can also be threatened by increased silt from bare eroded soils, and managing water run-off during construction is important.

Cement and concrete

Cement and concrete are very alkaline (pH ~12–13), and the common practice of disposing of excess concrete or the washings from equipment directly onto the soil can damage soils. Concreters also use stormwater systems for the washings, which may save soils from damage but shift the problem elsewhere. Ensure all contractors or employees know the appropriate way to handle cement and concrete.

Store cement properly under cover so it is protected from rain, run-off and wind that can wash or blow the dry powder onto soils or into stormwater drains where it can affect soil health or cause water pollution.

When concreting:

- Only mix or order the quantity of concrete or mortar that you can use so there is no wastage.
- Set up concrete mixers over a waterproof tarpaulin that will protect the soil from spills.
- Dispose of excess concrete (there should be little) only on a piece of ply, a tarpaulin, plastic sheeting or similar material. Never dispose of concrete directly onto the soil. Placing in small quantities makes it easier to handle and dispose of once set. Place in the waste bin for appropriate disposal.
- Scrape as much residual concrete as possible from equipment prior to washing, and dispose of as above.
- Minimise the amount of water used for washing equipment and set up a waterproof containment system for the wash water, e.g. thick builder's plastic held in a depression or a large plastic container. After settling, re-use the water for mixing more cement. Allow remaining water to evaporate as much as possible and dispose of the solid residue appropriately.
- Never allow the wash water from equipment to enter the soil or stormwater gutters.

Build less

With any garden, the less that is built, the fewer materials and the less energy will be needed. Building less does not mean building to a lower standard, but for some purposes specifications may be safely reduced. Concrete pedestrian paths do not have to bear the loads of driveways, for example.

It is worth asking if we really need some of the proposed landscape features, and if we do, do they need to be as big? Reducing the width of a path can save a considerable amount of concrete over a few metres length, for instance. Shade can be provided by trees rather than vines over a pergola, saving the need for timber, nails, bolts, post holders, and wood stains or paint. There would be no energy required for saws or nail guns, and the washings from paint brushes will be avoided.

Selecting appropriate materials

Selecting appropriate materials is partly about their embodied energy and water, but also the direct impact on the environment of its extraction (e.g. mining) and manufacture (e.g. pollution). Table A3 in the Appendix provides a list of landscape materials with their embodied energies, showing that some materials have more embodied energy than others. Where possible select local, renewable or recycled materials.

Plastics have high embodied energy and are not always recyclable. Metals are potentially recyclable although they also have high embodied energies and can corrode. Timber generally has lower embodied energy and may come from well-managed plantations, but it can rot, and things like decking timber are often derived from the south-east Asian rainforests that are rapidly disappearing. There are always factors to balance. The embodied energy of concrete per kilogram is around the same as timber, but because of its density and the larger volume required, the same area of concrete will have a much higher embodied energy than timber decking.

A better measure than embodied energy or water is an LCA, but there are few published studies for the range of landscape materials used.

It is also worth considering how suited a material is for the purpose. It is no good using a material with low embodied energy if it has a short life in the situation where it is to be used. In constantly wet places, stone or concrete may be a better option than timber which will rot and need replacing frequently. In dry locations, timber may last for decades.

Being willing to use non-standard materials, like porous paving, can provide long-term benefits for gardens, in this case by increasing the amount of water percolating into soils.

Minimising waste

Minimising waste during construction makes both good environmental and economic sense. Ordering just the right amount of material avoids leftovers. Wastage is also avoided when materials are bought in lengths or areas that span or cover exactly or just over the size required, eliminating or minimising off-cuts. Timber lengths are sold in increments of 300 mm, for example, so the closest length is easily available. Pipes are also sold in standard lengths, although the increments can be larger. It is often cheaper per metre to buy longer lengths, but if only one metre is required, a six-metre length leaves a lot of unused material.

The ultimate dimensions of paved areas can be governed by multiples of the lengths and widths of the pavers to be laid side by side, avoiding the need for cutting the pavers and producing useless off-cuts. No energy will be needed for masonry saws either. Sometimes existing structures like walls define the area, so selecting the best-fitting paver size may avoid the need for cutting.

Construction sites generally send most waste to landfill, but many materials can be recycled. Bricks can be used again on the site or sold on to a second-hand brick dealer. Pavers set on a compacted sand base are easily lifted and re-used. Concrete and timber off-cuts are recyclable. A trial in South Australia is recycling old black poly irrigation pipe from orchards so it may be possible one day to expand recycling to old irrigation systems from ornamental landscapes and from other parts of Australia. Packaging from materials, even food and drink containers from worker's lunches, are better sorted and recycled than sent to landfill.

Timber accounts for as much as 10% of landfill in Australia. Over 2 million tonnes of timber waste are generated from urban areas annually, and about the same in off-cuts from milling and other wood processing. Recycling timber, including off-cuts from builders and carpenters, is not yet well developed, but studies in various parts of Australia are helping to define the extent of this resource and opportunities to use it. *Wood Recovery and*

Recycling – A Source Book for Australia and New Zealand is a joint venture by the CSIRO and New Zealand's Scion (formerly Forest Research) to define the generated volumes of waste timber, recovery opportunities, materials handling requirements and sustainability.[4] These sorts of initiatives will stimulate wood recycling systems. It is worth remaining in touch with developments in your region.

There are companies that grind building materials on site for recycling including bricks, concrete and wood, claiming that up to 90% of a house can be re-used. Recycling on-site reduces the road transport needed to take the materials to landfill. Wood from house framing can be chipped for on-site erosion control, construction site walkways and landscape mulching. Masonry after grinding can be used as a base for paths and drives, or for backfilling. Plasterboard is made of gypsum (calcium sulphate) and can be used to improve the soil structure of some clays or to add soluble sulphur to sandy soils. Caution is required though, because not all soils will respond to the addition of gypsum or need it, and using painted materials for recycling (plaster is usually painted) risks contaminating soils with chemicals like lead in older paints and should be avoided.

Concrete is recycled by grinding, removing steel reinforcement and re-using the original aggregate with new cement. It is claimed that there is no noticeable difference in strength if up to 30% recycled old concrete-coated aggregate is used, and trials to test this are being undertaken by the CSIRO Division of Manufacturing and Infrastructure Technology. Since the early 1990s, hundreds of thousands of tonnes of recycled concrete have been used in base layers for road building in Australia. Sand from recycled concrete can also be used again in concrete.

There can be hundreds of empty pots remaining after planting a new garden and recycling is the best option, although recycling plant pots is problematic and not well developed in Australia at present. We discuss this further in Chapter 12. Landscape contractors regularly planting new gardens can divert the pots away from landfill for re-use if plastic recycling is not an option.

Minimising site damage and the impact of mechanised equipment

If possible and where regulations allow, locate service lines like water and gas in the same trench. The main pipelines for irrigation can, for example, run in the same trench as the conduit for the solenoids wiring, minimising soil damage by reducing trenching and not criss-crossing the site. Keep service corridor width to a minimum; this is especially important for larger scale projects where vehicle access tracks may be required.

Use mechanised equipment only when absolutely necessary. Human labour and manual tools are often as quick and as effective. If powered tools are needed, use the right tool and the lightest equipment appropriate for the job to reduce energy and fossil fuel use. Contractors often buy larger, more powerful equipment that will handle their biggest jobs, but for the most part they could use something smaller. A more economical and environmentally appropriate approach may be to purchase equipment suitable for most work and hire or subcontract someone else for the occasional big job. Larger equipment not only uses more fuel, but can cause more site damage, make it more difficult to protect existing important vegetation and compact and damage soils, requiring additional energy and resources to repair the damage. Smaller or manual equipment is often easier and more effectively used.

When purchasing machinery, ask about fuel consumption and the pollutants emitted under normal operating conditions. The wattage of electrically powered tools will give an idea of power consumption, but if possible find out its efficiency as well. Favouring brands and businesses that provide this information will strengthen their position in the marketplace. Energy-efficiency (energy star) labelling needs to be encouraged.

Ensure all powered equipment is well maintained so it runs efficiently.

Planning transport and deliveries to the site to minimise the number and distance of the trips can significantly reduce fuel use.

Preparing soil for planting

Healthy plant growth can be encouraged with good soil preparation prior to planting, and healthy growth means resource use is not unnecessarily increased when plants need to be replaced or require more intensive care.

Construction rubble needs to be removed, although hopefully the rubble has been confined during the works, leaving most of the site clear, protecting the soil and reducing work required prior to planting.

Using the soil already on the site rather than importing additional soil avoids damaging other landscapes from where precious topsoil is removed. Energy is saved by not transporting soil too, but there may still be a need to improve soil structure and fertility, both of which may increase the embodied energy of the soil.

Compacted soils are common especially on work sites, and if not remedied prior to planting, plant growth can be retarded through reduced oxygenation for the roots, slower entry of water into the soil, and the mechanical resistance of the closely packed soil particles to root penetration. Compaction may be restricted to the upper layers of the soil, but can extend into lower layers. Compacted soils can be identified by measuring soil density, for example by measuring bulk density (grams of soil per cubic centimetre) of undisturbed soil and comparing it with disturbed areas.

There is a temptation to use rotary hoes to break up the soil, but they work only on a very shallow surface layer, perhaps no more than 200 mm, and in wet clay soils the blades can leave a compacted surface underneath forming a barrier to drainage (Figure 11.3). On larger sites, ripping relatively dry soil with long tines shatters the soil, improving water infiltration and root penetration. For smaller sites hand digging with a spade or fork works very well, and this of course does not require fossil fuels.

The plants to be grown may not grow optimally without further improving the physical properties of the soil. Whether the soil is clay-based or sandy, structure can be improved with the addition of organic compost amendments. Clay soils benefit by

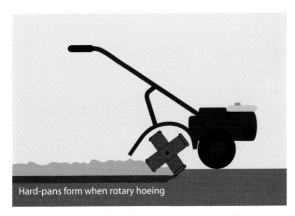

Figure 11.3 Rotary hoe blades can compact a layer of soil.

clumping the fine clay particles together into larger aggregates and creating bigger soil spaces. Sandy soils benefit by increasing the water- and nutrient-holding capacity of the soils. Some clay soils can also benefit from the addition of gypsum (calcium sulphate) which also clumps the particles together, but not all clays are responsive. Often the quantity of compost required is greater than the capacity of a site to generate organic composts so it will have to be brought in. Composts are available commercially, often with unique blends, but some are more standard in composition, like mushroom compost or composted pine bark. Check with the supplier the pH, salt content and manure content of their composts. Mushroom compost can have high levels of salt and, if fresh, a high pH, although microbial action brings the pH back to around neutral. Some composts have chicken or other manures added and can be high in phosphorus which may not suit some plants like *Banksia* and other members of the Proteaceae. If the composting process is not well managed, the compost may not reach the temperatures that kill weed seeds or diseases.

Composts are also vulnerable to re-infection if not stored well. Importing weeds or diseases can result in costly control measures, although using reputable suppliers should avoid these problems.

The low water-holding capacity of sandy soils can be improved by adding clay soil to the top layer to about 20% of the soil volume. If clay subsoil is used, additional phosphorus is most likely needed to compensate for the clay's ability to hold onto phosphorus making it unavailable to plants.[5] Water-holding crystals (acrylic copolymers, e.g. polyacrylamide and sodium polyacrylate) can increase the water-holding capacity of sandy soil too, expanding many times their original size as they absorb water, and decreasing in size as water is lost. They are not readily biodegradable and can remain effective in soil as a water reservoir for plants for long periods, perhaps even years. They are apparently non-toxic to plants, but the swollen crystals can cause health problems (such as internal obstructions in animals) if inhaled or ingested. Hence they should be kept out of waterways. Hydrophobic soils may also benefit from the use of wetting agents. We discuss this further in the section on water and soils in the following chapter.

Humates can also be used to improve the nutrient and water-holding capacity of sandy soil. The relatively young brown coal deposits of Victoria can be a rich source of organic humates, the variable complex mixtures of large organic molecules resulting from long-term weathering and oxidation of plant matter deposits. The humate molecules can be associated with hydrogen ions making the molecules acidic (humic acid); however, the Australian humates are instead often associated with magnesium and calcium (insoluble humates), or sodium and potassium (soluble humates). Soluble humates are the basis of commercially available products. These, when applied to sandy soils, form layers over the soil particles, giving increased attachment sites for plant nutrients so that less are leached away after fertiliser application. In clay soils, humates reduce the close packing of the fine clay particles, increasing water penetration and reducing the tendency of the soil to crack during hot dry summers. Humates also give soils a buffering capacity against the tendency to become more acidic.[6]

Zeolites, naturally occurring aluminosilicates found in the basaltic regions of eastern Australia, are another soil additive worth considering. The very high cation exchange capacity (CEC) of this mineral results from an open internal molecular structure. When added to sandy soils it can increase the soil's water and nutrient holding ability. Zeolite is milled and graded for horticultural uses, and growing trials have shown the addition of zeolite can increase growth of flowering plants and vegetables.

Biochar, also known as agrichar, is a type of charcoal produced from biomass via a no-oxygen burn process called pyrolysis that can also produce biofuels at the same time. It is the main component of the very fertile Amazonian dark earths (*terra preta*) that were probably created by the local indigenous people between 500 and 2500 years ago. Amongst other uses, biochar is used as a soil amendment. Being essentially a form of activated charcoal, it is largely inert, largely unaffected by microbial action, relatively long-lasting in soils and highly porous, serving both to retain water in the soil and provide large surface area for microbes. Adding biochar to soils has been shown to increase plant productivity. There is some evidence that increased plant growth can be attributed to using low-temperature biochar rather than high-temperature biochar, the former perhaps retaining organic matter beneficial to micro-organisms, such as mycorrhizal fungi. Biochar applied at the rate of 10 tonnes per hectare tripled the biomass (yield) of wheat and more than doubled the biomass of soybeans.[7] It is also suggested that biochar added to soils is a way to lock up carbon. Trials indicate that it can also reduce emissions of carbon dioxide and nitrous oxide from soils. Little biochar is produced in Australia, although a pilot plant exists near Gosford in NSW.

Freeing areas from weeds prior to planting sets the garden up for lower maintenance from the start. Glyphosate is often used as a broad-spectrum herbicide, although concerns have been raised about its potential for contaminating soil water after being banned in Denmark in the early 2000s. An alternative, although slower, method to using chemicals is soil solarisation where clear plastic is stretched across soil that has been dug, raked smooth and watered. The plastic is then sealed at the edges and left in place for a period of six weeks or more.

Tree roots generally grow in the top layer of soil, spreading horizontally with few roots penetrating deeply into the soil, so preparing the soil well for any planting will also benefit tree growth, and the planting hole needs only to be large enough to accommodate the root ball. If, however, the tree is to be planted into uncultivated clay-based soils that have not been amended, a hole two to three times the width and the same depth as the root ball needs to be prepared to encourage stronger root growth. Backfilling with the same soil maintains similar water holding characteristics as the surrounding untouched soil, but compacted clods need to be broken up for better aeration. Importantly, the walls of the hole in clay soils can become smooth and compacted through the digging process, potentially restricting root growth from the planting hole into the surrounding soil for some species. Rupturing this layer with a fork prior to planting and backfilling is suggested.

Landscapes sometimes need to be created on sites that have been excavated into the subsoil, as often occurs when buildings are constructed on sloping sites. It may be possible, with appropriate plant selection and some subsoil amelioration, to establish plantings in some subsoils. For others, however, the extended period of years that is usually required to develop a good topsoil needs to be compressed and so topsoil is imported. Taking time to investigate sources of good quality soil avoids plant establishment problems later, and it is worth asking if the soil complies with the Australian Standard.[8] The soil should be disease-free, of neutral pH, low in salts, and of a texture that will allow adequate aeration and good drainage but have reasonable water and nutrient holding capacity. Some landscaping 'soils' available are actually soil-less, combining composts with a sand base. They may benefit from the addition of clay to increase the soil's ability to hold water and nutrients, particularly as the compost decomposes. Deep ripping to fracture the subsoil before overlaying the imported soil promotes good drainage.

Installing irrigation systems

Irrigation systems need to be tailored to the plants grown, to the often irregular shapes of ornamental landscape plantings, and to the topography. They also need to reflect zoning for plants with different water requirements or differences in soil type. The system should be rapidly responsive to daily and seasonal weather variations. It also should be capable of easy modification, as the maturing of plants changes a garden's characteristics, intensifying the drying of soil when tree root zones or rain shadows from canopies extend, or where higher, denser foliage blocks the wetting patterns of sprinklers and microsprays. Drip and microspray systems are both suitable for these criteria, but because drip systems apply water more efficiently and they can be used at higher level water restrictions, they are preferred. Drip systems lose little water through surface evaporation or wind drift, and because only a small area of the soil surface is wetted, they have the added advantage of reducing weed growth.

Thoughtful planting design will group plants with similar watering requirements together, defining irrigation zones that can be managed differently:

- No irrigation (e.g. indigenous, many Australian, some exotic).
- Infrequent deep watering (e.g. many shrubs and trees).

- Regular deep watering (fruit-producing trees and shrubs).
- Frequent low application (e.g. vegetable beds).

In most situations, lawns would not be irrigated and indeed cannot be under many water restrictions. The less area irrigated, the less materials used, lowering environmental impacts like the high embodied energy of the plastic pipes used.

Manually operated zoned irrigation systems have the advantage of encouraging the operator to monitor plant health, weather and soil moisture conditions more closely, but watering times may suit the operator more than the plants. Automated irrigation systems give the freedom to water at any time including early morning, but gardeners may become complacent, monitor less, and end up watering more than is needed. From day to day, when irrigation is not needed, a rain switch can be used to suspend watering without losing the programming. Rainfall or soil moisture sensors can automate this and some large public landscapes use on-site weather stations. Rainfall sensors are cost-effective for home gardens. Seasonal adjustment of the programming is necessary to reflect the changing rainfall, temperature and evapotranspiration throughout the year. For each of the defined zones, a solenoid or manually controlled valve permits watering at different frequencies or for different periods.

Drippers come already installed at regular intervals (e.g. 30, 40, 50 mm) in 13-mm polyethylene pipe (poly pipe), or they can be placed where needed along standard black poly pipe either directly or via a 3-mm

Figure 11.4 Two drip irrigation systems: Top: individual drippers extend from the main line; Bottom: regularly spaced inline drippers.

microtube coming off the main line and usually mounted on small stakes (see Figure 11.4). In the first type, the distance between the emitters is less for sandier soils where lateral horizontal movement is low (e.g. 30 mm), and conversely greater for heavier soils where the radius of the wetting pattern from each emitter can be large (e.g. 50 mm).

Ideally, the driplines are laid in parallel rows spaced at the same distance as the drippers along the line, giving a grid pattern and fairly evenly wetting most of the soil. This encourages roots systems to colonise much of the soil. Exact layout will depend on plant placement, bed shape and other site constraints. When drippers are fitted onto microtubes coming from a main line, they can be placed closer together where root systems are thought to be more concentrated.

Drippers can be pressure-compensating for water supplies that have variable pressure to ensure water is emitted at a standard rate. These are useful for sites with variable topography where significant changes in elevation causes pressure changes along the line, or in large systems where friction between the tube and the water alters pressure along its length. When the pressure of the water supply hardly varies, non-compensating drippers are fine, but if the pressure does vary, so will the drip rate.

Drippers come in standard drip rates of 2, 4 or 8 litres per hour. Slower drip rates are better if the infiltration rate into the soil is low. Faster drip rates give wider wetting patterns.

Drippers can be laid on the surface or installed subsurface. In most domestic situations, surface installation is used although the lines are often covered with mulch.

There are design features that will help maintain well-functioning driplines over a long period keeping line replacement to a minimum together with the environmental impact that comes with that. Blocked drippers

Figure 11.5 A dripper that is responsive to soil moisture – dip stick®. (Photo: Sustainable Choices)

can quickly make parts of the irrigation system ineffective, but adequately filtering the irrigation water before it enters the driplines avoids one cause. Another cause of blockage is when water and potentially clogging particles are sucked into the line through the drippers as the line is draining after irrigating. Including, at the highest point of the system, an air or vacuum release valve that reduces suction through the drippers helps solve this problem. At the low point of each irrigation zone a flushing valve lets the line be purged of sediment periodically.

A recently released dripper made in Germany has a moisture-responsive fibre that contracts as it absorbs moisture from the potting media or soil, closing a valve as it does so, and stopping the flow of water to the dripper outlet. As the soil and therefore the fibre dries, it expands to open the valve again and reinstate the flow (see Figure 11.5).

Planting

Resources can be saved by using healthy, well-grown plants that need little after-planting care and that won't need to be replaced. These plants will be in pots large enough to allow a large root volume compared with shoot and leaf growth (definitely not pot-bound), growing in well-designed potting media that promote strong root growth (media with well-drained soil-less or low soil content), and from nurseries who practise high quality hygiene so are unlikely to pass on plant diseases or weeds. Among other things, look for weed-free pots kept on raised wire benches or, if on the ground, on deep layers of coarse well-drained gravel. Don't risk buying stock from nurseries where hygiene is poor. Controlling a new weed adds to the time and resources needed for maintenance, and once established in the soil, diseases like *Phytophthora cinnamomi* (cinnamon fungus) are difficult to eradicate and can restrict the plant species that can be grown.

There are many nurseries supplying plants indigenous to their local area and importantly, propagated from plants growing in the area. It is not uncommon for general nurseries to also have an indigenous section, but it is worth checking that they are propagated from local indigenous plants. Local councils may run their own indigenous nurseries or will have contact details for those in the area. The Indigenous Flora and Fauna Association has a listing of some indigenous nurseries.[9] The Association of Societies for Growing Australian Plants and its associated state bodies have a listing of nurseries specialising in Australian plants, some of which have local indigenous plants.[10] The Australian Plant Society is another good source of information about growing Australian plants.

Sourcing some species or cultivars in a garden design may be difficult, and alternatives with similar aesthetic and functional characteristics may be needed. Replacement plants ought to have comparable water, light and soil needs, and be able to serve similar functions such as being nectar-producing for local birds, insects or other animals. If at all unsure about why a particular plant was included in the design, consulting the designer will ensure the replacement plant has appropriate characteristics. Assess the replacement for its weed potential in your area.

Most plant stock in Australia is available as container-grown plants, and these are easily transported to the planting site from the nursery. Enclosed but ventilated transport vehicles reduce the risk of damage from direct sun or drying winds. Some deciduous plants, including trees and shrubs like roses, are available bare-rooted during winter. It is essential to avoid further root damage; plant as soon as possible or protect the roots with a material like moist composted saw-dust until they are planted.

Fertilising

Any additional nutrients added to garden soils means that materials have been brought in from elsewhere, adding to the garden's environmental footprint. Ideally plants selected will grow well without additional fertilisers, but even if they do need supplementing, fertiliser use can be kept low by applying the correct amount for each plant. Less fertiliser is usually better, and a high level, broad-scale application of fertiliser is rarely needed.

An important first step is to determine existing nutrient levels of the unmodified soil or, if amendments like compost have been added, the modified soil. Private analytical laboratories can test for a standard range of important plant nutrients and advise which nutrients may be limiting, but care is needed interpreting these results because they usually look at the levels needed to produce high yielding agricultural crops rather than maintaining a healthy ornamental landscape. Tests may reveal an overall low fertility, or low levels of one or more nutrients, these results then guiding the choice of fertiliser and the rate of application. It may be necessary to exclude adding some nutrients so their concentration will not become toxic or add to already toxic levels. pH, an important measurement in any soil testing, may also need to be adjusted as values outside an appropriate range can affect nutrient availability to the plant.

Debate about organic versus inorganic fertilisers still occurs. Both are able to raise nutrient levels, and it can be argued that both may have their place in plant nutrition. Kevin Handreck's *Gardening Down-Under* provides useful discussion on the topic. From a sustainability viewpoint, it is important to ask: what are the impacts of each? A great deal more research is needed in this area.

One study, conducted in Florida, reported on the use of chicken manure for growing commercial tomatoes. It concluded that the energy costs in transporting the manure from the chicken farm to the composting facility, creating windrows, turning the composts, lifting the composts and transporting them to the tomato farms would be less than using inorganic fertilisers only if the composting facility was within 30 km of both the chicken and tomato farms.[11]

Organic manures used for fertilising can vary in their nutrient content depending on the animal, their health, the feed used and the age of manure, and they may need to be supplemented with other organic or inorganic sources. Ultimately, there may not be enough organic fertilisers for all horticultural and agricultural needs. Organic manures may be important sources of nutrients other than the major nitrogen, phosphorus and potassium. They can also contribute to improving soil structure unlike inorganic fertilisers. An important environmental question to ask is what would be done with the manures if they were not recycled? How would they be disposed of and what environmental problems, such as eutrophication, may result? It is good to see them as an important resource.

Inorganic fertilisers can have high embodied energy, especially the nitrogenous ones, with urea having around 70 MJ/kg, superphosphate 8.5 MJ/kg and potassium sulphate 5 MJ/kg. (See Table A3 in the Appendix.)

The production of nitrogenous fertilisers begins with the fixing of nitrogen (N_2) from the atmosphere using the non-renewable fossil fuel natural gas (CH_4), producing anhydrous ammonia molecules (NH_3). The various nitrogenous fertilisers like urea ($CO(NH_2)_2$), ammonium nitrate (NH_4NO_3,) and ammonium sulphate $((NH_4)_2SO_4)$ can then be made. The process needs high energy and a non-renewable resource.

Phosphate rock is a sedimentary rock used to manufacture phosphorus fertilisers. Nauru, an island in the Pacific, had large deposits that have largely been mined out now. Current global reserves are estimated to be 34 000 million tonnes, deposited mainly in Morocco and the western Sahara, although deposits are also known on the ocean bed. In Australia, the largest deposits are in Queensland and, depending on demand and how much fertiliser is imported, they are expected to last until about 2040. Phosphate rock has a complex chemistry and can vary in exactly what high phosphorus minerals it contains. Fertilisers made from it include superphosphate, triple superphosphate, monammonium phosphate (MAP), and diammonium phosphate (DAP). Superphosphate is made when phosphate rock reacts with phosphoric acid, but higher phosphorus MAP and DAP are made by first reacting the rock with sulphuric acid to produce phosphoric acid, and then reacting this with ammonia. Natural gas is required to produce the ammonia and the sulphuric acid. So the inorganic phosphate fertilisers are dependent on two non-renewable resources, phosphate rock and natural gas.

Potassium is widespread in soils, but there are natural deposits in old sea or lake beds that are mined and processed to remove other compounds like common salt, producing the potassium fertilisers potassium sulphate (K_2SO_4), potassium chloride (KCl), potassium nitrate (KNO_3), and potassium and magnesium sulphate (K-Mag) $(K_2SO_4 2MgSO_4)$.

Controlled-release fertilisers are coated with wax, solidified oil or plastic. The coats are semi-permeable, allowing water vapour in to dissolve the nutrients, which are then released through the coating at a rate dependent on the ambient temperature and the coat's thickness. The controlled nature of the release usually lessens the amount of soluble fertiliser leaching away into groundwater or drainage systems. Eventually the coat should degrade through microbial action or exposure to light, although sometimes this takes some time. The coats of some products degrade into ammonium, carbon dioxide and water. The fertilisers coated with plastic are high in embodied energy.

Human urine is a good source of nutrients and its use for horticulture is being explored in Europe. Urine-collecting toilets, designed to capture urine and faeces separately, have been installed in parts of Sweden, Denmark and Austria, with the urine directed into large collecting tanks. In Queensland, a collaborative project involving the Department of Natural Resources and Water and based at the Currumbin Ecovillage is trialling urine separation toilets, and the potential for using the nutrients in the urine for fertilising soils.[12]

The solutes in urine have an NPK in the order of 10:1:1.3, although a study funded by the EPA in Denmark found urine to contain total nitrogen of 2.5 g/l, total phosphorus 0.17 g/l and potassium 1.2 g/l.[13] It has been suggested that urine may solve the phosphorus shortage expected within 50 years. One potential problem with using human urine is the presence of drugs in the urine of people on medication, which could have environmental consequences. Urine should be diluted to 1:10 before use as a fertiliser.

Different species tolerate different ranges of soil fertilities, so for any given soil some plants will thrive without any additional nutrients while others will be stunted and perhaps display signs of nutrient deficiency. Nutrient levels may be too high and toxic for other species; however, most Australian soils are not highly fertile and many Australian plant species will happily grow with low nutrient levels. Similarly many exotic (non-Australian) species can grow well with little or no additional nutrients. Indigenous species are adapted to their native soils and will need no or only low levels of additional fertilising to grow well.

Increasing soil fertility can have the unintended side effect of stimulating weed growth at the expense of the plants being established. It can also increase the need for irrigation to support increased plant growth. Excess fertiliser can be leached from the garden causing increased fertility in other habitats and may lead to eutrophication of waterways. Groundwater can be polluted.

Better plant growth has been shown to result from applying fertiliser to the soil surface after backfilling rather than incorporating it into the backfill or placing it in the hole where it can be leached down away from the roots.[14] The fertiliser can be placed over the area of root growth, extending beyond the existing root ball, the amount and the area being dependent on the existing soil fertility, the season and the expected growth rate.

Mulching

Mulches offer important benefits for newly planted gardens. The young undeveloped root systems of plants benefit from the mulches' ability to keep the root zone cool and to retain soil moisture, the evenness of conditions placing less stress on the plants at a time when they are more vulnerable.

Importantly during establishment, the competition from weeds for water and nutrients that is often experienced by new plants can be significantly reduced by controlling weed growth with mulches. Competition causes reduced growth rates of the desired plant, and it has been shown for *Betula pendula*, *Acacia mearnsii* and *Eucalyptus camaldulensis* that the critical period for maintaining a weed-free area around the plants is three months after planting, although for crop plants like fruit trees it has been suggested that a period of at least four years or even ongoing control is needed.[14]

Different kinds of mulches, their action and benefits are covered in more detail in the landscape maintenance chapter.

Monitoring the new landscape

Even with the greatest of care taken during the design and construction phases, without adequate follow-up monitoring and maintenance, the landscape may deteriorate very quickly, requiring additional natural resources and dollars to remedy. All landscapes are vulnerable in their early stages, even those that are well planned with suitable plants. Watering is necessary at planting to help settle soil around the roots, and may be needed from time to time as root systems are establishing. Planting can be scheduled to take advantage of natural rainfall or the cooler conditions that reduce evapotranspiration, but even at those times of the year soil moisture levels may be low enough to warrant irrigation. Using a wetting agent for non-wetting sandy soil may be helpful.

Newly installed irrigation systems cannot be relied upon to deliver the necessary amount of water to all plants, and they too need to be regularly checked for deficiencies or failure. Landscapes contain living things and benefit from regular inspection by an experienced horticultural eye, or someone who will respond to adverse changes.

Sustainable construction summary

There are a few key ideas guiding sustainable landscape construction.

- Protect the natural assets of the site and surrounding areas during potentially damaging construction activities to enhance the healthy establishment of the garden and its ability to integrate into the broader ecological systems.
- Carefully plan construction schedules and methods, the use of powered equipment, and the selection of materials to significantly reduce resource use and waste generation.
- Use quality materials and use the right materials for the job to provide the foundations for long-term landscapes, needing fewer additional resources and to retain high standards through their life.
- Monitor the plants, irrigation and drainage systems of the new landscape to identify problems before they become large and costly to remedy.
- Improve knowledge about constructing sustainable landscapes, and support the development of sustainable technologies and products that facilitate low impact landscape construction.

LANDSCAPE MAINTENANCE

> **KEY POINTS**
> Choose equipment with low energy needs.
> Maintain soil quality for healthy plant growth.
> Recycle the green wealth of the garden.
> Irrigate efficiently to maintain the plant quality required.
> Encourage a robust garden ecology for minimal pest problems.

Routine garden maintenance includes activities like mowing, pruning, weeding, soil care, fertilising, mulching, watering, composting, pest control and care of the hard landscape.

Low impact garden maintenance flows naturally from well-considered landscape design and construction. It includes sustainable practices like retaining or improving soil quality, regular mulching, using fertiliser sparingly according to plant nutritional needs, supplementing rainfall with needs-based watering, monitoring pests to guide the use of low impact controls, and using prunings, dead leaves and other organics as a compost and mulch resource for the garden, and more.

Maintenance for optimum sustainability requires an awareness of garden ecology; a sensitivity to the changing needs of different plants, and how they interact with each other and other organisms living in or passing through the garden. It understands the influence of the physical environment and seasonal changes on plant growth and health, and uses this knowledge to develop routine maintenance programs. Normal, healthy growth is clearly recognised. Conversely, for abnormal growth, the timing of an appropriate intervention is clear. Working with nature, its cycles and natural processes, gardens and landscapes can become largely self-sustaining, needing only low levels of our input.

Water, energy and material use are kept low so the garden has a small Ecological Footprint. The principles of re-using, recycling and local sourcing are followed.

There may be little need for materials to be moved in or out of a well-managed sustainable garden. Garden stakes for supporting vegetables can, for instance, be fashioned from the woody stems of plants growing in the garden rather than purchasing milled timber transported over long distances. Recognising and using the resources of a site reduces the need to import materials, including organic matter or water. It decreases the energy of transporting composts or mulch in, and wastes out for recycling or landfill. Kitchen scraps from houses or work places are a resource, as are chipped prunings and other organics from the garden itself, all adding to composts and mulches for the garden.

Maintenance equipment can be non-powered, or if powered, using renewable energy if the option is available, and the equipment is carefully selected for efficiency, matched to the task and well maintained.

Rainwater harvested from roofs or paving, and greywater from buildings significantly contribute to garden water needs as we have seen in the chapter on design, and these sources are recognised as the primary supplies, with mains water reserved as the backup. Checking and maintaining on-site water harvesting and irrigation systems is a critical and regular component of garden maintenance.

Chemicals, including fertilisers, pesticides, herbicides, and those in greywater are used cautiously and sparingly, and measures taken to avoid them accidentally affecting non-target species or the wider environment.

Different levels of maintenance are applied to different areas of the garden or to specific plants. For example, food-producing plants have higher maintenance inputs, so fertilising rates often need to be higher than for other plants to ensure high crop yields.

Changing gardens, changing maintenance

Maintenance requirements of gardens change as the various garden components pass through their life cycles. Working with these cycles can reduce the resource demands of the maintenance schedule.

New gardens require more monitoring and often greater levels of maintenance as the plants grow into a balanced garden ecosystem. Small, undeveloped root systems of plants in new gardens are much more vulnerable to dry periods until the root systems colonise a greater volume of soil, and monitoring for water stress is more critical. Irrigation frequencies may therefore be higher, depending on the season. The open areas of new garden beds give greater opportunity for weed colonisation until the garden plants establish root systems and canopies that compete more vigorously for nutrients, water and light. The interactions of garden organisms take time to establish too, and the lack of predators or parasites may allow garden pests to establish more readily. Young trees may need to be trained into shapes that increase fruit production or that promote strong branch attachments.

Declining gardens may need higher maintenance levels as ageing plants are replaced. Removing deteriorating trees for safety reasons, or those trees that no longer contribute to the beauty of the garden, can have a particularly significant impact as the micro-environment once created by its sheltering branches is radically altered. Light levels and heat may dramatically increase, while humidity may be much lower. Some previously sheltered plants may not thrive in the new conditions and others may flourish. Habitat for many animals may be lost and biodiversity decline. These changes will often mean increased monitoring and maintenance, and perhaps remedial changes to the garden (e.g. temporary shelters installed to help sensitive plants through the transition, or nesting boxes erected for displaced animals).

Understanding the seasons

The European concept of four seasons came to Australia with the first settlers in the 18th century, and they transferred the names winter, spring, summer and autumn to their new location even though the seasonal regime was clearly different in the generally warmer Australia. Even in southern Australia, there is no winter in the northern hemisphere sense, when most plants are dormant. Here 'winter' is a time of the year when many indigenous species like wattles and orchids flower as they take advantage of cool, moist conditions. The hot, dry 'summer' of southern Australia is the harshest season that plants need to endure, just as the plants of Europe, North America and northern Asia need to survive cold, below-freezing winter temperatures. Tropical northern Australia has even more marked differences and terms like wet and dry season are more appropriately used.

Indigenous Australians had their own way of categorising the seasons. Reinterpreting the seasons for Australia's regions may give

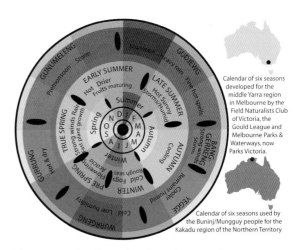

Figure 12.1 An alternative view of seasons for two regions of Australia.

horticulturists a greater insight into the cyclical changes in day length and climate, and, importantly, the changes experienced and expressed by plants (Figure 12.1). It may help refine maintenance schedules and identify those seasons which place the greatest stress on the garden. Irrigation needs, for example, can be affected not just by the long days and high temperatures of our summer, but also by the increased evapotranspiration during windy seasons like the spring in the south.

The harvest time of food plants can also define seasons and highlight the more sustainable periods for locally grown, in-season foods.

Equipment

Non-powered tools are more than adequate for many garden tasks, and should be preferred over powered tools. It is possible to maintain domestic-scale gardens without any

Figure 12.2 Non-powered horticultural equipment.

powered equipment at all, as was commonly done until the middle of the 20th century. Lawn edgers, push mowers, saws, hedge shears, loppers, secateurs, rakes and brooms all have their powered equivalents (see Figure 12.2), but it is worth asking: are they really needed? Powered equipment is more expensive, has more components that can fail, will usually have greater embodied energy, and requires energy to run, mostly from the non-renewable fossil fuels petrol or coal via electricity.

If, after consideration, powered equipment is thought to be more appropriate, select models that adequately perform the work required (e.g. don't use a rotary hoe to dig a 2 m^2 garden bed), that run using renewable energy or the least amount of energy from non-renewable sources, preference being given in the following order:

- Powered by renewable electricity (green energy from a supplier; photovoltaic panels; domestic wind turbines).
- Electrically powered (fossil fuels like coal or the less greenhouse gas producing natural gas).
- Fuelled using petroleum products (petrol, two stroke, diesel).

Electrical equipment uses less energy per hour than petrol driven equivalents (see Table 12.1). However, sometimes the greater torque gained from petrol engines, rather than electrical motors, is needed for heavier tasks, hence petrol-driven chippers are superior for large-tree work.

In most cases small machinery is suitable for gardens and larger for parks or commercial use.

Whatever equipment is chosen, it should be the right tool, the right size and have the right power for the job. Purchasing more powerful equipment does not necessarily do the work better and may use considerably more energy than smaller counterparts over its lifetime.

The pollution from small engines powering gardening equipment or outboard motors is often ignored. Detailed studies in Australia are only just being undertaken. It is estimated, however, that over the same operating period, the pollution from one brushcutter may be similar to 10 cars and a lawn mower the equivalent of four cars.[1] Further, on an annual basis, it is estimated that 4.1% of volatile organic compounds (VOCs) in Sydney's atmosphere are from lawn mowing, and up to 11% during weekends in spring or autumn. A Voluntary Outboard Emissions Labelling Scheme (VELS) exists but there is no equivalent for gardening equipment.

Maintaining equipment properly, whether powered or not, makes maintaining a garden easier for the operator, prolongs the life of

Table 12.1 Comparison of energy consumption of electrical and petrol powered garden tools

Equipment	Energy consumption (MJ/hour)	
	Electrical	Petrol
Chainsaw (30-cm blade)	5.8	26.1
Lawn mower (75-cm cut)	16.9	79.1
Lawn edger (small)	1.1	9.4
Blower (hand held)	5.0	14.2
Hedge trimmer	2.7	14.2
Chipper	19.8	67.1

the equipment, and ensures powered equipment is running efficiently so no more energy is used than needed and less pollutants are emitted.

Secateurs regularly sharpened using an oil stone and cleaned using fine wet and dry paper and steel wool will cut branches more easily, and by cutting more cleanly, they will be better for plant healing. Similarly, blades of other cutting equipment – shears, edge cutters, mowers, chippers – should be sharpened regularly.

The engines of powered equipment need to be maintained too. Four stroke engines need oil changes approximately every 50 running hours (the first service after purchase to be done after five hours use). Professionals and home handymen may feel comfortable doing the service or it can be taken to gardening equipment service centres. Correct disposal of the used oil is important, collecting it in a sealable container and taking it to an oil recycling centre.[2] Between services, the oil level should be checked regularly and topped up if low. Air filters also need replacing, perhaps once per season, more regularly if operated in dusty conditions. Long-term storage of fuel can change its properties and affect the operation of the equipment, so it is recommended to buy petrol to last no more than a month. Guidelines for maintaining the commonly used Briggs and Stratton engines can be found on their website (www.briggsandstratton.com).

It may be useful to know the energy contained in different fuels (see Table 12.2).

Maintaining soil quality

Soil is one of the most important resources in a garden, its quality being a major contributor to how well the plants will grow. A healthy soil is a complex material made from mineral particles, living organisms (plants, animals, micro-organisms, decomposing organisms), humus, water and air spaces. It needs to be valued and the qualities of these components protected to ensure our plants grow well. Maintaining a quality soil requires consideration of its structure, organic matter content, fertility and water-holding ability, and to tailor these to the needs of the plants we are growing.

It is usually better to work with the soil already present than to import additional soil, which is very often a subsoil, sometimes amended with compost. The imported soil may contain weed seeds or plant diseases, may have an imbalance of nutrients or may have an unsuitable pH range. The mining of soil damages other habitat, and non-renewable fossil fuels were used for its extraction, its blending with compost and its transport.

Soil structure

Soils supporting undisturbed natural vegetation like forests and woodlands tend to

Table 12.2 Energy contained in fuels[3]

Fuel	Quantity	Energy (MJ)
Coal (black)	1 kilogram	23–30
Coal (brown)	1 kilogram	10–15
Crude oil	1 litre	37–39
Petrol	1 litre	34.2
Natural gas	1 cubic metre	38–41
Propane	1 litre	25.5
Wood (dry)	1 kilogram	16.2

Figure 12.3 Pores between aggregates of minerals and humus in soil. (Photo: CSIRO)

be well structured. Structure refers to how a soil's organic (humus) and mineral components clump together, and how these small aggregates are then positioned to form open air pores through which water can percolate and drain (Figure 12.3). Aggregates tend to hold together when rain falls on them or with infrequent, light cultivation. Soils of poor structure are less likely to have aggregates, but be a mass of more closely joined particles. Aggregates can be created by adding organic matter or in sandy soils adding some clay, and the particles in certain clay soils will aggregate in the presence of gypsum (calcium sulphate).

Soil structure can be destroyed by prolonged periods of cultivation, cultivation when clay soil is wet (e.g. immediately after rainfall), erosion, compaction, depletion of organic matter from the soil or increasing levels of sodium ions leading to sodicity (see the section on recycling water in the design chapter).

Although we know that vehicle tyres can easily compact and damage soil structure in gardens, pedestrian traffic regularly passing over garden beds can also be a cause. Roots find it more difficult to penetrate the tightly packed soil particles, and the smaller pore sizes decrease the aeration needed for optimal growth. Water infiltration and percolation is reduced, lowering the amount of water reaching the root zone and increasing the surface water run-off perhaps causing topsoil erosion. Compacted soils may not drain well, remaining waterlogged for longer periods.

Compaction along regularly trodden routes through garden beds can be reduced by placement of spaced stepping stones or thick layers of coarse organic mulch. In vegetable beds or other places where maintenance is frequent, it is also critical to avoid compaction. Standing on moveable, small, light plywood boards, which fit between rows, spreads a person's weight over a larger area and reduces compaction. Two or three boards can be used like temporary stepping stones.

Regular cultivation will destroy the aggregates and the soil structure. It also hastens the loss of organic matter from the soil. Apart from that, each time the soil is cultivated, the fine roots of existing plants are damaged.

Water and wind erosion can quickly strip topsoil from a site, but can be minimised by maintaining plant cover, protecting root systems that bind the soil from damage, applying mulch and ensuring water can easily infiltrate. Non-wetting soils are more prone to water erosion and the use of wetting agents could be considered.

Organic matter

Returning organic matter to the soil supports the complex web of soil organisms, contributes to the maintenance of soil structure, increases aeration of clay soils, and improves the water and nutrient holding capacity of sandy soils. Much of the organic matter from a garden can be recycled back into the garden, either as partially composted mulch or dug into the soil after composting. Of all powered equipment, a chipper is one of the most useful by enabling more of the 'green wealth' from a garden to be retained within it. Managers of large public gardens will often have access to large chippers, but even domestic gardeners can purchase robust chippers able to take freshly pruned branches up to 3 to 4 cm in diameter for about the same price as a good lawn mower. If chipping is not an option, many local councils have regular 'green waste' collections, and sometimes make compost available for residents, giving them the ability to cycle the 'green wealth' back into their garden, albeit with an extra fuel cost due to the transport. Community 'green waste' collections ensure this important resource can be used in landscapes rather than being buried in landfill with rubbish and becoming a significant source of the greenhouse gas methane, which is formed during anaerobic decomposition (although sometimes the methane from landfill is used for energy generation).

Composting systems are easily set up, and collecting all the suitable plant material from the house and garden effortlessly becomes a part of daily routines. The kitchen vegetable and fruit scraps, vase flowers, paper and garden clippings can be routinely diverted into it. Having a designated compost bucket in the kitchen makes it easier, and there are bin systems that fit neatly under sinks (Figure 12.4).

Figure 12.4 Compost buckets can fit neatly under sinks.

Everything except thick branches, diseased plants, weeds with seeds or difficult to control runners (like couch or kikuyu) can be used. Composting can be done in commercially available bins, some having better designs for aerating the compost which aids decomposition, or heaps can be established (Figure 12.5).

A simple system can be made with four star pickets with chicken wire stretched between them, with one of the four sides able to open for accessing the compost (Figure 12.6).

Figure 12.5 Some compost bins, like the Aerobin™ illustrated, are better for aerating the compost which aids decomposition.

Figure 12.6 A simply made, inexpensive compost heap that can take large amounts of garden material.

These can be made quite large to contain masses of autumn leaves or prunings that can seasonally inundate composting systems. The materials needed for this system probably carry less embodied energy, being made from metal rather than plastic, although no studies have been done as yet. It is worth having two or three bins or heaps with one being added to, one completing the composting process and, if space allows, one being fully composted and available for garden use.

Decomposing organisms take from two months to a year to convert plant material to humus depending on the ambient temperature, moisture levels, aeration and balance of different types of organic matter. Mixing components high in carbon like twigs, wood chips, sawdust, dead leaves and paper, at about 30 times the amount of those high in nitrogen like food scraps, lawn clippings, hair, manures and urine, will provide the micro-organisms the balance needed to grow well. The process releases carbon dioxide and heat – hence the steam rising from disturbed compost heaps. The final product has an earthy smell, and the texture of a light soil. It also contains nutrients for plant growth, the levels being dependent on the organic matter added to the compost originally. Typically dried compost contains 1.4–3.5% nitrogen, 0.3–1.0% phosphorus and 0.4–2% potassium and some other nutrients in smaller amounts.[4]

Humus is the material left after most of the plant and animal material has decomposed in a well-managed compost bin or heap. It is composed of a heterogeneous mixture of large, complex molecules. Humus molecules improve soil structure by attaching to multiple mineral particles, forming aggregates. Humus also dramatically increases the surface area of a soil so it is able to hold on to more nutrients and water that can then be available for plants.

Related to humus are the humates found in Australian brown coal, some deposits being very rich in these complex mixtures of plant-derived organic molecules. Now commercially available, they can be applied to soils to increase the nutrient and water-holding capacity of the soil (for more details on humates, see the section on soil preparation in the construction chapter).

At times, a garden will not be able to produce enough compost. The regular cultivation and replanting of vegetable gardens speeds the depletion of organic matter from the soil, for example. The redevelopment of existing garden beds may also require above-normal inputs of organic matter, although some plants, including many indigenous or Australian species, grow very well in soils low in organic content. If additional compost is brought in, it is worth remembering that those organic materials are coming from elsewhere. Some commercially available composts are by-products of other industries like mushroom growing, but not all will be. SGA-approved nurseries are mindful that the components of composts should be from renewable sources as much as possible, and it is worth asking nurseries about the origin of their composts. A detailed LCA of composts is yet to be done.

Composting large areas usually requires plenty of space and heavy machinery; there are sometimes unpleasant smells generated. In November 2002, the Royal Botanic Gardens Sydney, with funding assistance from the EPA NSW, introduced a Vertical Composting Unit (VCU – better known as 'Sylvester the digester') that processes up to a tonne of green waste per day using natural composting processes but without artificial heating, air injection, chemicals or agitation. Material fed in at the top of the VCU unit takes about 14 days to be converted to pasteurised (over 70°C), odourless, unleached compost drawn from the base.

Soil fertility

(See also the section on fertilising in the construction chapter.)

Different plant species are adapted to different soil nutrient levels, so for just about any soil there will be species that can be grown without any additional nutrients being added. Fertilising then can be restricted to the root area of those plants needing additional nutrients, in amounts and at a frequency that each species needs.

Fertilisers provide chemicals to plants as simple inorganic salts that can be taken up directly by plants. They are sometimes packaged as controlled-release fertilisers. Fertilisers can be bought with stated levels of major nutrients (the NPK value) and minor nutrients listed as well. Prescribed nutrient mixes appropriate for particular kinds of plants are available.

Fertilisers can also be organic, like blood and bone, requiring micro-organisms to metabolise and release similar salts that become the plant nutrients. Fish can be used as a basis for fertiliser. A recent innovation is the use of the invasive European Carp which is minced and processed into a concentrated, balanced, organically certified liquid fertiliser. Their removal from river systems benefits native fish and other organisms of freshwater habitats because carp erodes river banks, causes silting, encourages blue-green algae blooms and feeds on native fish eggs. More than 150 tonnes of the fish are removed from our river system each year for use as fertiliser.[5] Fish-based fertiliser dry weight levels of NPK are in the order of 4–9% N, 1–4% P and 0.8–1% K. Seaweed-based fertilisers are also available with NPK dry weight levels in the ranges 0.5–1.5% N, 0.1–0.2% P and 0.1–1.9% K.[4] The sustainability of long-term seaweed harvesting needs to be more fully researched.

Using soil tests to determine the soil's nutrient levels and pH (which affects nutrient availability), and targeting those plants known to need higher soil fertility, as fruits and vegetables do, leads to the optimal use of fertilisers. Plants like trees may only need

Table 12.3 Energy requirements for production, packaging, transport and application of fertiliser nutrients in agriculture

Nutrient	Energy (MJ/kg)				
	Production	Packaging	Transport	Application	Total
N	69.4	2.6	4.5	1.6	78.0
P_2O_5	7.7	2.6	5.7	1.5	17.5
K_2O	6.4	1.8	4.6	1.0	13.8

Source: Helsel (1992).[6]

fertilising when young. Many Australian species can grow very well without fertilising, which may raise nutrients to toxic levels for some species, or have the negative effect of encouraging the establishment of invasive weeds which often thrive in soils of higher fertility.

How much fertiliser is needed depends on the type of fertiliser (i.e. rate of release), type of plant and environmental conditions. Controlled-release fertilisers are affected by ambient conditions of temperature and humidity. If excessively applied or applied in a season of little or no growth, fertilisers can be wastefully leached away before the plant can use the nutrients they contain. They may also contribute to pollution of water-bodies.

Applying fertilisers from the beginning of growing seasons provides food at the time the plant most needs it, but application may not be needed for each plant or during every growing season. Reducing applications can slow growth, but as long as plants remain healthy, this is not necessarily a bad outcome for ornamental plants as it may also reduce the frequency of irrigating and pruning.

Reducing fertiliser applications has other environmental benefits. Each time fertilisers are used, their embodied energy is added to the garden. Lowering fertiliser use lowers the energy needed for their production, packaging and transport. Large amounts of energy are needed to produce fertilisers, particularly nitrogen-based ones, as Table 12.3 shows (see also the fertilising section in the previous chapter). Nitrogen is extracted industrially from the atmosphere. Although less nitrogen for agricultural purposes is fixed this way than by growing legumes with their associated nitrogen-fixing bacteria, it is still a significant amount and the process uses fossil fuels as an energy source rather than the sunlight used by legumes.

Apart from the energy needed, extracting raw materials for phosphorus, potassium or other plant nutrients causes loss of natural habitat, adds to air pollution, may affect groundwater quality and lower water tables. The natural nitrogen cycle of the biosphere has been altered due to the large quantities of fertilisers produced, especially for agriculture. Fertiliser use for ornamental horticulture may not have anywhere near this impact, but using them sparingly in well-targeted ways contributes to an overall lessening of demand.

Commercially supplied fertiliser products should meet standards for ingredients, impurities and labelling as prescribed by chemical regulations.

Water and soils

Some sandy soils are naturally non-wettable but a potentially unhelpful outcome of adding organic matter to sandy soils is they can become non-wettable (hydrophobic) when organic matter is decomposed by micro-organisms producing water-repellent waxy compounds that coat the soil particles. Hydrophobicity is easily recognised when water forms beads on the surface of a dry soil sample for longer than one-and-a-half minutes (Figure 12.7).

In practical terms, water pooling on the soil surface for long periods without penetrating reduces the effectiveness of rain or irrigation. Even where the soil appears moist after prolonged rain, the surface can be scraped back to reveal powder-dry soil beneath.

It is still important to add organic matter to sandy soils for the other benefits it brings, and wettability can be increased by adding clay or by using wetting agents. When planting, the soil can be made wet by pummelling water in (Figure 12.8).

Soil wetting agents are detergents that are non-toxic to plants and degrade slowly. They help to penetrate the waxy layers in much the same way as a dishwashing detergent removes the fatty residues from a dinner plate, but wetting agents biodegrade much more slowly, being effective in the soil for much longer periods (months). They can be applied at a lower rate than kitchen detergents, and are not toxic to plants at the recommended application rates. Applying at higher concentrations than recommended can decrease plant growth. They should be applied no more frequently than every six months.

Figure 12.7 Water beads on the surface of non-wetting soils.

Soil wettability is very important, not only for plant health and growth, but to make the most effective use of limited water supplies in many areas of Australia. Studies currently underway at the Royal Botanic Gardens Melbourne in collaboration with the School of Geographic and Environmental Sciences, Monash University, are examining hydrophobicity and its relation to irrigation management.

The water storage of sandy soils may be increased by using organic matter, clay, or water storage crystals.

In heavy clay soils where drainage is potentially a problem, there are a number of strategies that can be implemented. In many cases these will not include the installation of subsurface drains as there are few situations where they are helpful. Lateral movement of water is extremely slow in soils and subsurface drains will effectively drain soils only up to about 300 mm from the drain.[7] Spacing the drains further apart will leave large areas of the soil undrained, resulting in a waste of the high embodied energy in the slotted plastic pipes that are usually installed. Spacing them

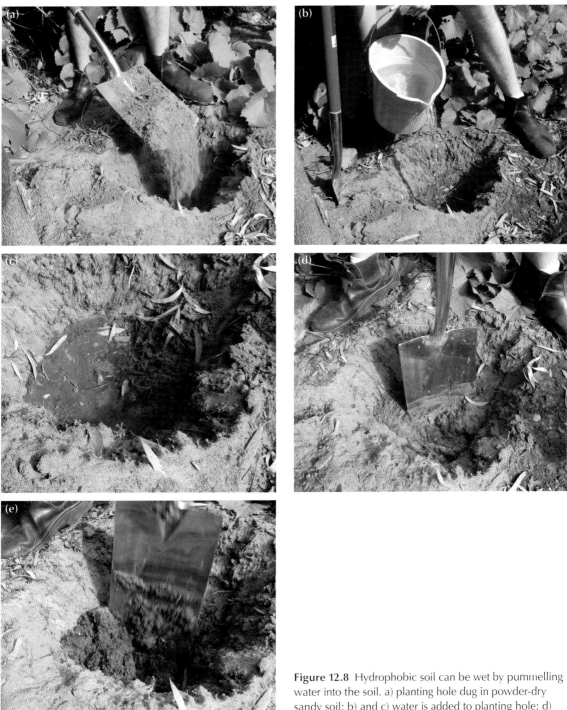

Figure 12.8 Hydrophobic soil can be wet by pummelling water into the soil. a) planting hole dug in powder-dry sandy soil; b) and c) water is added to planting hole; d) spade used to mix water into soil; e) soil is now wet through this mechanical action.

closer together will use even more resources, not to mention the cost involved, so their usefulness needs to be established before installation. Better strategies for many cases include:

- Avoiding water entering the soil during long or heavy rain periods by sloping the site with at least a 1:70 grade – a significant amount of the rainwater will run off these sites.
- Creating a better drainage base by:
  - ripping the soil to a depth of at least 400 mm, or
  - shattering compact subsoils using machinery such as vibra moles or high pressure air injection, or
  - applying gypsum to responsive clay soils (responsive soils can be identified by gently placing small, dry portions of soil about the size of a marble into deionised water and leaving it undisturbed for a number of hours. A cloudy halo will develop around soils that gypsum will improve, but the water around non-responsive soils will remain clear).

Mulching

Using mulches has four main benefits for the garden:

1. They reduce evaporation from the soil surface, conserving soil moisture for plant growth and reducing the need for irrigation.
2. They insulate the soil, keeping the root zone cooler during hot weather and warmer when it is cool.
3. Their presence dramatically decreases the growth of weeds, and reduces the time spent removing them during general garden maintenance.
4. Organic mulches decompose, adding humus and nutrients, and contributing to a healthy soil environment for plant roots.

Mulches may also reduce soil erosion at times when heavy or prolonged rain causes run-off or on slopes more prone to eroding. For some plant diseases, mulches can slow the disease cycle by reducing the likelihood of disease micro-organisms being splashed up on to leaves with the soil during rainfall or overhead irrigation. Mulches also control weeds and reduce the need for cultivation.

The term mulch is now applied to a number of inorganic materials and organic ones in various states of decomposition although it was probably originally applied more often to partially decayed plant material. Table 12.4 lists currently available mulches.

Table 12.4 Currently available commercial mulches

Type	Mulch
Organic	Pine bark mulch (fine, medium, coarse)
	Eucalyptus-derived mulch
	Wood chips
	Sugar cane
	Recycled, chipped, dyed, packing pallets
	Pea straw
	Hay (wheat straw, lucerne, etc.)
	Composts
	Seaweed
	Peanut hulls
Inorganic	Pebbles/stones
	Recycled glass beads
	Synthetic fibre sheets
	Sand

Table 12.5 Effect of mulches on evaporation from the soil surface

Mulch material applied 75 mm deep	Water evaporated in 2 days g/m^2	Reduction in evaporation compared with bare soil (%)
Bare soil	27.9	–
Dry lawn clippings	5.2	81
Lucerne hay	4.9	82
Oat straw	4.9	82

Source: Handreck & Black (2002).[8]

An appropriate mulch correctly applied can prevent up to 80% of water lost through evaporation from the soil (see Table 12.5). If the particle sizes of the mulch are many times larger than the soil particles, the continuity of the water film connecting soil water with the atmosphere can be broken, preventing the wicking effect that usually occurs. Finer mulches, like composts, will not be as effective. However, any mulch that keeps the soil cooler lowers evaporation, and mulches suppressing weeds reduce the loss of water via transpiration from these unwanted plants.

Creating a barrier to evaporation with plastic sheets certainly reduces moisture loss from soil, but plastic sheeting also prevents rain infiltrating the soil, hinders air diffusion to the oxygen-requiring plant roots, and during warm to hot weather can increase soil temperatures to levels that stress or kill the roots. They are not recommended for most situations.

Other mulches moderate soil temperatures so that there are fewer fluctuations from day to day and between seasons. The limited data available suggests that optimal root growth occurs between 20 and 35°C.[10] The various effects of different mulches on soil temperature is displayed in Figure 12.9.

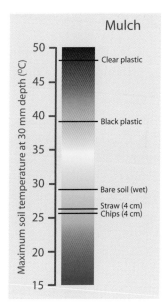

Figure 12.9 The effect of mulch type on soil temperature 30 mm below the surface.[9]

The prevalence of weeds is dramatically reduced by using mulches (Table 12.6), so the time spent removing them is low and the need for herbicides is avoided. Triggers for weed seed germination like light, increasing temperature or temperature fluctuations are not activated under the insulating, dark mulch. Mulch, particularly if composed of stone aggregate, pebbles or coarse organic mulches dry rapidly, providing a hostile environment for the germination of newly arrived seed. Some organic mulches such as

Table 12.6 Frequency of weeds with different mulches and depths

Mulch	Number of weeds/m^2	
	Mulch 50 mm deep	Mulch 100 mm deep
Bare soil	260	260
Sawdust	8	1
Pine bark	4	0
Wood chips	1	0

Source: Handreck & Black (2002).[8]

fresh tree bark can leach compounds that inhibit weed growth, but a cautionary approach is needed because they may also affect the growth of the desired plants.

Mulches may not be as effective controlling weeds that spread vegetatively, for example via stolons, as do couch and kikuyu. It is better to remove them using solarisation before applying the mulch (see section on soil preparation in construction chapter), or digging out by hand, which may not be entirely effective. This is a situation where herbicide may be required, but if most of the weed population is physically removed, small quantities of herbicide can be spotted on any regrowth that occurs.

Mulches improve soil structure by eliminating the need for cultivation, and through organic decomposition add organic matter to the soil. Mulch encourages soil fauna like worms and the benefits they bring to soil structure. In trafficked areas, the mulch particles spread the forces laterally, relieving the pressures on the soil itself. That is one reason why, during large public events in parks, mulch is temporarily used over lawns, but the concept can equally be applied more permanently as paths through gardens.

Mulching frequency is dependent on the type of mulch used and its resistance to breaking down. Pea straw decomposes quickly and will need to be replaced at least annually. Pine bark decomposes slowly, but it is worth checking annually to see if it needs topping up. When fresh plant-derived mulch is used on garden beds, micro-organisms compete with plants for soil nitrogen (known as 'nitrogen drawdown') and this may result in nitrogen deficiency symptoms in the plants (e.g. chlorosis). Using organic materials that are more fully decomposed or adding additional plant nutrients when placing the mulch reduces this problem. Mulch placed against tree trunks may result in collar rot and is best avoided.

Care is needed not to restrict air movement into the soil by applying mulches too deeply, or by watering so frequently that a saturated layer forms where the finer particles collect at the soil-mulch interface. Recommended mulch depth ranges from 2 cm for fine to 10 cm for coarse materials. The insulative properties of mulch may also increase frost damage as the heat radiating from its surface at night is not readily replaced from the warmer soil below, causing the air just above the ground to cool more rapidly. Dry, aerated organic mulches may become fuel for bushfires, so fire authorities recommend mulching at the end of one fire season so that it can decompose prior to the onset of the next. During drier seasons or drought, mulches may absorb all moisture from light, short periods of rainfall before it can percolate through to the root zone, so it is important to link irrigation scheduling to regular monitoring of the soil under the mulch.

Bringing any material into a garden will have environmental impacts, and it is helpful to be aware of at least some of them until fully informative LCAs for mulches can be done.

All mulches have embodied energy and contribute to greenhouse gases through their transport from their site of origin to the site of use. Processing can add more embodied energy, as does the milling and size-grading of pine bark, or the extraction and grading of pebbles. Pebbles may need washing (is the water recycled?). Pine bark may be partially composted by wetting and ageing to allow toxins to be metabolised, or the process may be controlled by adding nitrogen (high embodied energy) to massive windrows turned by powered equipment. Plantation pines, from where the bark comes, have their own environmental impacts, and the pine mulch product may be purchased in high embodied energy plastic packaging that will probably end up in landfill. When sourcing mulch materials:

- Use a mulch suitable for the situation (gravel or sand for rain gardens; organic mulches over garden beds, etc.).
- Ascertain the source of the material (ask for and select products derived from sustainably managed timber plantations, that are by-products of other crops or inorganic products mined with minimal habitat damage, pollution or siltation of waterways, etc.).
- Select products that come from local sources (e.g. peanut shells used locally in Queensland, sugar cane used locally in NSW and Queensland).
- Be aware of processing methods that may add significantly to non-renewable energy use and greenhouse gas production.
- Ensure high quality hygiene practises are employed so diseases are not spread via the mulches.
- Avoid ordering in small amounts to reduce multiple deliveries.
- Avoid wastage from over-ordering (calculate the number of cubic metres required by working out the area to be covered in square metres, and multiplying by the depth in metres, e.g. mulch required for a depth of 10 cm is $10 \text{ m}^2 \times 0.1 \text{ m} = 1 \text{ m}^3$).

Many suppliers may not have this information readily available, but it is worth asking so they can respond more fully to the needs of their customers and so you can identify those suppliers who are more aware of the issues.

Supplementing rainfall

Irrigation

Irrigation systems allow us to overcome variability in the frequency, intensity and duration of rainfall. When supplementing rainfall by irrigating, water conservation is paramount and this can be achieved by watering:

- Only those plants that need it.
- At a rate that will not cause losses through surface run-off.
- At a frequency that trains the plants to survive periods without rainfall.
- Using sufficient water to maintain the plants at a desired quality.
- For a period that will not cause losses through deep percolation below the root zone.

The most efficient way to do this is to create irrigation zones with similar watering requirements, using a system that delivers water to the roots of the plants with the least losses through surface evaporation or wind drift. Frequency of water application can be estimated by monitoring weather conditions, soil moisture and plant health.

Landscape irrigation is currently based mostly on intuition combined with some observations, but large water savings can be made if a more scientific approach is used. The Royal Botanic Gardens Melbourne, for example, has been able to halve its water use over 10 years from the mid 1990s even through an extended drought, by careful use of weather data. Similarly, residential gardens can reduce the amount of water that is applied yet still be healthy.

If we are to manage water more efficiently in residential gardens or public landscape, then we need to understand how water flows through the soil, the factors controlling this flow, and the volumes involved. We discussed the major garden flows earlier in the book; now we need to quantify these flows, assess them and prioritise for better water management. Geoff Connellan, a University of Melbourne lecturer specialising in irrigation, emphasises efficient irrigation practices and draws attention to the need for routine performance benchmarking for irrigation techniques and practices.[11] Costello and Jones introduce the classification of landscape plant species based on their water needs.[12] A brief overview is given here.

For any particular irrigation system, at a given time we need to know:

1. **How** to water: what sort of irrigation system is most appropriate?
2. **When** (and where) to water?
3. **How much** water is needed?

1. How to water – irrigation systems

Irrigation systems can vary from hand-held hoses and hose sprinklers through to installed sprinklers, microsprays and drip. They can be operated manually or automatically by an electronic timer. Restrictions limiting the use of mains water for irrigation are becoming standard, and if mains water can be used at all, it may be restricted to hand watering or to drip systems. Drip systems have application efficiencies as high as 95% because they avoid losses from wind and soil surface evaporation (see Table 12.7).[14] They can be used in all soils although spacing of the outlets will depend on soil type, being much closer in sandy soils (about 30 cm).

Irrigation efficiency can provide a measure of performance. It is affected by the system design, its maintenance, and effective management using observation-based irrigation scheduling. The scheduling will be done based on the soil type, prevailing weather conditions, the specific requirements of the plants being irrigated and the amount of water in the root zone. In agriculture, *water use efficiency* can be related to water productivity (crop per drop) but in turf or landscape, irrigation is generally referred to as *irrigation efficiency* and simply reflects how much of the water applied is available to the plants.

Table 12.7 Attainable application efficiencies for different irrigation systems

Application type	Application efficiency range (%)	Average application efficiency (%)
Residential sprinklers	10–85	75
Microspray	70–85	80
Surface drip	70–90	85
Subsurface drip	70–90	85

Source: Dukes MA.[13]

2. When to water

Watering *frequency* will depend primarily on the soil texture and its water storage capacity as well as the current and expected weather conditions, the water demand of the plants, and our assessment of an acceptable level of plant quality.

Plants have different water requirements and we have different expectations for plant quality. Knowing when to water requires knowing the water demand of the plants we are growing and the quality we require of them, as well as knowing the water content of the soil. Ornamental plants can often be water-stressed with minimal effect on their visual quality, but this kind of stress should be avoided for high-yielding food production plants.

The likelihood of rainfall should also be taken into account, along with the expected evapotranspiration, resulting from the expected temperatures and wind. The water levels in the soil may be estimated roughly by observing plant quality/stress, or measured using soil moisture instruments (although accurate ones are very expensive), or by weighing soil samples before and after oven-drying. The water demand of particular plants may be estimated using a crop factor, a number that reflects how freely or otherwise a species will transpire, although these do not often exist for ornamental garden plants.

For watering purposes the soil 'tank' is full when at *field capacity*, and empty at the *permanent wilting point*. Overfilling the tank will result in deep percolation and/or run-off. The point where the soil is fully replenished is called field capacity or the *refill point* (RP).

How much water should be allowed to leave the soil tank before we top it up by watering? For maximum crop production, the soil tank is maintained at close to full, but for plants that can and need only to survive drought, watering may be minimal. As a general rule, watering is carried out at 50% depletion of field capacity, but the precise level will depend on the plant water demand and the acceptable plant quality.

In irrigation terminology, watering is needed when water levels have fallen to about 50% field capacity and the *frequency* of watering is reduced to a minimum by watering the soil to field capacity.

The Royal Botanic Gardens Melbourne uses rainfall, solar radiation, air temperature, wind speed and relative humidity data recorded by an on-site automatic weather station to estimate the evapotranspiration rate and determine when irrigation should occur for different plant categories.

3. How much water?

The purpose of irrigating is to fill the soil between the soil surface and the bottom of the root zone to field capacity, avoiding run-off and deep percolation. For each soil type and irrigation system it is important to be aware of how much water needs to be applied and the rate of application.

Measuring soil water availability

As water decreases in the soil, the more firmly the remaining water is attracted to the soil

> **INFO BOX 12.1: A GENERAL IRRIGATION GUIDE**
>
> Water:
>
> - In the morning when wind and temperatures are at their lowest, avoiding evapotranspiration before the water has a chance to infiltrate into the soil (the increased humidity resulting from overhead watering in the evening can encourage disease).
> - Less frequently but deeply enough to encourage deep root systems that support plant growth through longer periods without rainfall or irrigation (avoiding overwatering where water extends well below the root zone).
> - Sandy soils by applying less water more frequently than clay soils.
> - Ahead of hot weather.
> - Regularly monitor the weather, particularly for temperatures and wind, and irrigate more frequently when either or both of these are higher.
> - Check the 4-day forecast and avoid watering if good rainfall is expected.

Table 12.8 Sample soil water tensions for maximum yields

Plant	Pressure (kPa)
Shallow-rooted vegetables in sand	25–35
Plants in loams	35–50
Plants in heavy soils	50–60

10 kPa = 0.10 bars = 10 centibars = 100 millibars = 1.45 psi.

particles and the more suction pressure is required to remove it by the plant. This can be measured by instruments like tensiometers which indicate the amount of water available to plants. Water is applied when a predetermined trigger level of suction pressure is reached (Table 12.8).

Sophisticated modern soil probes like the one depicted in Figure 12.10 give a greater insight into water movement through the soil by providing real-time data that can be used to determine the best times to water and how much to apply. Already used for commercial crops, tools like this will increasingly be valuable for efficiently managing the irrigation needs of large public landscapes. Currently a collaborative soil moisture study at the Royal Botanic Gardens Melbourne with the University of Melbourne and Sentek Pty Ltd is recording the movement of water through the soil after irrigation or rain to increase our understanding of water behaviour in the soil and its effect on plant growth.

Water demand of the plants

Irrigation scheduling is partly dependent on the rate of transpiration from the various kinds of plants growing in the garden, some species losing water more rapidly than others. This, in turn, is dependent on the total leaf area and how the plant's stomata function. Estimates of evapotranspiration can be made by measuring evaporation (from a Class A pan) and multiplying by a crop factor (or a similar crop coefficient) which represents the rate of transpiration for a particular crop. The lower the crop factor, the lower the transpiration rate. Crop factors are available for commercial crops. They are not available for many ornamental plants although an evaluation of plant water needs based on field observations has been completed for 1800 plants cultivated in California.[15] Ornamental plantings tend to

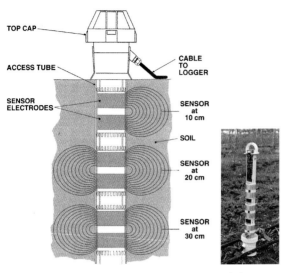

Figure 12.10 Sentek's Enviroscan probe currently being used in studies at the Royal Botanic Gardens Melbourne. (Images: Sentek Pty Ltd)

be of mixed species so approximate crop factors for plant groupings such as broad-leaf ornamentals are useful (Table 12.9). For mixed ornamental landscapes, landscape coefficients that take into account the species, the density of planting and the microclimate are potentially more useful.

Table 12.9 Approximate crop factors for landscape planting

Crop factors	
Cool season grass (lush)	0.8–0.85
Cool season grass (moderate growth just acceptable)	0.65–0.7
Warm season grass (lush)	0.55–0.7
Warm season grass (moderate growth just acceptable)	0.25–0.4
Broad-leaf ornamentals (establishing)	0.7–0.85
Broad-leaf ornamentals (strong growth)	0.6–0.7
Low growth, moderate drought tolerance	0.2–0.4
Minimum for desert plants	0.05–0.2
Vegetables and fruit trees	0.7–0.85

Source: Handreck & Black (2002).[8]

Delivery rate

Water is best applied at just below the infiltration rate to bring the water level in the soil to field capacity over the depth of the plant root zone. So the delivery rate will be governed by the infiltration rate and the rate at which the irrigation system can deliver the water.

Water infiltrates into the soil at different rates depending on the soil type, ranging from about 5 mm per hour for clay loams, to 7 mm per hour for sandy loams, to about 20 mm per hour for sand.[4] Sloping soils can have significantly lower water infiltration rates. If water is applied to the soil more quickly than it can infiltrate, then there will be surface run-off.

Drip irrigation systems are graded for their delivery rate with dripping rates of 2, 4 and 8 litres/hour for example. Sprinkler system delivery rates can be measured simply by capturing water in ice cream containers or fruit cans placed within the wetting pattern and measuring the water depth after a period of time the sprinkler has been on. Or, if a predetermined amount of water needs to applied, see how long this takes, by monitoring the containers until they contain water to the depth required. Timers can be adjusted accordingly.

The delivery rate for both sprinkler and drip systems can also be measured using the property's water meter, noting the reading before and after the irrigation system has been on (Info Box 12.2).

Excess water not held in the soil by capillarity will move mostly downwards as a wet front.

> **INFO BOX 12.2: CALCULATING IRRIGATION SYSTEMS WATER DELIVERY RATES**
>
> 1. Read water meter (443.512 kL).
> 2. Turn on irrigation system for 1 hour without any other water being used during this period.
> 3. Read water meter (444.002).
> 4. Calculate the water used (0.490 kL = 490 L).
> 5. Calculate the area being watered by the irrigation system (4 × 10 = 40 m^2).
> 6. Divide the litres used by the area, to give the delivery rate of the irrigation system in litres of water/m^2/hour, which translates to the equivalent of rainfall measure of mm/hour (490/40 = 12.25 L/m^2/hour).

Table 12.10 Water-holding capacity of different textured soils

Soil texture	Water held (mm H$_2$O/m soil)		
	Field capacity	Wilting point	Available water
Sand	90	20	70
Loamy sand	140	40	100
Sandy loam	230	90	140
Sandy loam plus organics	290	100	190
Loam	340	120	220
Clay loam	300	160	140
Clay	380	240	140

Source: Handreck (1979).[16]

When there is no longer any free water, the wet front will stop and additional water added at the surface will flow directly to the water front. If there is sufficient water, the front will pass into the water table; it may meet a hard pan or flow to deeper levels through cracks in the substrate.

Subsurface bedrock, layering (soil profile) and compaction

Soils are rarely uniform in texture but generally have a *layered* profile. Loams are often found above clays. Infiltration in such a soil will be initially quite rapid, but the clay will slow penetration and after an extended period of rain, permeable topsoils may become waterlogged even when there are dry, deeper layers below. Soils prone to these conditions may require the clay layer to be fractured, although applying gypsum will work for some clay soils. (Note that drainage systems are often not very effective and if they are effective are very expensive in resources and cost to install).

Following a heavy watering or rainstorm, drainage of water through the soil will usually stop after a few hours in sand, and up to seven days in clay soils. The moisture that remains is being held by the soil particles, the soil is at field capacity, and the water will remain in the soil unless evaporation and/or transpiration occur. The volume of water being held by the soil at field capacity depends on the soil structure and texture (see Table 12.10). It is clear that a coarse soil with a high sand content will dry out quickly and require regular watering or rainfall to provide plants with enough water. Loamy soils have good moisture-retaining ability, and if good structure and organic matter are present, they will also drain freely. Although clay has the greatest capacity for water storage, it can become sticky and difficult to work. Clay being slow-draining, run-off is more likely and lack of aeration may affect the growth and health of some plants whose roots are adapted to freely draining, well-aerated soils.

Soil management

Major watering efficiencies can be made by improving water storage in the soil, increasing the infiltration rate and reducing evapotranspiration. To help achieve efficiencies:

- Aim for a soil like a crumbly loam with mix of large and small pores. Treat clay with compost and/or gypsum if clay is responsive. Treat sand with clay and/or compost.
- Release any surface compaction and break up impervious lower layers.
- Use surface mulches to reduce evapotranspiration.
- Use environmentally sound soil wetting and water-absorbing agents (acrylamide crystals, vermiculite crystals or granules).
- If garden is sloping use terraces, roughen the surface, mulch, direct drainage to the garden.
- Don't dig wet soil (particularly clay soils).
- In general, the less soil disturbance during dry periods the better.

Seasonal changes – long-term monthly trends

The amount of water lost by transpiration in a well-watered garden approaches the amount of water lost by evaporation. A very rough guide to the water demand in your area can be gained by using Bureau of Meteorology data (see Figure 12.11). Monthly data of rainfall and evaporation smooth out daily fluctuations for an area and can indicate strong seasonal trends to help guide irrigation planning.

All major Australian cities have evaporation levels greater than rainfall for a large portion of the year (Figure 12.11). In southern cities, June and July, and for some, May and August too, are the only months where rainfall exceeds evaporation, and evaporation can be many times higher in the summer months. Brisbane has a different pattern, with evaporation and rainfall similar for the first half of the year, and evaporation overtaking rainfall in the second half. Darwin has a different pattern again, with evaporation being fairly even through the year, and rainfall peaking in January and almost non-existent in June, July and August.

Like the major cities, almost all Australian regions have evaporation exceeding rainfall for at least part of the year indicating that without careful plant selection, most places may need irrigation, and the more that evaporation exceeds rainfall, the larger the volume of water that may be needed. In a climate where rainfall exceeds evaporation year-round it may never be necessary to irrigate and the choice of plants will be less restricted. There are few places like that in Australia.

Plant quality

The amount of water we use also depends on the *plant quality* we require. For ornamentals this will be a standard of appearance, the degree of lushness and general health. For crops, it will be primarily yield. In general, garden plants need enough water to keep them alive and fresh with decorative appeal,

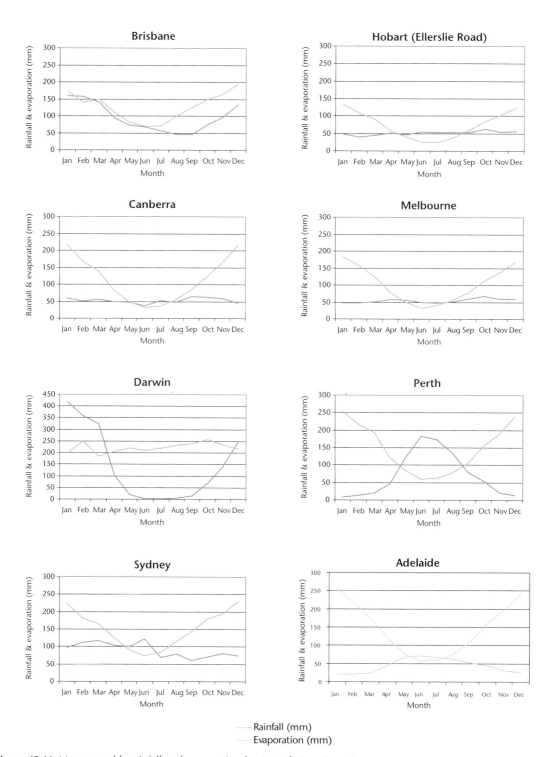

Figure 12.11 Mean monthly rainfall and evaporation for Australia's major cities.

and food plants usually need larger amounts of water to increase the quantity and quality of the fruits and vegetables. Home gardeners will need to use less water on most of their garden, except food plants, than nurserymen or market gardeners who need high yields, rapid growth and a lush appearance to attract buyers and maximise profits. Commercial suppliers may add excess water so that they are sure water is never limiting plant growth. Plants bought from garden centres may need a period of adjustment to a reduced watering schedule in gardens.

The quality of plants adapted to a particular climate can be maintained at a high level by irrigating with an equivalent of only 25% of the water lost through evaporation (as measured by a Class A pan). This may be used as a very rough guide to managing watering requirements during the months when watering is needed. Indigenous plants should survive on rainfall alone, although applying up to 10% of the evaporation rate may help improve their quality. Additional water applied will still be taken up by the plants which will transpire water up to 80% of the evaporation rate, but it will also mean the plants will grow faster, needing more pruning, and lawns more frequent mowing. For better yields of vegetables and fruit trees, up to 70% of the evaporation rate may need to be applied when there is full groundcover on unmulched soil, and 30% of the evaporation rate when there is 20% cover on mulched soil.

The water requirements of agricultural crops and plants grown in production horticulture are quite well known. The demand for water conservation, low costs and maximum water productivity are now an incentive to develop similar guidelines for the water requirements of garden and landscape plants.

Controlling growth

Even in well-planned gardens with good plant selection, maintaining the size and shape of the plants is a routine part of maintenance. Lawns need mowing, edges trimmed, shrubs clipped, trees trained into shape and pruned to encourage fruit, and dead wood removed. What and how this is done can improve energy efficiency and add to resource recycling within gardens.

Mowing lawns using non-powered mowers is ideal and, when done regularly before the grass is too long, is not difficult, and can be handled even by some people in their 80s. Keeping the blades sharp and adjusted helps. Manual edge cutters are also available. If powered mowers or edgers are needed, use models with low fuel consumption. Prepare the lawn by removing all obstacles like garden furniture and hoses prior to starting the mower so the mower will be on only for the minimum time. Mow only when necessary, which is usually less than every week for most of year. Once per fortnight is often adequate even when the grass is growing rapidly. Clippings can be collected for composting.

Almost all prunings are suitable for recycling back into gardens either as mulch or compost for digging into the soil. Ideally recycling this green wealth is done on-site if prunings can be chipped, but council green waste collections can also be used. There are a number of chippers on the market; some

Figure 12.12 Chippers help to recycle the green wealth of a garden.

powered electrically and others by petrol engines. The latter, in general, are able to chip larger branches meaning that more of the green wealth of a garden can be easily recycled (Figure 12.12).

Plants that show the potential for self-seeding and becoming garden or environmental weeds should be removed or at the very least strictly managed so that flowers are removed to prevent seeding. *Dietes grandiflora*, for example, has been found to self-seed in sandy soils in Melbourne and is better not grown or used only when maintenance schedules always include the removal of old flowers before the fruit matures (Figure 12.13).

When plants are placed in the wrong location without the space they require for growing naturally, constant pruning will be needed, such as trees under power lines or in inappropriately small gardens. Often lopped in these situations, the subsequent growth will refill the space, the new branching will be weakly attached below the cut, and as the growing branches become

Figure 12.13 *Dietes grandiflora* flower and seedlings that have self-sown in a Melbourne garden.

heavier, they are more likely to break off (Figure 12.14). Trees in some cases may be able to be pruned and trained correctly keeping a strong branch structure, but in many cases it is better to start again with a plant of appropriate size and habit. Well-chosen plants will not only look better, but will need less resources invested in their maintenance.

Vegetation management needs to support the solar passive principles of buildings, or the optimal functioning of solar-assisted appliances, ensuring minimal shadowing over windows or solar panels. In southern areas, this is particularly important during winter months when the sun is weakest,

Figure 12.14 Lopped trees result in weaker more dangerous branching, and higher maintenance needs.

Figure 12.15 Shadowing of solar features needs to be regularly assessed and pruning undertaken where necessary.

lowest in the sky, the days shortest and the need for heat the greatest (Figure 12.15). Yearly autumn maintenance programs can assess whether recent plant growth results in altered shading patterns that compromise solar efficiency, and pruning undertaken where necessary. Related to this are plants growing over windows, darkening rooms and increasing the use of artificial lighting and energy consumption.

Green waste – recycling the garden's green wealth

Green waste (sometimes called green organic material or green organics) is the term used to collectively describe plant material which has been discarded as non-putrescible waste. It includes tree and shrub cuttings and prunings, grass clippings, leaves, natural (untreated) timber waste and weeds (noxious or otherwise). In the last decade, green waste processing and technology has become mainstream. Many councils now have special collections of green waste and dedicated bins

Figure 12.16 Council greenwaste collection helps recycle a community's green wealth.

for prunings (Figure 12.16). Giant chippers and mulchers grade, macerate and shred green waste into a form that can be recycled as compost or mulch.

Green waste processing reduces the amount of waste that goes into landfill, and provides a useful resource.

In some cities, the support for the green waste recycling is so high that the amount of compost being produced is greater than the market is currently able to absorb. It is sold through retail nurseries, and has been used in large-scale projects such as roadside plantings.

Residential green waste from council kerbside collection contains grass clippings and small prunings. Some councils also collect bundled larger branches. It is taken to regional processing facilities where it is chipped and composted. In terms of green waste, it is preferable that residents recycle most if not all the waste on their own property, and councils encourage this by selling compost bins and worm farms to residents, often at cost price.

There are a number of companies and methods used for recycling the materials from council collections, including windrows and sophisticated systems that process the compost in enclosed chambers where environmental conditions can be more tightly controlled and composting accelerated (Figure 12.17).

Chipping, sieving and composting the green material in windrows is one method, guided by Australian Standards (AS4454-1999 Compost, soil conditioners and mulches) to ensure diseases and weeds are eliminated.

Chemical control of weeds

Herbicides have environmental implications relating to their extraction, processing, manufacture, packaging and transport. They also have high embodied energies (see Table 12.11).

Except for public landscape maintenance where it may be impractical not to use herbicides, there should be little need for their use. Even in public landscape other horticultural practices like mulching should minimise their need. However, controlling perennial weeds like couch or kikuyu can be difficult without using herbicides like the commonly used glyphosate. Its use can be minimised by physically removing as much of the plant as possible, and applying the glyphosate to any regrowth that occurs while it is still small.

Glyphosate, which has no pre-emergence (it cannot act on weeds before their leaves appear above ground) or residual activity, is the most widely used herbicide in Australia (15 000 tonnes per annum).[22] The pre-emergent herbicide Simazine (3000 tonnes per annum) is also commonly used. Atrazine (3000 tonnes per annum) is a pre- and post-emergent herbicide. Both are used for broad-

Table 12.11 Embodied energy in herbicides

Herbicide	Embodied energy (MJ/kg)
2,4-D	85.0
Alachlor	277.9
Atrazine	187.6
Diuron	268.6
Fluazifop-butyl	517.8
Glyphosate	453.9
Paraquat	459.9
Trifuralin	150.0

Source: Helsel (1992).[6]

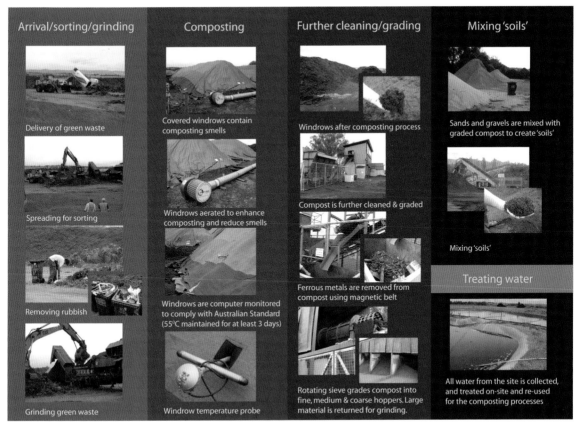

Figure 12.17 Municipal green waste processing by Australian Native Landscapes, Coldstream.

leaf weed and some grass control in forestry and agriculture. In the second draft report by the Australian Pesticides and Veterinary Medicines Authority released in 2004, it was noted that non-agricultural uses of Atrazine may 'pose a possible risk to the environment'.[17] The report found that levels in waterways are generally below the upper recommended limit, but they can be exceeded during storms. The report also noted that the jury is still out concerning the effect of Atrazine on amphibian development. Plant-cell-growth-disrupting chemicals less frequently used include 2,4-D and its derivatives, MCPA, paraquat dichloride and, to a lesser extent, diquat.

A cautionary note is needed for glyphosate. In Denmark, tests by the Denmark and Greenland Geological Research Institution (DGGRI) indicated that glyphosate had contaminated groundwater at levels higher than the allowable 0.1 $\mu g l^{-1}$ standard for drinking water and its use was banned in 2003. Glyphosate applied at the recommended rate leached through to the groundwater before microbial action was able to degrade the chemical. It has widely been assumed that it degrades rapidly. A confirming study showed the amount of leaching of glyphosate and its breakdown product amino-methylphosphonic acid varied in different soil types and was also dependent on the amount of rainfall, leaching being greater where water percolation through the soil was higher in the first months after application.[18] Amino-methylphosphonic acid was often detected more than 1.5 years after application.

Sometimes it is not the main active ingredient that may cause environmental problems. In 1995, Bidwell and Gorrie published a report for the Western Australian Department of Environmental Protection raising concerns about the effect on tadpole health of surfactants found in some glyphosate formulations. The report made recommendations that would avoid the contamination of waterbodies.[19]

Managing pests and diseases

Gardens with healthily growing plants and a rich variety of plant species that are supporting a wide diversity of micro-organisms, arthropods (insects, mites, spiders) and other organisms, provide the checks and balances that make it difficult for pests and diseases to take hold. These gardens generally need little pest control on our part, and although some damage to garden plants may occur from time to time, it is usually limited.

The keys to managing pests and diseases are to:

- Monitor garden plants for damage.
- Identify its cause(s).
- Know at what point intervention is required (if at all).
- Select a control method appropriate for the pest and for the environment.
- Use the control at the optimal time(s).
- Assess the control's effectiveness.
- Continue to monitor.

Remembering that small amounts of pest or disease damage usually do not detract from the overall beauty of a garden, we can focus on those plants where applying pest control may be critical for a plant's survival, its ornamental qualities or its productivity as a food plant. There are also routine maintenance practices that we can adopt to minimise the harbouring and spread of pests and diseases. Many control methods do not rely on using synthetic chemicals, but if they are required, their use can be limited and well targeted.

Insects are the animals that cause the most pest problems for plants, and fungi cause the most microbial diseases.

Common pests and diseases and their control

Table A8 in the Appendix is a guide to pest control without using chemicals. Judy McMaugh's *What Garden Pest or Disease Is That?* is useful for the identification of many garden pests of both native and exotic plants and includes organic, non-chemical control methods.

General guide to controlling pests and diseases

Disease-resistant cultivars have been developed for a number of horticultural crops including fruit trees and ornamental plants, and it is always worth giving preference to these in gardens to minimise the need for pest control.

Choosing the right place for a plant in a garden with varying micro-habitats can be an important contributor to a plant's ability to resist pests and diseases. Environmental conditions such as soil type and fertility, sun exposure and humidity can all affect susceptibility. Some site modifications to improve conditions are possible, like improving soil structure and fertility.

Just as humans are healthier with good nutrition, providing adequate levels of plant nutrients ensures robust plant growth and less vulnerability to pests and diseases. This does not mean over-fertilising plants. Toxicity is possible if nutrient levels are too high, resulting in poorer plant growth and, potentially, death, not to mention the wastage of excess fertiliser leaching away. Nutrition of garden plants is also aided by removing the competition of weeds, as is the availability of water.

Managing how and when irrigation water is applied can affect the establishment and progress of diseases. The increased moisture around foliage resulting from overhead watering can provide the right conditions for fungal diseases to flourish on susceptible plants, particularly if watering occurs in the evening when cooler temperatures slow evaporation. Early morning irrigation and/or drip systems delivering water directly to the soil are better practices in many cases.

Encouraging good organisms – the natural predators or antagonists of pests and diseases – in the garden or around crops can keep pest problems at a low level. Not using broad-spectrum insecticides, for example, ensures the good bugs are not killed along with the bad.

Garden hygiene can reduce the presence and transfer of pests and diseases. This includes removing and destroying diseased or infested plants or plant parts as soon as possible, keeping the garden free of weeds that may support pests (mulching helps with this), and sterilising garden equipment, including secateurs, between using on one plant and the next (for example, with methylated spirits).

Hygiene includes quarantining diseases out of gardens. Cautious purchasing and exchange of plant material, soils, composts or mulches from reliable disease and pest-free sources can save the cost of eradicating later, or the restriction of having to live with a disease that cannot be eliminated once established. Hygiene may need to be extended to footwear or car tyres if they have been in areas with soil-borne diseases like *Phytophthora cinnamomi* (cinnamon fungus).

The traditional rotation of vegetable crops, even including fallow periods without any plants, lowers the risk of disease populations building up in soils.

Seasonally recurring pest and disease problems can be managed by incorporating preventative measures into garden maintenance. For example, to reduce the incidence of brown rot (*Sclerotinia* sp.) damage to stone fruit, yearly pruning of the trees can retain open canopy centres, providing better air circulation and faster drying of the fruit after rain, conditions less favourable for the disease. The planting of some crops can be timed to avoid crop development and maturation during the peak pest time.

Horticulturists correlate pest cycles with Degree Day calculations, giving greater accuracy in timing the implementation of pest control. Used in large public landscapes (for controlling elm leaf beetle, for example), the method takes into account the effect of weather conditions, including temperature, on the growth rate of pest populations.

Insects and plants begin seasonal growth when a temperature threshold is reached. Growth then increases as temperature rises until an upper threshold is reached when it slows or stops. Seasons with larger numbers of warm days will have pest populations appear earlier than those with cooler days. The warmth of a season can be measured in Degree Days which are calculated in a number of ways, the simplest (although not necessarily the most accurate) being:

$$\text{Daily Degree Days} = \frac{\text{Temp.}_{max} + \text{Temp.}_{min}}{2} - \text{Temp.}_{base}$$

Each day the Degree Days are added to the previous cumulative total for the season. Base temperatures vary for different pest species, and the number of Degree Days can be used to predict various stages of development (egg, larvae, crawler, nymph, adult). The control method is applied at the developmental stage when it will be most effective (e.g. larvae). Minimum to maximum Degree Day ranges for different pests have been published.

Companion planting has been used by gardeners for many years, but it is now gaining more scientific attention. University of Sydney's Associate Professor Geoff Gurr (Faculty of Rural Management) believes that ecological engineering may offer useful low environmental risk, economical solutions to controlling pests, and is assessing systems that have been developed elsewhere for their suitability in Australia.[20] In Orange, NSW, borage planted with potatoes provided increased nectar supplies for wasps that parasitise potato moth caterpillars allowing the wasps to live longer and attack more caterpillars. Similarly, *Trichogramma carverae*, used as a biocontrol for the grapevine pest, light brown apple moth, lived longer in vineyards when sweet alyssum (*Lobularia maritima*) was planted as a companion plant. Importantly, the alyssum was not a food source for the light brown apple moth.

These examples clearly show the benefit of thinking of crops and gardens as ecological systems, and the importance of providing an environment that supports the continued presence of beneficial organisms. It may mean that low levels of the pest species need to be maintained for the continued presence of its predator, each contributing to a balance that leaves the garden plants largely intact.

Biological control methods

Predators or parasites of particular pests may be specially bred and control-released to attack the pest organisms (microbes, viruses, bacteria, nematodes or weed plants). Although people may be more aware of examples on a large scale, like the successful release of *Cactoblastis* moth reducing populations of prickly pear cactus, it also has relevance to gardens and public landscapes where beneficial organisms can be released for controlling pests. Large-scale introductions require considerable research and trialling to ensure there are minimal negative environmental consequences, and can therefore be slow and expensive to implement, but will possibly become more widely used in the future. The release of blackberry rust has had only moderate success, reminding us that truly successful cases are relatively few. Ideally, a balance is required between the pest and its predator.

Beneficial organisms

Beneficial organisms are those that are natural competitors, predators or parasites of organisms that cause damage to the selected garden plants. It is possible to encourage the presence of beneficial organisms by providing food plants for them (e.g. nectar), appropriate environmental conditions for their reproduction and survival, and by avoiding use of non-specific chemical controls.

For some plant pests, beneficial organisms can be purchased for release in greenhouses, amongst crops or in gardens. The Australasian Biological Control Association was formed in 1992 to foster cooperation between commercial suppliers of beneficial organisms for biological control, to promote the use of Integrated Pest Management (IPM) and lessen the use of pest control chemicals.[21]

What is Integrated Pest Management?

IPM uses natural pest control methods as an initial line of defence, such as: physical barriers, resistant cultivars, optimal growing conditions, cultural practices, hygiene, companion planting, encouragement of beneficial organisms in the garden or glasshouse, the introduction of populations of beneficial organisms, etc. Chemical pesticides are only used when absolutely necessary, and then only those chemicals least likely to harm beneficial organisms, humans or the environment. Organic growers have few choices as they are restricted to products that are naturally derived botanical pesticides. Other growers have more options, but should still be strongly guided in their chemical choices by a concern for the environment.

IPM tolerates low levels of pest organisms in balanced garden or glasshouse ecosystems, and management aims to retain this balance. This requires accurate identification of pests, understanding their life cycles, knowing their vulnerable stages, identifying what other plants in the vicinity support their populations, monitoring their population levels (with the eye and with traps, including pheromone traps), assessing pest damage to garden or crop plants, monitoring the levels of beneficial organisms, keeping good records, and knowing when to be more vigilant (using Degree Days, or associating with the stage of the crop or other indicator plant).

When it is assessed that further control methods are needed, the aim is: to bring the pest population down, but not to eliminate it; to restore a balance with beneficial organisms, and if possible, to do this by additional cultural practices or biological controls, and lastly, by using least harmful chemical pesticides.

Synthetic chemicals

Synthetic pesticides, like synthetic fertilisers, carry environmental implications, not just in the garden and surrounding environment but also through the raw materials and resources (such as fossil fuels) used for their extraction, processing, manufacture, packaging and transport (see Table 12.12).

Chemical solutions to pest problems often appear more convenient, more efficient and cheaper than other solutions, so there is a tendency to see a chemical solution to every biological problem when there may be alternatives. This tendency probably stemmed from the wide use of synthetic chemicals in agriculture. Home gardeners adopted a similar approach to nutrient, pest and weed problems.

Herbicides and pesticides are chemicals that are manufactured to kill. Rarely are they specific to their target organism. Apart from their occasional unexpected or unpredicted effect on people, herbicides can also poison soil micro-organisms, mycorrhizal fungi, earthworms, insects, birds and animals. Herbicides settle on plant tissue, soil or water where they will accumulate unless they degrade or are transported elsewhere. Most problems arise due to the herbicide's mobility and to their long-term effects. In general, in the developed world the pesticides and herbicides used are safer, more selective, less persistent and with improved methods of application over those used a decade or two ago. However, for the home garden, the environmental cost of their manufacture alone is good reason to reduce their use to a minimum.

The potential environmental and health effects of synthetic pesticides, herbicides and fertilisers are many and varied. They include:

Table 12.12 Embedded energy in fungicides and insecticides

Pesticide type	Pesticide	Embodied energy MJ/kg
Fungicides	Benomyl	396.9
	Captan	115.0
	Maneb	99.0
Insecticide	Carbofuran	453.9
	Cypermethrin	579.8
	Malathion	228.9
	Methyl parathion	160.0
	Parathion	138.0

Source: Helsel (1992).[6]

- Impacts on human health – following label recommendations should be safe but there needs to be training in application.
- Impact on the environment – often high energy and water input during manufacture and transport.
- Potential for development of resistance in target organisms, especially with continued application of the same chemical, potentially leading to increased dosages, destruction of natural predators and resurgence of pests or formerly harmless insects becoming pests.
- Build-up of toxic residues in foods, soils and animals.
- Lethal effects on non-target organisms (for example chlorinated hydrocarbons are fat-soluble and passing through the food chain building up in fatty tissues by bioaccumulation, sometimes becoming dramatically more concentrated as they pass up the food chain (biomagnification). There may be unpredicted environmental consequences. For example, broad

spectrum pesticides may kill natural predators of the undesired pest.

Our pervasive use of synthetic chemicals often goes unnoticed and may have unexpected impacts on gardens, their plants and the broader environment. However, the scale of impact of synthetic chemicals used on gardens is very small compared with agriculture.

When used as directed, chemicals may not have detrimental environmental effects and can provide benefits. However, a cautious approach to their use is advisable in view of unfortunate experiences of the past. The need for their continued use is greater in agriculture and municipal areas than in home gardens. When they are used, great care is needed to ensure the correct concentration, rate of application, time of application, weather conditions (e.g. no wind, rain-free period), and importantly, that chemicals are used only in those situations for which they are prescribed and where natural ecosystems, including waterways, will not be contaminated. Long-term effects may take a while to emerge.

Chemicals should be used only as a last resort and with as complete knowledge as possible of all their potential consequences on living systems. In the light of actual or potential risks to human health and the environment, gardeners and professional horticulturists are urged to adopt a precautionary approach by avoiding synthetic chemicals if possible. If they must be used then use the minimum necessary, of the least toxic ones, with short persistence in the environment, being selective and clearly targeted to the pest organism and with little likelihood that pests will develop resistance. Always take care in disposing of any excess, which should be kept to a minimum. We can also take extra care with the application method, using wicks rather than sprays for example. In public areas, there should be ample community consultation and information.

Chemicals and the environment – information and monitoring

The most comprehensive report on pesticide use in Australia was produced by the Australian Academy of Technological Sciences and Engineering in 2002.[22] It excluded veterinary and urban use. The Executive Summary commented:

'At present there is no detailed publicly available information in Australia, on individual pesticides, either nationally or by regions … [A] database on agricultural and veterinary chemical use would allow government, industry and the wider community access to use data, giving the ability to recognise changes in use patterns, determine what is causing any observed trends, and relate them to changes in productivity, the environment and any perceived health risks, allowing sound scientific conclusions'.

A major recommendation of the report was the establishment of a comprehensive and integrated pesticide use reporting system, including a cost–benefit analysis of the value of pesticide use in production systems and the value of any regulatory changes.

On environmental effects the report continues:

'Relatively speaking, little is known of the effects of pesticides on Australian species in their natural habitats. More information is also required on the effects of newer pesticides on birds and termites in their natural range. Although the risk of off-target herbicide damage to commercial crops, especially vineyards, is well established, off-target damage to native plants and trees needs further attention. More emphasis needs to be given to monitoring the biological effects of pesticides on organisms and ecosystems rather than just testing concentration effects in individual species … A comprehensive integrated national environmental monitoring program should be implemented. The recommended National Adverse Health Effects register should be broadened to become a National Adverse Pesticide Effects Register, recording acute incidents where pesticides have had an adverse effect on the natural environment'.

And, on the economics of pesticide use:

'there is justification on economic policy grounds to have government intervention in a pesticide regulation system'.

A national approach to the management of chemicals

A Working Group of the National Chemicals Taskforce, set up by the Environment Protection and Heritage Council (EPHC), is currently developing a proposal for a national environmental risk management framework for chemicals which should address some of the issues raised above.[23]

The Australian National Pollutant Inventory is a publicly accessible database containing information on emissions of 90 substances from more than 3800 facilities around the country. In a positive sign, data on emissions to Australia's air, land and water during 2005–06 showed that 47 of the 90 NPI substances reported by industrial facilities decreased compared to the previous year.

In addressing the problems posed by fertiliser use in Australia, environmental precautions are gradually being put in place. National Land and Water Resources Audits assess the loss of nutrients from farming systems. Farming industries now have a Nutrient Management Code of Practice. In 2002 the Fertilizer Industry Federation of Australia Inc. (FIFA) endorsed an eco-efficiency agreement with Environment Australia. This was in line with the World Business Council for Sustainable Development's aim at achieving eco-efficiency. Eco-efficiency is reached by the delivery of competitively priced goods and services that satisfy human needs and bring quality of life, while progressively reducing ecological impact and resource intensity throughout the life cycle, to a level at least in line with the Earth's carrying capacity.

FIFA has also produced Nutrient Management Codes of Practice specific to particular industries and regions. One example is the potential damage to the Great Barrier Reef by fertiliser nutrient discharged into the sea from the cane-growing industry.

Sustainable Gardening Australia's chemical rating system

SGA, in consultation with Burnley College (University of Melbourne), developed a chemical rating system for grouping garden chemicals according to their effect on the environment. The system covers pesticides, herbicides, fertilisers and potting media. Chemicals that have the lowest ratings have the least environmental impact, and are promoted in SGA Certified Sustainable Garden Centres. Scoring chemicals using a list of numerically weighted questions, the final score is used for directing products into high, medium or low environmental impact level categories. The ratings relate primarily to toxicity and mobility in the environment, but also include other factors like the type of packaging, whether the product has a child-proof lid, and method of delivery; aerosol formulations, for example, due to the small droplet size, are more easily absorbed into the lungs by inhalation.

Communities and chemical use

Science provides information on the chemicals available, but communities can determine the acceptable level of risk, monitor the types of chemicals and their use in their locality, and encourage appropriate legislative checks and balances with rigorous screening and safety checks.

In Canada, many cities have already passed laws banning pesticides and the entire province of Quebec has restricted their use and sale. The European Union is following suit with demands for chemical manufacturers to put all their products through a very rigorous, extremely expensive testing procedure.

Providing food

(See also the section on food in the design chapter.)

Food gardens have the best success when frequently monitored. There are many books that can help with the general maintenance tasks of vegetable and fruit gardening, but as with all gardening, use the minimum resources needed for healthy productive growth.

Depending on local climate, species and cultivar, fruit trees and shrubs may not need high levels of maintenance, high volumes of irrigation water or large amounts of fertiliser to have high yields. Pruning can be used to decrease fruit number thus increasing fruit size and quality, or to open the canopy to more light or to reduce susceptibility to disease. There are good references to help with pruning specific crops. Poorly performing or disease-prone plants can be replaced with those needing less resources to maintain them.

Regular harvesting of vegetables depletes soil nutrients and lowers soil structure quality. Returning nutrients and organic matter (compost) to the soil is therefore important, and will usually require more material than can be produced from within the garden. When additional composts are needed, use local sources like green waste processors, working to the Australian Standard for composts.

Harvest produce at its peak and minimise wastage. Maximum yields may result from harvesting the crop frequently over a period of time; grapefruits for example, can be harvested over six months in some cases (see Info Box 12.3).

When crop production is greater than what can be readily consumed at the time, preserve

> ### INFO BOX 12.3: HINTS FOR HARVESTING
>
> - Continual harvesting to extend season: broad, bush and runner beans, capsicum, cucumbers, eggplant.
> - Harvesting early: partially developed broad beans can be eaten like runner beans, capsicum can be eaten green.
> - Picking just prior to eating: avoids the need for energy-requiring refrigeration, and it tastes best then too.
> - Storing energy-free: carrots and parsnips can be left in the soil for long periods; avocados can be left on the tree for many months.
> - Preserving: bottled, pickled or dried foods keep without energy-powered equipment.

or exchange the excess. Excess production can be avoided by staggering planting times so fewer vegetables are maturing at once. This works well for lettuces and carrots.

Maintaining special features

Potted plants

Potted plants can make attractive features in gardens or provide conditions for growing species that cannot grow in the existing soil. However, they can be hungrier for resources, requiring regular repotting, and more frequent watering and fertilising than plants in the garden soil because the volume of roots is restricted. Consider too that imported decorative ceramic pots come with greater environmental impacts from transport, and plastic and concrete pots have high embodied energy. They all contribute to CO_2 emissions. The main constituent of potting media is often composted pine bark, which requires energy for harvesting, transporting and processing, and it is packed in high embodied energy plastic bags. Lower numbers of potted plants is worth considering.

Rain gardens

Rain gardens are relatively low maintenance. In general, the plants are selected to form an appropriate cover, surviving well in local conditions and with periodic inundation. It is important to:

- Keep overflow drains clear of any obstruction.
- Top up mulch if required, or consider using inorganic mulch (like stones) which doesn't float and cause drain obstruction as organic mulches can.
- Filter out, or divert, very silty water which can reduce the infiltration rate.
- Periodically check the effluent water for contamination (e.g. heavy metals) if there is a risk the catchment water is polluted. This may indicate the rain garden can no longer absorb the pollutants and will need to be replaced (likely to be after many years if at all).

Green roofs

Properly designed and installed, many green roofs need minimal management once established. *Intensive* green roofs, especially those accessible as outdoor recreational roof gardens, may require more maintenance than *extensive* ones due to the style of plantings and the increased structural complexity required for their greater weight (more growing media, larger plants, planters, other garden structures). To maintain green roofs:

- Ensure the green roof has a warranty and consider including a maintenance schedule for the establishment period.
- Weed, irrigate and fertilise the plantings when necessary during establishment. Beyond this period, extensive green roofs with well-chosen plant species may only require irrigating during extended drought.
- Inspect and maintain extensive roofs at least twice a year, and intensive roofs more frequently. Include:
  - checking the drainage system.
  - checking roof membranes for leaks, especially at seams.
  - topping up the growing media where required, removing weeds, replacing dead plants and fertilising.
  - assessing the success of plant combinations and adjusting the proportions of different species if necessary.
- Inspect and maintain intensive roofs similarly but more frequently. They may need
  - ongoing and higher levels of watering and fertilising.
  - pruning, perhaps mowing, sweeping and raking, especially when they are regularly used recreationally.

Green walls

Ideally, climbers that form green walls will have a slow to moderate growth rate and light stem frameworks to reduce the weight the support frames need to carry. Frequency of pruning away from windows, roofs, guttering and other building features depends on the species or cultivar grown, and the growing conditions. Consider the following:

- Lower fertilising and irrigation rates to slow growth.
- Checking fittings of frames to ensure they are not corroding or damaging the masonry they are attached to.
- Pruning the climbers regularly to lower the weight on frames and to maintain an attractive, healthy screen of vegetation over the whole framework.
- Maintaining single-storey green walls will be easier. Higher screens may need work-safe platforms or travel towers for each maintenance cycle, adding to costs and energy use.

How sustainable are lawns?

Quality lawns are established in well-prepared soils, require regular rainfall or watering, need frequent fertilising, must be kept free of broad-leaf weeds to favour the grasses, are aerated on a regular basis, and need weekly or fortnightly mowing during the warmer months. Lawns like this are rare in Australia, but when they do exist they are resource-hungry features. Even poor quality lawns mown once a fortnight during the six months or so of more rapid growth will use scarce energy.

Lowering demand for materials

Recycling and re-using pots

Plastic plant pots usually end up in landfill. To reduce or avoid wastage we can wash,

Figure 12.18 Bioplastic plant pots. (Photo: Plantic™)

sterilise and re-use the pots, we can recycle their constituent plastic, or we can avoid the problem by using an alternative environmentally friendly material. Recycling pots has been difficult, but systems are being introduced in some parts of Australia.

As pot turnover is fairly rapid, some manufacturers are now using plastics that are relatively short-lived; these photo-degrade, deteriorating and becoming brittle after exposure.

Another approach for pots is to use biodegradable materials. At present, this is a developing area for the nursery industry, and just starting to be used in the retail industry. Caution is needed when assessing biodegradable materials, as it is possible to use conventional plastics and have a very small (less than 10%) biodegradable additive that is sufficient to allow partial breakdown and a claim to biodegradability. It is important to ensure that products are truly biodegradable.

Biodegradable pots and tubes are made from materials like sphagnum peat, wood fibre and bamboo. Biodegradable bioplastic made from corn, potato and tapioca starch is beginning to be used (Figure 12.18).

Maintain structures and fittings

Regular maintenance of structures, fittings and furnishings extends their useful life, sometimes indefinitely. Less frequent replacement reduces resource use.

Sustainable garden maintenance summary

Some key ideas to guide sustainable garden maintenance:

- Being careful with how, when and where irrigation occurs can dramatically decrease water use and make it easier to manage with the water harvested on-site.
- Choosing, maintaining and using equipment well can lower (or eliminate) the need for fossil fuels.
- There are alternatives to synthetic non-specific chemicals for keeping pests at acceptable levels in gardens.
- Recycling the green wealth of a garden helps maintain soil quality and a healthy garden for pleasure or food.

The way we maintain gardens can lower environmental impact significantly, not least because garden maintenance occurs over many years. Even low impact tasks, repeated regularly, can have a large cumulative impact. The best maintenance is responsive to the needs of the garden, accounts for the changes that occur from day to day, season to season and over the years, and uses methods and equipment with least environmental impact.

SUSTAINABLE GARDENS, LANDSCAPES AND LIVES

Sustainable living

Since the 1960s, a road map for a sustainable future has been coming gradually into focus. Based on overwhelming scientific evidence, it is now apparent that achieving a fossil-fuel-free green economy will require a sustainability transformation that must gather momentum at all levels of the sustainability management hierarchy.

The global human population is expected to increase from the current 6 billion to reach about 9 billion by 2050. Over this period we need to find ways for the average global citizen to live well on less than half the present global average Footprint to ensure no more than 'one planet consumption'. Currently 20% of the world's population is estimated to consume 80% of global resources and, as the non-industrialised world aspires to 'catch up' with high-consumption Western lifestyles, human consumption will continue to take a heavy toll on its life-support system, the biosphere.

The world's resources are unevenly distributed across the globe. One simple outcome of sustainability accounting is to consider equitable shares per person for energy use, food, materials and water. That is, to improve the standard of living of relatively poor nations and to moderate the high consumption levels of affluent countries. A simple challenge to us as individuals is to consider what are acceptable levels of consumption that provide quality of life, a sense of well-being and protection for the biosphere.

In the non-industrialised world it will mean raising living standards by reducing the rate of population growth while at the same time adopting sustainable technologies (effectively bypassing the environmentally harmful technologies of the West) and exploring ways to increase the Earth's biocapacity without putting further pressure on natural biodiversity and resources.

Response times of human populations are slow so, even with falling birth rates, populations will continue to expand as life expectancy increases. Infrastructure, too, will last many decades so the present population and infrastructure will shape resource consumption for most of this century. Current projections based on moderate population growth indicate a doubling of human consumption of the Earth's bioproductivity by 2050 and this gives us a measure of the challenge ahead.

We need more thorough environmental and resource accounting at all action levels to identify effective ways to decouple the connection between economic growth and environmental deterioration. The task for sustainability science is to provide ways of reducing consumption with the least 'pain' and to provide tools that relate consumption patterns meaningfully to environmental impact. Sustainability science itself requires greater integration of expertise from many disciplines and areas of life from politics, social science, urban planners, engineers, agriculturists, horticulturists, etc. Scientists, especially in biologically related fields, can explore ways to increase the Earth's biocapacity by protecting and restoring ecosystems to maintain biological productivity and ecosystem services while avoiding risky technologies, and without putting further pressure on biodiversity.

As individuals the challenge is to reduce the resource intensity of our activities. Our determination for economic growth must be matched by an even greater determination that this should be sustainable; that the global eco-footprint must be less than the Earth's capacity for renewal; and that well-being should not depend on material accumulation, and that quality of life also comes from community service and activities, intellectual, recreational and cultural pursuits and personal development.

We need sustainable cities, buildings, construction techniques and transport systems.

As nature and resources diminish, their value will inevitably increase and 'nature' will become a more market-based commodity. Environmental costs are already being built into the economy using market reform (e.g. carbon, water and nitrogen trading and taxes, conservation banking and economic acknowledgement of eco-assets).

We can experiment more with resource use at the individual and local levels (vegetable gardens, water tanks and solar panels) rather than expecting massive infrastructure projects (dams, power stations, industrial agriculture) to fulfill our needs. In this way we accept greater personal responsibility for our resource use while becoming more self-reliant.

The ideas of embodied resource use, resource intensity and resource productivity will gather importance. Our direct use of energy (grid) and water (mains) as it appears on household bills is small compared to the energy and water embodied in the goods and services we use. Green purchasing (ethical consumerism) can play an important role here.

Dematerialisation and zero waste with continued emphasis on 'reduce, re-use, recycle' and cyclical materials metabolism, assisted by green design and manufacture, can change the way we manage the physical objects around us.

Most of our individual environmental impact occurs through the resource use associated with food consumption. A diet of reduced (or no) meat and dairy and the encouragement of sustainable organic agriculture or permaculture production methods combined with home/community grown food has the potential to make major environmental savings and this carries with it the potential to reclaim land for nature.

As global water management becomes more focused on the green and blue components of the water cycle and implementation of demand management, individuals can make a big difference by harvesting their own water and plumbing it into toilets and laundry.

In the final analysis sustainability transformation is about engagement and willpower. Our response to climate change is a good sustainability indicator for environmental stewardship. Sustainability accounting by the world's foremost climate scientists has indicated that there is a 90% probability that the enhanced greenhouse effect is due to human activity and also that high income nations must reduce their greenhouse emissions by 60–90% from current 2006 levels by 2050 to avoid CO_2 levels rising above 475 ppm (which would produce a temperature rise of more than 2°C, a dangerous level of climate interference) and further, that the current rise must be arrested by 2015 to avoid disproportionate future effort.

Is this possible? What can we all do now to help? As individuals we can head towards carbon neutrality by using public transport, buying 100% green power, installing solar panels, and encouraging renewable energies and efficiencies in all walks of life.

Sustainable horticulture

The following general principles underlie much of the philosophy concerning environmentally aware horticulture.

Ecological and systems thinking
(see Chapter 2)

Permaculture and other movements related to sustainable horticulture encourage holistic thinking – an awareness of cultivated urban space as an (eco)system of processes, interactions and resource flows that connect with the wider environment. The goal of management is to integrate gardens with the wider environment to encourage a healthy, functioning 'whole'. And, in general, the more diverse the system the more resilient it is.

Self-sustaining systems
(see Chapters 1 to 8)

The biosphere is a closed system except that it is powered by the input of sunlight. A system is 'closed' when it can persist with no (or very few) inputs and outputs; it is then self-sustaining. It is possible for a small farm to be effectively self-sufficient with the only demands on ecosystem services being made and satisfied on-site.

Urban spaces are 'open' systems with inputs and outputs of energy, water, materials, chemicals and organisms. Inputs and outputs connect with both the human economy and the natural economy and have significance for everything from local ecology and biodiversity to global biogeochemical cycles and global ecosystem services. Even though they are open systems they can still aspire to a cycle of production and consumption that is in tune with the wider environment.

Production and consumption
(see Chapters 1 to 8)

The ecology of urban space is a microcosm of the Earth's biological cycles of production and consumption driven by sunlight – but on a smaller scale. Horticultural consumption uses resources from both the natural economy and the human economy. Natural resources are more 'renewable' and include sunlight, rainwater, carbon dioxide, plants, nutrients and organics. From the human economy there is non-renewable mains water, hard landscape, garden products and equipment, chemicals, fossil fuels and some soft landscape. All these items from the human economy contain embodied energy and water.

Urban space can be made more sustainable in exactly the same way that we make our lives more sustainable – by modifying its 'consumption pattern'. Some urban space consumption comes from the goods and services of the human economy and this can be managed in the ways suggested in this book. But greater environmental gains are to be made through management of horticultural primary production.

All living organisms depend on the primary production of plants. However, the primary production of the human economy, agriculture, does not provide food energy to the biosphere, only to humans, and in the process it takes a heavy toll on nature's ecosystem services. Crop and livestock production now takes up 30–40% of the planet's total ice-free land surface and around 25% of global net primary productivity. In Australia food makes up, on average, 48% of the total individual Ecological Footprint (it also makes up more than 36% of our water Footprint and 28% of our emissions Footprint). This has placed a huge demand on ecosystem services not only by removing land from nature but by using up resources like water (70% total global human water use, 67% total Australian water use) and fossil fuels.

Smart management of horticultural primary productivity can make the most of nature's renewable resources by directing horticultural primary production towards environmentally friendly objectives.

In other words, when answering the question 'if we are to become more sustainable, what should we be growing in the urban space that is available to us?' we can see three major opportunities:

- Urban agriculture.
- Indigenous biodiversity (land for nature).
- Carbon sequestration.

These objectives can, of course, be pursued in combination.

Urban agriculture (see Chapter 7)

Urban landscapes can reduce the environmental impact of agribusiness by taking over some of its primary production and, where possible, supporting the cultivation of crops and raising of livestock. This approach could range from growing fruit and vegetables in the backyard to converting under-utilised urban space, degraded and other land to community-based urban agriculture. Food produced this way would, in principle, release more agricultural crops for export and the relief of global poverty. Inevitably there are sustainability trade-offs: even community-based urban agriculture would require some resource inputs like water and organics.

Indigenous biodiversity, land for nature (see Chapter 8)

Sustainability from increased biodiversity and ecosystem services can be secured by regaining for nature some of the land lost to the human economy. This can be done by growing indigenous plants in gardens, along creeks and in public space. Growing indigenous plants is a way of linking back into the natural ecology of an area and developing a renewed sense of place.

Carbon sequestration
(see Chapter 4)

There is also the capacity for carbon sequestration in the woody tissue of trees, although this would make only a small contribution relative to the major reforestation and plantations that are needed to alleviate climate change.

Of course, sustainability values must be balanced with other values we choose for our gardens.

Figure 13.1 summarises the various inputs and outputs of a garden that link into the human and natural economies.

Horticultural sustainability accounting
(see Chapter 3)

Chapter 1 emphasised the onset of a period of sustainability accounting for all human activities: to manage we must measure.

Horticultural sustainability accounting (like all sustainability accounting) is in the very early stages of development and likely to become much more sophisticated in future.

The environmental impact of a site can be calculated by assessing the combined effect of its individual inputs and

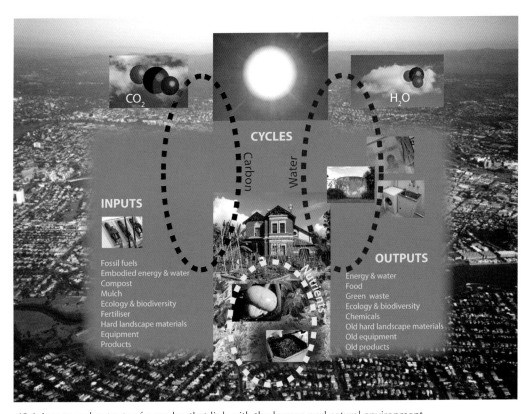

Figure 13.1 Inputs and outputs of a garden that link with the human and natural environment.

outputs (its consumption pattern). This can be done as a sustainability budget or audit. A sustainability budget compares the sustainability losses of horticultural consumption and the sustainability gains of horticultural production.

These losses and gains will depend, in turn, on the design of the site and the way it is managed. Both of these factors can be guided towards sustainability goals and need to be included in any sustainability assessment of urban space.

Some major goals and limitations of an audit are discussed below and in Info Box 13.1.

An audit for landscape can rank major elements of horticultural primary production (its biocapacity) and resource consumption (the Footprint) of a site, indicating the major physical inflows and outflows. Inputs and outputs are given individual sustainability rankings which together give an overall sustainability ranking. The rankings given will depend on the quantities involved and the conditions at the particular locality. For example, the sustainability of mains water use will depend on how much is being used relative to the available supply. An ideal complete audit would include quantitative estimates of all the inputs and outputs over a given time.

Various strategies, alone or in some combination, can be employed to increase sustainability. Try to:

- Aim for a self-sufficient closed system by reducing inputs and outputs to a minimum while making the site as self-maintaining as possible.

INFO BOX 13.1: A PROPOSED SUSTAINABILITY AUDIT

- It is a measure of inputs and outputs according to selected sustainability categories, and includes an assessment of their collective and individual impacts.
- It includes measures of comparability; while giving a ranking of sustainability at a particular site it also allows for the comparison of sites.
- It provides management guidelines based on the results (and prioritised where possible).
- It takes account of absolute results – though any audit will encourage good sustainability practices, priority must be placed on the absolute results achieved. For example, in a drought-prone region a garden that is designed to survive on natural rainfall alone is completely water-sustainable unlike one that uses mains water (no matter how small the volume of mains water used or how smart the way the mains water is managed).
- It rests on on-balance decisions – some activities affect the use of several resources and on-balance decisions then become necessary and they may well depend on local circumstances (shade trees cool houses saving the need for air conditioners but may also increase garden water demand).
- It is applicable to local circumstances – for example, water may be plentiful at some localities and therefore not a sustainability issue.
- It is flexible in content and level of detail – audits will be constructed around the needs of the user: landscape professionals need more detail than the home gardener.

- Find a balance where consumption inputs and production outputs have maximum positive environmental impact and minimum negative environmental impact.
- Maximise sustainable horticultural production (from the renewable natural resources of sunlight, carbon dioxide and rainfall) while minimising consumption from the human economy.

The design, construction and maintenance of landscapes

It is through site-responsive design that we can bring all these ideas together so the environmental impact from using water, energy and materials in gardens is low, natural biodiversity is encouraged and food is harvested. Good designs promote healthy self-sustaining garden ecologies. These complement sustainable communities which are encouraged and supported through the facilities and infrastructure provided, and through sensitive urban design. A summary guide to sustainable landscape design is found in Table A9 in the Appendix which can be used as reminders of the considerations that can be made at each stage.

During the construction process take care of healthy soil, the remaining vegetation and the surrounding environment. Minimise waste and construct gardens for the long term using quality, low environmental impact materials that last. A summary guide to sustainable landscape construction is found in Table A10 in the Appendix.

Gardens are monitored and maintained for a healthy garden ecology, remembering that how we maintain can significantly reduce resource use over its long life. A summary guide to sustainable landscape maintenance is found in Table A11 in the Appendix.

Where does this lead us? What can a sustainable garden look like?

… Karkalla …

Karkalla

The landscape designer, Fiona Brockhoff, and her partner, David Swann have created a beachside garden on the Mornington Peninsula which demonstrates beautifully the fusion of excellent garden design with sustainability. The garden, known as Karkalla, is approached along a gravel drive, the undulations reflecting the tertiary dunes of the site and gently reminding the visitor to 'slow down', as they enter another world.

Figure 13.2 Fiona Brockhoff.

Figure 13.3 Indigenous plants used in the ocean facing courtyard. Water tanks are a design element of the house and garden.

Figure 13.5 Local limestone is used sculpturally.

Locally sourced fine granitic gravel links the outdoor living areas, and rounded formally clipped plants, some trained as standards, combine with other plantings to define spaces. *Alyxia buxifolia*, Sea Box, *Correa alba*, White Correa, *Allocasuarina verticillata*, Drooping Sheoak, and *Austrostipa stipoides*, Prickly Spear Grass, all indigenous to the area, become feature plants through their placement and shaping and yet remain as habitat for fauna. Local limestone is used for

Figures 13.4 Indigenous plants become striking visual elements in the garden with careful shaping.

The garden contains a carefully selected, limited range of indigenous, non-indigenous Australian and exotic plants, these maintained formally close to the house, and further away, their forms sculpted by natural forces rather than human, and merging with surrounding indigenous vegetation of the adjacent Mornington Peninsula National Park. *Leptospermum laevigatum*, Coast Teatree, *Leucopogon parviflorus*, Coast Beard Heath, *Pimelea serpyllifolia*, Thyme Rice Flower, and *Melaleuca lanceolata*, Moonah, contribute to this landscape and habitat for indigenous fauna.

Figure 13.6 The garden near the house blends into the surrounding indigenous vegetation.

Figure 13.7 Copper bird baths, designed by Fiona, and sculpture decorate the garden.

Figure 13.9 Sculpture by Victorian sculptor Jonathan Leahey (steel, 2004) extend the garden from its formal core.

Figure 13.10 Gravel sourced from a nearby quarry has a sand-like quality, low embodied energy and being porous is good for rainfall infiltration.

Figure 13.8 *Koonya Beach Columns* by New Zealand sculptor Chris Booth (limestone, 1997).

Figure 13.11 Glass fragments collected from beaches and placed on the gravel surface complements the pattern of leaves of *Alyxia buxifolia*.

Figure 13.13 David Swann developed the distinctive patterning in the limestone walls.

Figures 13.12 Limestone walls extend through the house into the garden. The stone is local.

The garden requires very little maintenance. The shrubs are pruned a couple of times a year, the gravel generally stays clear of debris and is raked infrequently. Irrigating the ornamental plantings rarely occurs after the initial watering-in when planting. Sandy soils without irrigation limit weed establishment and growth. The orchard is drip-irrigated with greywater. All green waste is returned to the garden as compost and organic mulch.

walls and path construction, being a traditional building material for this part of the Peninsula. The limestone walls extend from the house interior into the garden, emphasising the almost seamless flow between these spaces. Recycled pier timber, reminding us of the coastal location, is used for retaining walls, and to define walkways within broad sweeps of gravel. Locally sourced and recycled materials ensure the garden has a strong sense of place, blending with its surroundings and, importantly, reducing the garden's Footprint.

Figure 13.14 The limestone path at the top of the vineyard is constructed in a traditional way used in the region since the early days of European settlement.

Figure 13.15 Service areas can be attractive and functional. Herbs are planted under the washing line. Recycled pier timber placed within the gravel defines a walkway.

Figure 13.16 Olive trees are pruned for views, their trunks reminiscent of the gnarled older olives in the Mediterranean, and of Snow Gums in Australia. The fruit is harvested for pickling which avoids the seed spreading and naturalising.

Figure 13.17 The kitchen garden produces throughout the year. Enclosed by a fence of Teatree branches, the raised rectangular beds are retained by pier timbers.

Figure 13.18 Woven willow edging further protects the flowers and vegetables.

Powered equipment is rarely used and synthetic chemicals are avoided.

Rainwater is harvested from roofs, the four silver corrugated iron tanks becoming sculptural elements as well as functional ones. Greywater is re-used for irrigating the orchard. A dry composting toilet (Rota-Loo) reduces household water use and provides high quality compost for soil improvement.

The landscape also provides food from the kitchen garden, orchard, herb area (under the clothes hoist) and egg-laying hens. Care of the hens and the collected eggs is shared with the neighbours, and produce from the kitchen garden is exchanged. A small Pinot Noir vineyard produces around 300 bottles of wine in a good year.

The garden is creatively designed – interesting, comfortable and functional, providing a relaxed and stimulating setting for coastal family living with a low environmental cost.

Karkalla is open to the public through the Australian Open Garden Scheme.

Figure 13.19 Compost bays recycle vegetable matter from the garden.

Figures 13.20 The hen house is within the kitchen garden. Water is collected from the roof.

13 – SUSTAINABLE GARDENS, LANDSCAPES AND LIVES

Future developments

There is the need for much more empirical information to help gardeners and landscapers design, construct and maintain landscapes; to choose the most sustainable methods and materials; and to audit landscapes more accurately.

LCAs and studies like them are needed for many more products and would help immensely. The real environmental impact of using many landscaping and horticultural methods and products is yet to be identified.

Designers and those constructing landscapes will also benefit from design software that easily:

- Compares the embodied energy in alternative products that can serve the same function.
- Gives a running total of embodied energy as different elements and materials are included in a design.

Developing software tools like this, however, will largely be dependent on completing the LCA studies that necessarily underpin it.

Example trends

In these early days of sustainability we often don't have access to the technology that may reduce resource use or it may be prohibitively expensive. However, the need for sustainable solutions is fuelling the rapid growth of ideas, research and technological development, and industry is responding with smarter, less resource intense products. Who would have thought just five years ago that a walk around suburban streets would find garden solar lights a frequent feature? Improvements may still be needed to increase their lighting efficiency, often being more decorative than functional, but as with many products refinements come as the market develops.

Global research into the future of urban lighting identified energy management as an urgent priority particularly with the increasing use of light.[1] Urban areas with good lighting, although energy consuming, can encourage energy saving walking and public transport use. So cutting back lighting is not necessarily the way forward, although light pollution has been identified as an increasingly important issue needing innovative design solutions to minimise its effects. Technological improvements for the future include the further development of LEDs, and visionary ideas about wireless lighting allowing lights to be easily moveable as lighting needs change.

A design collective, Civil Twilight, won the 2007 Metropolis design competition for its concept of Lunar Resonant Street Lights. Dimming under full moons and increasing light output as clouds build up or under new moons, they claim gigawatts of power could be saved in cities. The technology, if developed, could be applied to exterior home lighting too, when constant light is needed through the night for safety or other reasons.

Energy efficiency of many appliances has already increased, and renewable energy options are more accessible with power companies offering green energy alternatives or accommodating electricity being fed back into the grid from photovoltaic roof panels. Micro-hydropower, using the energy inherent in gravity fed mains water reticulation systems, could potentially contribute to the electrical power grid of cities and towns or possibly provide electrical energy more locally for parks and gardens, community street lighting or houses.

A solar powered public rubbish bin, trialled in New York in 2005, compacts the rubbish when it reaches a certain volume, allowing the bin to hold four times the amount of standard bins. In the trial the frequency of bin emptying was reduced by about 70%, which reduced fuel use by the trucks picking it up. The problem of overflowing litter was also lessened.

Water storage alternatives are now giving people a wide choice of systems that suit individual needs. Water storage can serve other functions like giving a building greater thermal mass. Paul Morgan, a Melbourne architect, made a feature of a water tank inside the living area of his Cape Schanck home, the body of water passively cooling in summer (Figure 13.25).
The tank provides water for some household needs and irrigation.

Evaporative losses from open water bodies can now be reduced by using a silicon-based liquid (Aquataine) that forms a molecular mono-layer across the surface of the water and therefore a barrier. It is claimed to be safe for drinking water and can be used in swimming pools.

Hundreds of kilometres of black poly pipe are replaced yearly when orchards and their irrigation systems are improved. The irrigation company Netafim and Yandilla Park have trialled a mechanical retrieval system for removing black poly pipe from orchards in the Riverland for recycling in Adelaide rather than ending in landfill – a system, if successful, that could be applied to urban landscape irrigation systems too.

Figure 13.21 A water tank in the living area passively cools Paul Morgan's house at Cape Schanck. (Photo: Peter Bennetts)

Plant tolerances will be more thoroughly researched and documented in readily accessible databanks, making plant selection for regional differences in climate and soils easier. Complementing this will be breeding and selection programs with greater emphasis on those qualities leading to low environmental impact when ornamental plants are grown, including reduced water and nutrient needs, and greater disease resistance. Already low water demand turf grasses have become commercially available and further research is continuing. Related to this, is the continued development of food crops from the Australian native plant pool; species that are adapted to Australian conditions and requiring much lower resource input to be productive.

Urban landscape design and horticulture will become increasingly ecologically based with greater understanding of the interrelationships between organisms, of how urban plantings can enrich biodiversity and also contribute to the linking of refuges of natural habitat.

It is up to us

Urban horticulture has a valuable role to play in the overall management of environmental sustainability. Most of its environmental impact is small – this applies to:

- Energy use.
- The amount of carbon that can be sequestered.
- The volume of water used (although this can be quite large as a proportion of urban use).
- The relative quantity of materials used.

However, parks and gardens can act as catalysts and models for the encouragement of sustainable lifestyles and help to make our lives more ecocentric.

Importantly, it is the contribution that can be made by plants themselves, the primary producers of the suburbs, that must be taken more seriously by increasing the food productivity of our urban landscapes and re-establishing indigenous flora and fauna.

The biosphere is at the same time both complex and simple. As a balanced life-support system it is the most unfathomable, intricate and miraculous product of the universe that we know. And yet, from its very beginning over 3 billion years ago, life has been sustained by the Sun's energy captured in carbon compounds built up in plants from carbon dioxide and water to be passed through the food chain. Humanity now controls the global distribution of these simple chemicals – and how we manage them will determine both our future and that of all life.

APPENDIX

Table A1 Hypothetical weekly energy use for a medium-size (150 m^2) Australian garden

Energy use	Energy use rating	Energy use (MJ/week)
INFLOWS		
Power tool fuel		
Handheld blower	¼ hr @ 14.2 MJ/hr	3.55
Mower	¼ hr @ 16.9 MJ/hr	4.2
Chainsaw	¼ hr @ 5.8 MJ/hr	1.45
Whipper-snipper	¼ hr @ 2 MJ/hr	0.5
Total		9.7
Chemicals		
Fertiliser		Negligible
Pesticide	3 apps on 75 m^2 @ rate of 1 kg/ha and ~ 263 MJ	0.025 Negligible
Mains water	Embodied energy of treatment and infrastructure	Negligible
Organics		Negligible
Mulch (off-site) and compost		Unknown (small)
Food scraps	2 kg peelings etc.	?
Increase in 'permanent' vegetation such as a growing tree	Each kg gain in fresh weight equates to about 4.65 MJ energy	5
Human labour	1 day light work	4
Hard landscape		
Energy costing of area with pool, decking and paving, labour, construction	12 218.5 MJ – assuming life span of 15 years = 814.6 MJ/yr	15.7
Manufacture of tools and machinery		Unknown
OUTFLOWS		
Organics garden produce		
Vegetables Asparagus Onion Potatoes Carrots Cucumber Lettuce Spinach	Annual production (kJ) ½ kg = 247 2 kg = 2920 10 kg = 29 260 2 kg = 2672 ½ kg = 209 ¼ kg = 160 1 kg = 138 Total c. 35kJ	0.0006 MJ
Green waste	Processed on-site (possible use of chipper)	?
Hard waste		?
Total*		c. 30
*Solar radiation over garden***		Total solar 2268

* this is extremely low relative to other energy use, e.g. a home of 4 people uses about 32 kWh (115.2 MJ) per day.
** In Melbourne with a mean annual solar radiation of about 4.2 kWh/m^2/day total solar energy input per week is:
150 (area of garden in m^2) x 4.2 (kWh/m^2/day) x 3.6 (conversion to MJ) x 7 (week) = 2268 MJ

Table A2 Energy content of vegetables and fruit*

	cal/item	cal/portion	cal/kg	J/item	J/portion	J/kg
Apple	44	–	440	184	–	1839
Asparagus	–	–	260	–	–	1087
Aubergine	–	–	150	–	–	627
Avocado	150	–	–	627	–	–
Banana	107	–	650	447	–	2717
Beetroot	–	–	380	–	–	1588
Carrot	–	–	320	–	–	1338
Celery	–	–	80	–	–	334
Cucumber	–	3	100	–	12	418
Dates	12.5	100	2350	53	418	9823
Grapefruit	100	–	–	418	–	–
Grapes	2.4	55	620	10	230	2592
Kiwifruit	34	40	500	142	167	2090
Lettuce	–	4	150	–	17	627
Mango	40	–	–	167	–	–
Melon	–	14	280	–	59	1170
Mushroom	–	–	120	–	–	502
Olives	6.8	50	800	28	209	3344
Onion	–	–	350	–	–	1463
Orange	35	40	300	146	167	1254
Peach	35	–	300	146	–	1254
Pear	45	–	380	188	–	1588
Pineapple	50	40	400	209	167	1672
Plum	25	30	390	105	125	523
Raisins	5	–	–	21	–	–
Peppers	–	–	180	–	–	752
Potatoes	–	–	700	–	–	2926
Radish	–	–	130	–	–	543
Spinach	–	8	80	–	33	138
Strawberries	2.7	10	300	11	42	1254
Sweet potatoes	–	–	900	–	–	3762
Sweetcorn	–	70	700	–	293	1225
Tomato	–	30	200	–	125	836
Watercress	–	5	200	–	21	836
Yam	–	–	1100	–	–	4598
Zucchini	–	–	200	–	–	334

Source: www.weightlossforall.com/free-food-calorie-counter.htm.
* the value of a food to the body as a fuel may be less than the heat value obtained experimentally by 'burning' the food outside the body in a bomb calorimeter because some of its energy may be 'lost' during digestion and metabolism.

Table A3 Materials in landscape categories ranked in order of increasing embodied energy per kilogram

Category	Material	kilojoules/kg
Fertiliser	Fertiliser (potassium magnesium sulphate)	5000
	Fertiliser (potassium sulphate)	5000
	Superphosphate	6266
	Fertiliser (superphosphate)	8500
	Fertiliser (ammonium sulphate)	58 000
	Ammonium	64 023
	Fertiliser (liquid UAN – liquid urea ammonium nitrate)	65 000
	Fertiliser (ammonium nitrate)	67 000
	Fertiliser (urea)	70 000
Irrigation, drainage	Plastic – PVC	80 000
	Plastic – general	90 000
	Plastic – polyethylene, HDPE and LDPE (poly pipe)	103 000
Metal	Steel reinforcing	8900
	Steel recycled	10 100
	Wire rod	12 500
	Galvanised steel	38 000
	Aluminium recycled, anodised	42 797
	Copper	100 000
	Copper pipe	169 592
	Aluminium	170 000
	Aluminium anodised	202 522
Miscellaneous	Aggregate river	20
	Sand	37
	Aggregate – general	100
	Cement – mortar	2000
	Concrete – precast, steam cured	2000
	Aggregate crushed stone	2155
	Fibre cement sheeting	4800
	Cement – dry powder	5600
	Aggregate – local dimension granite	5900
	Glass	12 700
	Aggregate – imported dimension granite	13 900
	Glass – tinted	14 900
	Glass – float	15 900
	Glass – laminated	16 300
	Glass – toughened	26 200
	Bolts – steel	61 712
	Rubber – natural latex	67 500
	Nails – steel	78 880
	Adhesives and sealants – epoxy and resins	81 200
	Rubber – synthetic	110 000
	Plastic – polycarbonate glazing	158 224

Category	Material	kilojoules/kg
Paint	Water based	88 500
	Oil based	98 100
Paving	Concrete – ready mix	1000–1600
	Concrete paver	1200
	Concrete – *in situ*	1900
	Asphalt paving	3400
	Paving and quarry tiles	20 919–53 096
Softworks	Earth – dry loose	399
	Earth – dry packed	399
	Earth – stabilised	700
Stone	Local	790
	Uncut	1035
	Granite cut	5886
	Imported	6800
Timber	Softwood air dried sawn	300
	Rough cut	378
	Air dried sawn hardwood	500
	Air dried dressed softwood	1160
	Kiln dried sawn softwood	1600
	Kiln dried sawn hardwood	2000
	Kiln dried dressed softwood	2500
	Kiln dried sawn softwood	3400
	Plywood	10 400
	Glue-laminated	11 000
	MDF	11 500
	Masonite	24 200
Walls – masonry	Concrete blocks	1500
	Bricks – clay	2500
	Concrete blocks – aerated	3591
	Concrete blocks – autoclaved, aerated	3600
	Bricks – clay glazed	7200

Source (blue): 'Building Materials Energy and the Environment: towards ecologically sustainable development', B Lawson 1996 RAIA Canberra; and 'The Embodied Energy of Living', G Fay Treloar, EDG GEN20 RAIA. via www.greenhouse.gov.au/yourhome/technical/fs31_3.htm or www.yourhome.gov.au/technical/fs31_3.htm
Source (tan): Embodied energy coefficients, Centre for Building Performance Research, Victoria University, Wellington, NZ
Source (black): 'Sustainable Landscape Construction: a guide to green building outdoors', J William Thompson & Kim Sorvig, Island Press Washington DC & Covelo California, 2000, ISBN 1559636467
Source (grey): Energy information handbook, Appendix C, Energy information document 1028, Institute of Food and Agricultural Sciences, University of Florida 1991

Table A4 Materials that are used in landscaping (largely from Landscape Industries Association of Victoria (2004) A guide for landscape works schedule of rates 2004/05, published by the Landscape Industries Association of Victoria, Hawthorn East, Victoria)

Category	Product
Hardworks	Concrete • Expansion joints • Reinforcement • Formwork
	Walls • Bricks • Concrete blocks
	Retaining walls • Concrete blocks • Timber
	Mortar
	Paving • Asphalt • Brick • Concrete • Clay • Gravel
	Crushed rock
	Pebbles
	Stone
	Step treads and risers
	Pool coping
	Edging
	Timber
	Fencing • Star pickets and wire • Timber
Drainage, irrigation and water features	Subsurface drains • Flexible coil • Slotted PVC • Filter sock • Gravel, sand

Category	Product
	Surface drains • Stormwater PVC • Grates (galvanised steel, iron, plastic) • Trench (plastic)
	Irrigation • PVC pipe • Poly pipe • Drippers • Controller • Solenoids • Wiring
	Water features • Fibre glass pond • Pond liners • Filter • Pump • Lights
Lighting	Path lights/step lights/bollards Wall lights Pond lights Transformers Low voltage cable
Softworks	Topsoil
	Compost
	Fertiliser
	Soil conditioners • Gypsum • Water crystals
	Mulch • Organic • Non-organic (pebbles etc.) • Hydromulching • Straw and bitumen
	Staking and tying
	Turf

Table A5 Energy use in food production to consumption in USA 1980

Activity	Total gigajoules of energy use in USA	% total energy use in USA
TOTAL ENERGY USE OF USA	80 000	100
CROP PRODUCTION:		
Primary crop production	2160	2.7
Total agricultural production	3086	3.9
Pesticides	34	0.04
Total crop production	**5280**	**6.7**
FOOD PROCESSING AND DISTRIBUTION:		
Food processing	2719	3.4
Drink processing	662	0.8
Distribution (to and by retail)	1558	1.9
Cars used for food retail business	732	0.9
Consumer shopping	1894	2.4
Food preparation and cooking at home	3051	3.8
Food preparation and cooking in restaurants etc.	1666	2.1
Garbage and sewerage disposal	323	0.4
Construction and maintenance of food related buildings	1173	1.5
Total food processing and distribution	**13 778**	**17.2**
TOTAL FOOD RELATED ENERGY	**19 058**	**23.9**

Source: Tables 7.5 and 7.6 combined, page 172, section 7, Chapter 7, 'Energy in plant nutrition and pest control'. In: *Energy in World Agriculture*. Volume 2. (Ed. Zane R Helsel) Elsevier, 1987.

Table A6 Potentially weedy plants

Plants that grow in a broadly similar climate and are therefore likely to naturalise	
Plants resistant to drought and therefore likely to grow well if naturalised	
Cacti and succulents, grey foliage plants, herbs	
In many parts of Australia we must be wary of plants from South Africa, California, the Canary Islands and the Mediterranean	
Genera with species that produce fruits that are known to be spread by birds and animals: *Pittosporum, Cotoneaster, Crataegus, Prunus, Pyracantha*	
Plant families with many representatives known to be weedy:	
Iridaceae	many South African species with known weed potential
Poaceae	many are seedy annuals (about a third of Australia's grasses are introduced)
Asteraceae	many are seedy annuals
Plants which produce prolific propagules (that are often easily transported):	
Seeds	Poaceae, Asteraceae
Stem fragments	*Tradescantia*
Leaf propagules	*Bryophyllum, Crassula tetragona*
Tubers	*Helianthus tuberosus* (Jerusalem artichoke)
Corms	*Watsonia*
Bulbils	*Oxalis, Cyperus rotundus* (nutgrass)

Table A7 Commodity impact for Melbourne, 1998–1999

Rank	Commodity	Impact (gha/person)	% total ecofootprint
1	Clothing	0.68	14.41
2	Retail trade	0.55	11.66
3	Accommodation, cafes and restaurants	0.54	11.40
4	Beef products	0.46	9.73
5	Electricity supply	0.20	4.20
6	Raw sugar, animal feed, processed seafood and other products	0.19	3.96
7	Dwelling ownership	0.14	2.97
8	Dairy products	0.13	2.86
9	Carpets, curtains, tarpaulins, sails, tampons, other textiles	0.13	2.71
10	Bread, cakes, biscuits, bakery	0.10	2.19
11	Petrol	0.10	2.09
12	Footwear	0.08	1.60

Source: Victorian Department of Sustainability and Environment (2006) *Melbourne Atlas*. Department of Sustainability and Environment, Melbourne.

Table A8 Garden pests and diseases and their control

Pest or disease group	Pest	Causes and symptoms	Control
Insects	Scales	Scales are sap-sucking insects which are easily seen on the leaves and sometimes branchlets of susceptible plants. There are many species, some classed as hard and others soft. Hard scales produce a characteristic hard waxy shield under which they shelter, and soft scales without the shield. Sugary secretions caused especially by soft scales are ideal for the growth of fungi, like sooty mould which in severe cases can block light from reaching chlorophyll, reducing photosynthesis. Ants also like the sugary secretions and can 'farm' the scale to promote this food source by moving the young crawlers to fresh sites.	Scales can often be kept in check by other insects including wasps, lacewings and ladybirds. Controlling ants can be important for reducing the spread of scales and so they do not deter natural predators of the scale. Small numbers of scale are easily picked off by hand. White oil, a product from mineral oil, can be sprayed over scale to suffocate it. White oil has a low toxicity; however, it does rely on the fossil fuel industry.
	Aphids	Sap sucking aphids can build up population numbers rapidly and colonise particularly fresh shoots or flower buds causing wilting or distortion of growth. Some aphid species attack the roots. Honeydew excreted from the digestive tract of aphids is a good food source for fungi including sooty mould. Aphids can carry plant viruses.	Aphid numbers build up most quickly when temperatures are in the high teens and low twenties and when it is humid. Some aphids, like the rose aphid and green peach aphid, are parasitised by naturalised wasps, with some of these parasitic wasps available from biological control companies (e.g. Biological Services). Ladybirds, lacewings and wasps can also control aphid numbers. Small infestations of aphids may be able to be squashed by hand or hosed off, but regular monitoring and squashing will be needed. Eco-oil is an emulsifiable botanical oil concentrate sold for the organic control of aphids containing canola and essential oils (www.ocp.com.au).
	Moths (e.g. codling moth, cabbage moth)	Caterpillars feed on leaves (cabbage moth) or fruit (codling moth).	In the case of codling moth avoid growing susceptible trees as vigilance is required to control the pest. Fruit damaged by the moth should be destroyed after removal. Site and tree should be kept clean, removing dead wood, flaking bark etc. Trees can be banded to act as a barrier for larvae. Pheromone traps can be used. For cabbage moth Dipel® (*Bacillus thuringiensis*) is an effective control. *Trichogramma*, a small parasitic wasp, can be used to control moth pests.
	Fruit fly (Queensland and Mediterranean)	Damaged results from the laying of fruit fly eggs under the fruit skin which also introduces fruit-rotting micro-organisms. After hatching, the larvae bore through the fruit's flesh.	Infested fruit must be collected from the ground or tree and burned, boiled or solarised in sealed plastic bags in the sun to kill larvae. Traps can be used to monitor for adults. Chemical control may be needed. *Clivia* species are recognised host plants for Queensland fruit fly, the larvae having been found in the fruits.

Pest or disease group	Pest	Causes and symptoms	Control
	Azalea lace bug	Azalea lace bugs are sap sucking insects found on the underside of leaves of *Rhododendron* (including azaleas). Yellow spots appear on the leaves and for heavy infestations, leaves become reddish-brown or bronze.	Select cultivars that are less prone to attack. Plants that are growing well are less susceptible to attack. Regular monitoring of the plants can detect the vulnerable nymph stage which can be suffocated by good coverage with horticultural oil (particularly of the underside of leaves) in a weekly spraying program (Crawford D (2004) Finding ways to control azalea lace bugs. *Australian Horticulture* 102(8), 8–9). Beneficial organisms do exist overseas, but not in Australia.
Arachnids	Two-spotted mite (red spider mite)	Sap sucking two-spotted mites are found on the underside of leaves where they feed. They are very small, being just visible with the naked eye, but they create a tell-tale woolly webbing underneath the leaf and a mottled silvering or yellowing of the upper leaf surface. Dry, warm conditions favour two-spotted mites.	Watering leaves can provide the less favourable moist conditions. Remove affected leaves or plants. Persimilis or Chilean predatory mite is a biological control agent for two-spotted mite. Eco-oil is an emulsifiable botanical oil concentrate sold for the organic control of two-spotted mite containing canola and essential oils. Note: Not all mites are plant pests and some are beneficial for plant health by preying on pest organisms, predatory mites being used in some biological control. Spiders, also arachnids, can be beneficial predators.
Nematodes	Bulb nematode	Leaves arising from bulbs may be distorted and with abnormal ridges. The fleshy bulb leaves are eaten, and become discoloured as they rot.	Only plant bulbs from a reliable source known to be free of nematodes. Don't plant susceptible bulbs in soil infested with nematodes. Destroy infested bulbs. Small numbers of nematodes infesting bulbs can be killed by exposing cleaned, lifted and stored bulbs to hot water at 43°C for 4 to 4.5 hours. Soil solarisation can be used to kill nematodes. (In summer, clear plastic is placed over watered soil, and its edges sealed. In a sunny hot position the process takes 5 or more weeks depending on the amount of hot weather.) Note: Only a few nematodes are pests of plants and some can be used in biological control systems.
Molluscs	Slugs and snails: common garden snail (introduced from Europe)	The common garden snail eats leaves, vegetable or flower seedlings, sometimes bark even, using its rasping tongue, leaving ragged holes and leaf margins. Snails leave telltale silvery trails.	Snails like the moist conditions of gardens, offering both food and shelter during warmer, drier weather. Controls can include reducing sheltering spots in the garden, identifying sheltering spots so the snails can be picked out and squashed, providing sheltering spots that we routinely check for snails (like flat boards in cool, damp areas near where they are likely to be), patrolling regularly in the evening, particularly after rain, when the snails are out, providing barriers around vulnerable plants like seedlings using crushed egg shells or diatomaceous earth. Copper bands are effective barriers due to the electric current formed as the snail's slime reacts with the copper (copper bands are commercially available to gardeners in the UK). Snails have natural predators like birds, frogs and lizards. Note: There are many Australian species of land snails, but they are not garden pests.
Birds	Blackbirds (introduced)	Fruits eaten.	Rat traps set on the ground in some localities almost exclusively trap blackbirds, but care is needed if other ground-frequenting birds or pets are present. The native blue wren, for example, can also be caught and traps should not be set if they visit the garden. Birdnetting, if installed well, can protect developing fruits.

Pest or disease group	Pest	Causes and symptoms	Control
Mammals	Possums	Fruits eaten; foliage of same plant repeatedly eaten threatening its survival.	Possums, protected animals, can be difficult to deter once they get a liking to a plant. Part of the control strategy can be to disrupt their normal pathways by banding trees with metal or plastic bands, placing flexible floppy plastic on fence tops, having the electricity authority place rotating pipe on wires, and pruning trees away from each other. Quassia chips and chili powder can make short-term deterrents, but need to be regularly reapplied. Ultrasonic equipment can be used effectively when the sound is directed to the plant being visited. It needs to operate only from early evening to dawn when the possums are active so using a timer is useful. Not all equipment sold produces the wavelengths needed to deter possums. Bird Gard (www.birdgard.com.au) supplies equipment that has effectively deterred possums by using complex and varying wavelength patterns producing sounds both audible and inaudible to humans. Some nurseries and home gardeners have found energised fence guards useful (like an electric fence that gives a mild electric shock).
	Rats	Fruits eaten.	Rat traps can be effective, avoids using chemicals and can be used multiple times.
Fungi	Powdery mildew	Powdery, diffuse white growth on the plant's surface. Growth is slowed and leaves eventually wither and die.	Use resistant varieties when available. Encourage good ventilation by spacing plants and pruning to keep canopies open. Remove affected parts or whole plants and burn to slow its spread. Avoid watering foliage. Sterilise secateurs etc after pruning infected plants. Peter Crisp at the University of Adelaide trialled a number of potential controls and bicarbonates were amongst the more successful along with 1:10 full-cream milk (trialled on grapes leaves and in earlier studies in Brazil, on zucchini). Eco-carb, with the active ingredient potassium bicarbonate, is sold as an organic control for powdery mildew (www.ocp.com.au).
	Black spot	Black spots develop on leaves, especially during humid weather. Leaves eventually yellow and drop off.	Choose cultivars that are resistant to black spot. Reduce humidity around leaves by heavy winter pruning, followed by a light summer pruning that opens up and aerates the foliage. Do not water leaves. Remove infected and fallen leaves, and burn to break the cycle. Sterilise secateurs etc after pruning infected plants. Bicarbonates have potential for controlling black spot.
Viruses	Passionfruit – woody fruit	Fruit walls become thick, woody and distorted. Leaves are mottled with yellow.	Viruses are difficult to control as they are often, as in this case, carried from plant to plant by aphids. Pruning equipment like secateurs can also carry infected sap from one plant to the next and should be sterilised between plants using 70% methylated spirits or household bleach diluted to about 0.5% available chlorine (about 1 in 10 depending on the concentration of sodium hypochlorite). Severely infected plants should be removed and destroyed by burning.
Bacteria	Apricot bacterial canker	The common bacterium, *Pseudomonas syringae*, gains entry when plants are damaged. Gum exudes, cankers form, branchlets may die, fruit and leaves may have dark spots.	Prevent damage to the tree. Prune during growing season so wounds can heal and seal. Regularly sterilise pruning equipment using 70% methylated spirits or household bleach diluted to about 0.5% available chlorine (about 1 in 10 depending on the concentration of sodium hypochlorite). Remove all infected branches and burn. Rain and overhead watering can help spread the disease.

Source: Lenzen M (2004) The 1998–99 ecological footprint of the population within the Port Phillip and Westernport Catchment. Sydney University; cited in: *Melbourne Atlas* 2006. See www.acfonline.org.au/custom_atlas/index.html for a more recent consumption profile of Melbourne.

Table A9 Summary guide providing sustainability prompts through the design process

Stage	Considerations
Choose a site appropriate for your needs if possible	Check that the: • soils are healthy • site is already with useful vegetation • site is suitable for growing proposed plants • project will not damage important natural environment including any nearby indigenous vegetation
Survey and analyse site; draw site plan	Note: • soil types including location of rocks that could impede plant growth • water including drainage lines, high dry spots and low collection points • vegetation to be retained • aspects (N, S, E or W) especially for hot/sunny or cool/shady spots • services or other features that may restrict planning • surveying in itself can cause damage to existing vegetation and soils • locals or contractors that have worked in a region for a long time can be a source of valuable knowledge
Develop and refine the design	Using garden design principles consider: • style of garden (some have more impact than others) • site uses (playing, outdoor eating, drying washing, growing food plants, service areas etc.) • overlapping uses for the same space (e.g. carparking space used as play space during the day; shade tree producing fruit) • relating to indoors (outdoor living areas like decks for eating adjacent to kitchen and family room etc.) • relating to surrounding landscape (other houses, natural areas) • placing plants in the best soils and locating buildings and hard landscape on the worst • storing rainwater collected from roofs (choose the optimum tank size for your climate and needs) • channelling surface water onto garden beds • using permeable paving to increase water reaching the soil • re-using water (greywater) • reducing water and energy hungry lawn areas • designing gardens without need for irrigating • installing efficient drip irrigation systems if irrigation is needed • ensuring the landscape design complements and certainly doesn't hinder the 'green' qualities of buildings • building less to reduce resource use • reducing the area of paving with high embodied energies, leaving more area for plants and their roots to grow healthily • designing to accommodate standard available sizes to reduce the need for cutting materials like pavers • locating services such as irrigation pipes and cables in the same trench to reduce site damage and energy used by trenching machines • using plants rather than masonry to stabilise slopes • using plants like trees and climbers to shade walls and windows from hot sun • installing a green roof to keep buildings cool in summer and warm in winter • placing plants according to their cultural needs, e.g. drought tolerant in drier areas, moisture loving in lower damp areas • whether informal hedges could replace high maintenance formal hedges • producing food in the garden • encouraging the indigenous biodiversity • ensuring the garden is not detrimental to the health of the surrounding environment

Stage	Considerations
Assess and select the materials to be used for their low environmental impact	Consider: • products and materials with lower embodied energy and water • equipment needing lower or no ongoing energy and water input • generating your own renewable energy (solar, wind) to offset garden and household use • using materials that are renewable and produced with minimal environmental impact and if possible, certified under recognised authorities • re-using or using recycled materials • using non-renewable materials sparingly and from 'best practice' sources • sourcing form local quarries, manufacturers and suppliers where possible
Create planting plan	Incorporate in the design: • plants ecologically suited to the area • plants needing little supplementary water in the area • plants requiring minimal fertiliser to grow well • disease resistant plants requiring preferably no chemical control • food plants • plants that support indigenous biodiversity • plants that will not become weeds • trees that provide shade for summer or sun for winter or in warmer latitudes shade all year • species and cultivars with low maintenance needs • species that are ultimately the appropriate size for the position • grouping of plants with similar cultural needs – especially water
Construct the garden	• develop low environmental impact specifications for contractors • choose contractors who practice low environmental impact construction methods

Table A10 Summary guide to constructing gardens sustainably, providing prompts and considerations that can be made

Construction area	Considerations
Construction	Plan to: • use landscape designs that are sustainable and suitable for the site • construct in an appropriate sequence • construct for the long term • use materials that can be easily recycled in the future Choose: • contractors trained in sustainable methods
Protect site and local assets	Consider: • working with natural processes, not against them • contributing to regional habitat conservation • restoring or improving damaged sites • supporting local suppliers to encourage a strong, diverse local economy which can also help reduce transport needs • restricting the use of heavy landscape equipment, and when used, it only affects a minimum area of site soil Protect: • healthy sites, especially with indigenous vegetation, and/or healthy mature trees • root systems of existing vegetation to be retained • site soil from compaction, erosion, pollution with chemicals • local waterways from pollution
Energy	Consider: • using green power • using materials with low embodied energy Remember to: • respect the sustainable energy features already existing on the site • use the lowest powered equipment suitable for the work
Materials	Consider: • building less • reducing resource use • using green purchasing • using low environmental impact materials • using materials appropriate for the situation • selecting quality materials that last for the long term • purchasing from sustainably run companies • sourcing locally produced materials manufactured to high standards (e.g. compost, mulch produced under Australian Standards with high hygiene, appropriate pH etc.) Consider site soil: • assess and improve quality if necessary (organic matter, pH, wettability (hydrophobicity) etc.) • test soil for nutrient content and fertilise only where necessary to bring nutrient levels up to that required by the plants to be grown (fertilise only in those areas where needed) • choose fertilisers with lower environmental impact (include their embodied energy) • avoid importing soil Source plants: • of high quality, that are well grown so they establish strongly in the new site and will not introduce diseases or weeds • that are indigenous from reliable sources which know and grow local provenances

Construction area	Considerations
Equipment	Consider: • installing low energy, low water equipment like pumps and water features
Water	Install irrigation systems: • only to parts of the garden that need irrigating • in zones that irrigate plants with similar water needs • that efficiently deliver water to plant roots (e.g. drip)
Waste	Consider: • producing minimal waste and pollution • recycling where possible and providing bins for sorting waste types • disposing of waste appropriately Contain and dispose of appropriately: • building rubble waste • chemical waste including herbicides, wash-up water from concreting etc.
Monitoring	After completion, ensure the new landscape is well monitored for: • adequate irrigation of the establishing plants • general plant health

Table A11 Summary guide to maintaining gardens sustainably, providing prompts and considerations that can be made

Maintenance area	Considerations
Change	Remember gardens change: • with time • with seasons and maintenance needs to reflect this
Equipment	If the tool can do the work, preference is given in the following order: • non-powered • powered by renewable electricity (green energy from a supplier; photovoltaic panels etc.) • electrically powered (fossil fuels like coal or the less greenhouse gas producing natural gas) • fuelled using petroleum products (petrol, two stroke, diesel) Consider when selecting equipment: • efficiency • size appropriate for job • maintaining equipment well for optimum efficiency
Soils	Remember: • healthy soils mean healthy gardens • protect structure from compaction and over-cultivation • avoid water and wind erosion • maintain or increase organic matter levels especially by using the garden's composted green wealth
Fertility	Fertilise: • after soil nutrient tests indicate a deficiency • only where needed (not needed for all plants) • at low rates to keep ornamental plants healthy • at higher rates for food plants • when plants are able to use the nutrients during growing seasons • using in preference fertilisers that have low environmental impact, including embodied energy Remember: • soil pH can affect nutrient availability
Mulches	Consider: • using mulches appropriate for the situation • using mulches from well-managed sustainable sources, processed with low environmental impact and high hygiene standards • sourcing locally
Soils and water	Remember: • to manage soils to have high water infiltration and water-holding capacity • to mulch to reduce evaporative loss from soils and via competing weeds • that non-wetting soils can decrease the effectiveness of rainfall or irrigation
Irrigation	Water: • only those plants needing additional water (e.g. using zones of plants with similar watering needs) • using efficient delivery methods (e.g. drip) • at a rate to avoid run-off and a period to avoid drainage losses below the root zone • at a frequency so plants adapt to periods without rainfall or irrigation • with an amount of water that maintains plants of a suitable quality • in response to weather conditions (not repetitively at set times) • appropriately for the season

Maintenance area	Considerations
Pruning	Consider: • most green waste from a garden is green wealth that can be recycled back into the garden as compost and mulch • using council green waste recycling if green waste cannot be recycled on-site • garden plants should enhance not hinder energy efficiency of buildings Remember: • prune well for healthy long-lived plants (especially important for trees)
Weeds	Remember: • herbicide production has environmental impacts (including the embodied energy they contain) Consider: • use of herbicides only for those situations where other control methods are impractical
Pests and diseases	The keys to managing pests and diseases are: • having a healthy garden ecology • selecting plants for disease resistance • providing the correct growing conditions for healthy growth • using appropriate companion plants • using routine maintenance practices, including good hygiene, that reduce the likelihood of pest and disease problems • monitoring garden plants for damage • remembering that small amounts of pest or disease damage do not reduce a garden's beauty • identifying accurately the cause of damage • knowing at what point intervention is required (if at all) • selecting an appropriate control method for the pest and for the environment • using the control at the optimal time(s) • assessing the control's effectiveness • continuing to monitor If considering the use of chemicals, use: • cautiously • the least toxic • the minimum for effective control • those well-targeted to the pest • those that are unlikely to promote resistance in pests • application methods that avoid unwanted spread (e.g. wicks rather than sprays) • only enough for the job but if there is excess dispose appropriately
Materials	For materials consider: • using less • re-using • recycling • using materials easily recycled or made from low environmental impact materials • maintaining well to reduce frequency of replacement

ENDNOTES

Chapter 1
1. *The Age*, 3 June 2006.
2. For an artist's visual representation of consumption statistics, see: www.chrisjordan.com/current_set2.php.

Chapter 2
1. Gott B, in Aitken & Looker (2002) *The Oxford Companion to Australian Gardens*. Oxford University Press, South Melbourne. pp. 1–2.
2. Macknight CC (1976) The voyage to Marege. Melbourne University Press, Carlton; cited in: Jumping the garden fence. 'Invasive garden plants in Australia and their environmental and agricultural impacts'. (RH Groves, R Boden & WM Lonsdale) CSIRO Report for WWF Australia, 2005.
3. Murray D (2003) *Seeds of Concern*. UNSW Press, Sydney. This is a scientifically referenced and rigorous explanation of plant genetic engineering – its potential and its dangers – by South Australian plant geneticist David Murray.
4. Harper M (2007) GM foods can be dangerous. But you do the research. *The Age*, 14 December, p. 17.
5. Matthews JM (1996) Aspects of the population genetics and ecology of herbicide resistant annual ryegrass. Australasian Digital Theses Program, The University of Adelaide. Thesis. See: http://search.arrow.edu.au/main/results?subject=Lolium (Acc. 3 June 2008).
6. Gaynor A (2002) Market gardens. In: *The Oxford Companion to Australian Gardens*. (R Aitken & M Looker) Oxford University Press, South Melbourne.
7. El-Hage Scialabba N (2007). Organic agriculture and food security. www.fao.org/organicag; and Badgley C *et al.* (2007) Organic agriculture and the global food supply. *Renewable Agriculture and Food Systems* **22**, 86–108. Cambridge University Press, cited in: *New Scientist*.
8. Holmgren D (2002) *Permaculture: Principles and Pathways Beyond Sustainability*. Holmgren Design Services, Victoria.

Chapter 3
1. McNiell JR (2003) Resource exploitation and overexploitation: a look at the 20th century. Cited in: Goudie A (2006) *The Human Impact on the Natural Environment*. 6th edn. Blackwell, Carlton.
2. Goldie J (2005) Population – the great multiplier. In: *In Search of Sustainability*. (Eds. J Goldie, B Douglas & B Furnass) CSIRO Publishing, Collingwood.
3. Goudie A (2006) The human impact on the natural environment. Blackwell, Carlton.
4. Adapted from FAOSTAT figures given in 2000.
5. UN Population Reference Bureau (2006). www.prb.org.
6. See real-time updating Population Clock: www.abs.gov.au.
7. Australian Bureau of Statistics 2007. cat. no. 3218.
8. See: www.footprintnetwork.org.
9. WWF *Living Planet Report* 2006 © text and graphics 2006 WWF.
10. Lenzen M (2004) 'The 1998–99 Ecological Footprint of the population within the Port Phillip and Westernport Catchment'. University of Sydney; cited in: *Melbourne Atlas*. See: http://www.dse.vic.gov.au.
11. www.acfonline.org.au/consumptionatlas/.
12. ACF (2007) Consuming Australia. Based on data collected and analysed by Centre for Integrated Sustainability Analysis, University of Sydney, and available at www.acfonline.org.au.
13. Adapted from data cited in Australian Conservation Foundation Australian

Consumption Atlas, www.acfonline.org.au/consumption_atlas/index.html.July 2007.

14 Foran, Lenzen & Dey (2005) 'Balancing act'. Available at: www.isa.org.usyd.edu.au/publications/balance.shtml. This is a detailed input–output sustainability (triple bottom line) analysis of the Australian economy. In four volumes it analyses 135 economic sectors of the economy in relation to their financial (three indicators), social (three indicators) and environmental (four indicators) intensities. The environmental indicators used were: greenhouse gas emissions, primary energy use, managed water use and land disturbance. Each related to $1 of final demand. For each of the 135 economic sectors the report presents a table of the estimated intensities. An analysis like this provides a measure of the social and environmental implications of financial flows across the economy.

15 Yale Center for Environmental Law and Policy. www.yale.edu/esi.

Chapter 4

1 Moll G & Ebenreck S (Eds) (1989) *Shading Our Cities. A Resource Guide for Urban and Community Forests.* American Forestry Association, Island Press.
2 Moore GM (2007) Managing trees in a changing environment. *Genus* **19**(4), 10–13.
3 International Energy Agency 2006. Key World Energy Statistics 2006.
4 www.greenhouse.gov.au/workbook/index.html.
5 World Resources Institute. www.wri.org.
6 United Nations Environmental Program/GRID/Arendal, www.grida.no/climate/vital/13.htm
7 FAO (2005) FAO, global forest resources assessment. Rome.
8 Millennium Ecosystem Assessment 2005.
9 Adapted from data in: Groombridge B and Jenkins MD (2002) *World Atlas of Biodiversity.* University of California Press, Berkeley.
10 http://svc237.bne113v.server-web.com/calculators/treecarbon.htm.
11 National Association of Forest Industries. Wood Waste Bioenergy Information Sheet No. 3. See: www.nafi.com.au/bioenergy_factsheets/WWFS03.pdf.
12 Myeong S, Nowak DJ & Duggin MJ (2006) A temporal analysis of urban forest carbon storage using remote sensing. *Remote Sensing of Environment.* **101**(2), 277–282.
13 Nowak DJ & Crane DE (2002) Carbon storage and sequestration by urban trees in the USA. *Environmental Pollution* **116**(3), 381–389.
14 Nowak DJ (1993) Atmospheric carbon reduction by urban trees. *Journal of Environmental Management* **37**(3), 207–217.
15 Groombridge B & Jenkins MD (2002) *World Atlas of Biodiversity.* California University Press, California.
16 Adapted from ABARE (2006) Australian energy consumption and production, 1974–5 to 2004–5.
17 *The Heat Is On. The Future of Energy in Australia.* Energy Futures Forum 2006. Data from ANUCLIM and map by CRES, ANU.
18 Adapted from Australian Greenhouse Emissions Information System. www.greenhouse.gov.au/inventory.
19 Dupont I & Pearman G (2006) 'Heating up the planet: climate change and security'. Paper 12. Lowry Institute for International Policy, Sydney.
20 Australian Greenhouse Office. www.greenhouse.gov.au/gwci/households.html.
21 ABS (2008) Australia's Environment: Issues and Trends, 2007. Cat. No. 4613.0.
22 Lenzen M (1998) Energy and greenhouse gas cost of living for Australia during 1993/4. *Energy* **23**(6), 497–516. Cited in: Foran & Poldy (2002).
23 Adapted from data supplied by Australian Greenhouse Office for 2004.
24 Adapted from Thomson & Sorvig (2002) *Sustainable Landscape Constructions.* Island Press, Washington.
25 www.dest.gov.au/sectors/science_innovation/publications_resources/profiles/water_for_

our_cities.htm; and Priestly, T. CSIRO Land & Water. pers. comm.

Chapter 5

1. UNEP (2002) Vital Water Graphics – an overview of the state of the world's fresh and marine waters. UNEP-Grid Arendal.
2. Adapted from figures in WWF *Living Planet Report* 2006.
3. Millennium Ecosystem Assessment 2005.
4. Australian Bureau of Meteorology.
5. www.waterfootprint.org/WaterFootprints.htm.
6. Hoekstra & Chapagain (2007) The water footprints of nations: water use by people as a function of their consumption pattern. *Water Resource Management* **21**(1), 35–48.
7. Australian Bureau of Statistics (2006) Water Account Australia, 2004–05. Cat. No. 4610.0.
8. National Heritage Trust (2002) Australia's Natural Resources 1997–2002 and beyond. National Land and Water Resources Audit, ACT.
9. Cullen (2005) Water: the key to sustainability in a dry land. In: *In Search of Sustainability*. (Eds. J Goldie, B Douglas & B Furnass). CSIRO Publishing, Collingwood.
10. www.vicwater.org.au.
11. WSAAfacts 2004. Water Services Association of Australia (2004). www.wsaa.asn.au.
12. ABS (2008) Australia's Environment: Issues and Trends, 2007. Cat. No. 4613.0.
13. *The Age* 3 May 2007.
14. National Land and Water Resource Audit (2002) Australian Catchment, River and Estuary Assessment. See: audit.ea.gov.au/ANRA/coasts/docs/estuary_assessment/Est_Ass_Contents.cfm.

Chapter 6

1. Schutz H, Moll S & Bringenzu S (2006) Globalisation and the shifting environmental burden. Material flows of the European Union. Wuppertal Institute on Globalisation. See: www.wupperinst.org/globalisation.
2. UNEP VitalWaste2 (2006). See: www.vitalgraphics.net/waste2/index.html.
3. Adapted from Australian Bureau of Statistics (2006) Australia's Environment: Issues and Trends, 2006.
4. www.sita.com.au/media/21643/plastic.pdf
5. WCS Market Intelligence (2001) 'Industry and market report – Australian waste industry'. WCS Market Intelligence, Sydney.
6. City of Whitehorse pamphlet, May 2007.
7. Millennium Ecosystem Assessment (2005).
8. See: www.ephc.gov.au/pdf/EPHC/chemicalsmgt_supdoc.pdf.
9. Cohen D (2007) Earth audit. *New Scientist* (May), 35–41.

Chapter 7

1. Rome Declaration of World Food Security. World Food Summit 1996. See: www.fao.org/docrep/003/w3613e/w3613e00.htm.
2. http://faostat.fao.org.
3. Imhoff *et al.* (2004) Global patterns of human consumption of net primary productivity. *Nature* **429**, 870–873.
4. See FAO: www.fao.org/worldfoodsituation/wfs-home/en (Acc. 2 June 2008), and related links.
5. FAO (2006) The state of food and agriculture 2006.
6. Ho M-W *et al.* (2008) 'Food futures'. ISIS-TWN Report. Sustainable World 2nd report. www.i-sis.org.uk/foodFutures.php (Acc. 10 March 2008).
7. Erlich R, Wahlqvist M & Riddell R (2007) Regional foods: Australia's health and wealth. Rural Industries Health and Development Corporation.
8. Halweil B (2002) 'Home grown. The case for local food in a global market'. Worldwatch Paper 163. See: www.worldwide.org/pubs; and Halweil B (2003) The argument for local food. Worldwatch May–June. See: www.worldwide.org/pubs.

9   Gaballa S & Abraham AB (2007) Food miles in Australia: a preliminary study of Melbourne, Victoria. CERES. www.ceres.org.au/index1024X768.htm.

10 Hamilton C, Deniss R & Baker D (2005) 'Wasteful consumption in Australia'. The Australian Institute Discussion Paper 77. See: www.tai.org.au/documents/dp fulltext/DP77.pdf.

Chapter 8

1   Millennium Ecosystem Assessment 2005.

2   See: www.environment.gov.au/cgi-bin/sprat/public/publicregisterofcriticalhabitat.pl.

3   www.environment.gov.au.

4   Lindenmayer D (2007) *On Borrowed Time*. CSIRO Publishing, Collingwood; and Fischer J, Lindenmayer D & Manning AD (2006) 'Biodiversity, ecosystem function, and resilience: ten guiding principles for commodity production landscapes'; and NK Wong *et al.* (2007) 'Establishment of native perennial shrubs in an agricultural landscape'. *Austral Ecology* **32**(6), 617–625.

5   Bisgrove R & Hadley P (2002) 'Gardening in the Global Greenhouse: The Impacts of Climate Change on Gardens in the UK'. Technical Report. UKCIP, Oxford. www.ukcip.org.uk (Acc. 9 Jan. 2008). See also comments at www.rhs.org.uk/climate (Acc. 6 Feb. 2008).

Chapter 9

1   Snape D (2002) *The Australian Garden – Designing with Australian Plants*. Bloomings Books, Melbourne.

2   Ferguson BK (2005) *Porous Pavement*. Taylor & Francis, Boca Raton, Florida.

3   See: www.lanfaxlabs.com.au.

4   Handreck K & Black N (2001) *Growing Media for Ornamental Plants and Turf*. 3rd edn. University of New South Wales Press, Sydney.

5   Handreck K (2008) *Good Gardens with Less Water*. CSIRO Publishing, Collingwood; and Handreck K (2001) *Gardening Down-Under*. 2nd ed. Landlinks Press, Collingwood.

6   Mobbs M (2001) *Sustainable House*. Choice Books, Marrickville.

7   Simpson J (1998) Urban forest impacts on regional cooling and heating energy use: Sacramento County case study. *Journal of Arboriculture* **24**(4), 201–214.

8   Dept of Transport Energy & Infrastructure (2008) Heating and cooling. Dept of Transport Energy & Infrastructure, SA. See: http://www.dtei.sa.gov.au/energy/energy_action/household/saving_energy/heating_and_cooling.html.

9   Reardon C (2001) Your home, your future, your lifestyle. Commonwealth of Australia, Canberra.

10 Turner L (2005) Solar panels buyer's guide. *ReNew* **91**, 47–55; and Reardon C (2005) Your home. The Australian Greenhouse Office, Canberra.

11 Cribb J (2006) Mineral processing: green steel. *Solve* **7**, 12–13.

12 After the European Commission LCA Tools Services & Data. See: http://lca.jrc.ec.europa.eu/lcainfohub/introduction.vm#.

13 Collins M, Fisher K, Fripp E & Nuij R (2005) Study and assessment of available information for a pilot project on a teak garden chair. Environmental Resources Management, Oxford for the European Commission. See: http://ec.europa.eu/environment/ipp/pdf/teak_chair_final_report.pdf.

14 Biever C (2005) White light at the end of the tunnel. *New Scientist* **184** (2479/80), 12.

15 See: www.greenhouse.gov.au/local/strategic/chapter7b.html.

16 George Wilkenfeld & Associates P/L (2004) Analysis of the potential for energy efficiency measures for domestic swimming pool and spa equipment. National Appliance and Equipment Energy Efficiency Committee and the Australian Greenhouse Office, Canberra.

17 Crawford H (2006) A review of forest certification in Australia. Prepared for the

Forest and Wood Products Research and Development Corporation.

18. Personal communication: Merrin Layden, FSC Australia. See: www.fscaustralia.org; and Emily Blackwell, Woodmark, Soil Association Bristol, UK. See: www.soilassociation.org/forestry.

19. Thornton JD & Johnson GC (1997) Revised CSIRO natural durability classification: in-ground durability ratings for mature outer heartwood. Forestry and Forest Products, CSIRO, Clayton.

20. Cookson L (2005) Safety of timber treated with CCA preservative. CSIRO Forestry & Forestry Products. See: www.ffp.csiro.au/TI-CCAFactSheet.asp#17.

21. Calkins M (2006) To PVC or not to PVC? *Landscape Architecture* **96**(3), 94–101.

22. Harrison J (2003) The TecEco project: the factors of cement and climate change. In Search of Sustainability Online Conference. See: www.isosconference.org.au.

23. Reardon C (2005) Your Home: design for lifestyle and the future. Australian Government, Canberra.

24. Ewan JM (2004) Conspicuous reconsumption: recycling concrete at Steele Indian Park. *Landscape Architecture* **94**(4), 134–137.

25. Standards Australia (2002) Guide to the use of recycled concrete and masonry materials HB 155–2002. Standards Australia, Sydney.

26. See: http://earthobservatory.nasa.gov/Newsroom/NPP/npp.html; http://gaim.sr.unh.edu/Products/News/Summer98/index.html; http://www.columbia.edu/itc/cerc/seeu/atlantic/restrict/modules/module06_content.html.

27. Netzel M, Netzel G, Tian Q, Schwartz S & Konczak I (2007) Native Australian fruits – a novel source of antioxidants for food. *Innovative Food Science and Emerging Technologies* **8**, 339–346.

28. Thompson K, Austin KC, Smith RM, Warren PH, Angold PG & Gaston K (2003) Urban domestic gardens (I) putting small-scale plant diversity in context. *Journal of Vegetation Science* **14**(1), 71–78.

29. Rich C & Longcore T (Eds) (2006) *Ecological Consequences of Artificial Night Lighting*. Island Press, Washington DC.

30. Hitchmough J (2004) Selecting plant species, cultivars and nursery products. Chapter 2 in: *Plant user handbook: a guide to effective specifying*. (Eds J Hitchmough & K Fieldhouse) pp. 7–24. Blackwell Science Ltd, Oxford.

31. Hitchmough J & Dunnett N (1996) Sustainable planting schemes. In: *Landscape Design*. Vol. 251, pp. 43–46.

32. Dunnett N & Clayden A (2000) Resources: the raw materials of landscape. Chapter 10 in: *Landscape and sustainability*. (Eds JF Benson & MH Roe) pp. 179–201. Spon Press, London.

33. Hitchmough J & Fieldhouse K (Eds) (2004) *Plant User Handbook: A Guide to Effective Specifying*. Blackwell Science Ltd, Oxford.

34. Seddon G (1999) Mission statements. In: *The Nature of Gardens*. (Ed. Peter Timms) Allen & Unwin, St Leonards.

35. See the Royal Botanic Gardens Melbourne Resource and Information Pack on Garden Plants as Environmental and Agricultural Weeds together with the Weed Strategic Plan 2004–2008: www.rbg.vic.gov.au/gardening_info/weed_strategy.

36. See: www.austep.net.au.

Chapter 10

1. After Virginia Department of Forestry (2005) Rain gardens – a landscape tool to improve water quality. Virginia Department of Forestry, Charlottesville.

2. Victorian Stormwater Committee (1999) *Urban Stormwater: Best Practice Environmental Management Guidelines*. CSIRO Publishing, Collingwood.

3. Melbourne Water (2005) *WSUD Engineering Procedures*. CSIRO Publishing, Collingwood.

4. Dunnett N & Clayden A (2007) *Rain Gardens*. Timber Press, Portland, Oregon.

5. Labrecque M (1995) Street trees in Montréal. American Association of Botanic Gardens and Arboreta Conference, Montreal, 25 July 1995.

Chapter 11

1. Thompson JW & Sorvig K (2000) *Sustainable Landscape Construction*. Island Press, Washington DC.
2. See: www.ecospecifier.org.
3. See: www.landscapingaustralia.com.au. In Victoria: www.liav.com.au; NSW: www.lcansw.com.au; Queensland: www.qali.asn.au; South Australia: www.lasa.org.au; Western Australia: www.landscapewa.com.au.
4. See: www.woodrecycling.info.
5. Handreck K (2001) *Gardening Down-Under*. 2nd edn. Landlinks Press, Collingwood.
6. Perry G (2007) The Story of Australian Humates. Omnia Specialities Australia Pty Ltd, Mulgrave, Victoria. See www.australianhumates.com.
7. Dr Lukas van Zwieten, Senior Research Scientist, DPI, NSW.
8. Soils for landscaping and garden use AS 4419-2003. Standards Australia, Sydney.
9. See: www.iffa.org.au.
10. See: http://aspag.org.au.
11. Florida Energy Extension Service & Helikson H (1991) 'The energy and economics of fertilisers'. Energy Efficiency and Environmental News, University of Florida. http://www.p2pays.org/ref/13/12141.pdf.
12. Natural Resources and Water Queensland (2007) Urine separation and collection to save water and feed gardens – an Australian-first demonstration project. Natural Resources and Water Queensland, Brisbane; Beal C, Gardner T, Ahmed W, Walton C & Hamlyn-Harris D (2007) Closing the nutrient loop: a urine separation and reuse trial in the Currumbin Ecovillage, Qld. On-site '07 Armidale.
13. Handreck K (2006) *Gardening Down-Under*, p.132. Also see study by Kolby E & la Cour Jansen J (2001) Exploitation of the nutrient content of urine at Svanholm Gods organic farming collective. http://glwww.mst.dk/publica/projects/2001/87-396-6.htm.
14. Hitchmough J (1994) *Urban Landscape Management*. Inkata Press, Sydney.

Chapter 12

1. Department of the Environment and Water Resources (2007) Comparative assessment of the environmental performance of small engines – outdoor garden equipment. Department of the Environment and Water Resources, Canberra.
2. See: www.oilrecycling.gov.au for your nearest local oil recycling centre.
3. Australian Bureau of Agricultural and Resource Economics (ABARE) (2008) Energy in Australia 2008. ABARE, Canberra.
4. Handreck K (2001) *Gardening Down-Under*. 2nd edn. Landlinks Press, Collingwood.
5. Ryan R (2006) Double win for environment. *Australian Horticulture* **104**(1), 34.
6. Helsel ZR (1992) Energy and alternatives for pesticide and herbicide use. In: *Energy in World Agriculture* (Ed. C Fluck). Elsevier, New York.
7. McIntyre K & Jakobsen B (1998) Drainage for sportsturf and horticulture. Horticultural Engineering Consultancy, Kambah, ACT.
8. Handreck K & Black M (2002) *Growing Media for Ornamental Plants and Turf*. 3rd edn. University of New South Wales Press, Sydney.
9. Adapted from Handreck K & Black M (2001) Growing media for ornamental plants and turf. University of NSW Press, Sydney.
10. Hitchmough J (1994) *Urban Landscape Management*. Inkata Press, Sydney.
11. Connellan GJ (2002) 'Efficient irrigation: a reference manual for turf and landscape'. Burnley College Report, University of Melbourne; Connellan GJ (2004) 'Performance benchmarking of turf and landscape irrigation systems'. *Proceedings of Irrigation Australia National Conference, Adelaide 1–13 May, 2004*. Irrigation Association of Australia, Sydney; Connellan GJ (2005) Evaluating and benchmarking the performance of urban irrigation. Irrigation Association of Australia, Sydney.
12. Costello LR & Jones KS (2000) Water use classification of landscape species (WUCOLS

III). In: *A Guide to Estimating Irrigation Water Needs of Landscape Plantings in California.* Sacramento, California Department of Water Resources.

13. Dukes M and Jacobs JM (2007) Types and efficiency of Florida irrigation systems. Appendix 8 in: Revision of AFSIRS crop water simulation model summary. Special publication SJ2008-SP19. (Eds JM Jacobs and M Dukes). St Johns River Water Management District, Palatka, Florida.

14. Rogers DH, Lamm FR, Alam M, Trooien TP, Clark GA, Barnes PL & Mankin K (1997) Efficiencies and water losses of irrigation systems. Cooperative Extension Service, Kansas State University, Manhattan.

15. University of California Cooperative Extension & California Department of Water Resources (2000) A guide to estimating irrigation water needs of landscape plantings in California. California Department of Water Resources, Sacramento.

16. Handreck K (1979) *When Should I Water?* Discovering Soils no. 8. CSIRO Division of Soils, Canberra, in association with Rellim Technical Publications.

17. Australian Pesticides & Veterinary Medicines Authority (2004) 'The reconsideration of approvals of the active constituent atrazine, registrations of products containing atrazine, and their associated labels'. Second draft final review report including additional assessments, October 2004. Australian Pesticides & Veterinary Medicines Authority, Canberra.

18. Kjaer J, Olsen P, Ullum M & Grant R (2005) Leaching of glyphosate and aminomethylphosphonic acid from Danish agricultural field sites. *Journal of Environmental Quality* **34**, 608–620.

19. Chemical Review Section National Registration Authority for Agricultural and Veterinary Chemicals (1996) Glyphosate. NRA Special review Series 96.1. National Registration Authority for Agricultural and Veterinary Chemicals, Canberra.

20. Gurr GM, Wratten SD & Altieri MA (Eds) (2004) *Ecological Engineering for Pest Management – Advances in Habitat Manipulation for Arthropods.* CSIRO Publishing, Collingwood.

21. See: www.goodbugs.org.au.

22. Radcliffe JC (2002) Pesticide use in Australia. Australian Academy of Technological Sciences and Engineering, Parkville, Victoria.

23. Supporting Information – Towards Ecologically Sustainable Management of Chemicals in Australia is a document available on the web at www.ephc.gov.au/pdf/EPHC/chemicalsmgt_supdoc.pdf.

Chapter 13

1. Philips E (2007) 'City People Light'. See: www.lighting.philips.com; Gordon A (2007) Lighting of the future. *Green Places* **40**, 14–18.

INDEX

Aboriginal gardening 13–5
Acacia 175
Acacia baileyana 209
Acacia mearnsii 249
Acacia species 199
acid rain 109, 130
acidification 131
Adelaide Botanic Gardens 213
Aerobin™ 258
Agenda 21 129–30
aggie pipes 93
agribusiness 18, 23–4, 26
    environmental demands of 23
agrichar *see* biochar
Agricultural Revolution vii
agricultural waste 52
agriculture 9, 11, 13–5, 17–8, 22, 26, 28, 116, 118, 124, 292
    alternative approaches 22
    development of 17–22
    genetic engineering 18, 22
    Green Revolution 17–8
    organic 26, 28, 292
    second Green Revolution 18
    sustainable 22, 26, 28, 116, 118, 124
agro-ecosystem 116
agro-industry 17
AIDS 16, 34
air pollution 130, 261 *see also* garden services
air travel 49, 62, 65–6, 100
albedo 56
algal blooms 81, 94, 96, 109
alien species 15, 131 *see also* environmental and economic weeds
Allocasuarina luehmannii 193
Allocasuarina verticillata 299
allotments 124
aluminium smelting 100
almond 198
Alyssum 175
Alyxia buxifolia 299, 300
Alternative Technology Association 175, 211, 230, 231
Amazonian dark earths 242
Anetholea anisata 199
animal welfare 122
aniseed myrtle *see Anetholea anisata*
anthroposophy 24
apple 198
apricot 198
Aquataine 305
aquifer 34, 73–9, 84, 88, 91

arable land 4, 34, 113–4, 117, 124
ArchBar *see* Green Screen®
astrological agriculture 24
atrazine 278, 279
Atriplex semibaccata 151
audit 36, 65, 87, 105, 111, 132, 231, 286, 296, 304
    energy 65
    garden sustainability 296, 304
    sustainability ix, 87
    waste 105
    water 111, 132
Aurora residential estate 221, 222, 223, 224
Australasian Biological Control Association 283
Australia
    Ecological Footprint 38
    energy and emissions 60–6
    food 118–26
    forestry 193–4
    land use 63
    population and projections 36
    rainfall distribution and availability 78, 80–4
    solar radiation levels 61
    water use 82–6
    wind resources 61
Australian Academy of Technological Sciences and Engineering 285
Australian City Farms and Community Gardens Network 215
Australian Conservation Foundation 41
Australian Cooperative Research Centre for Greenhouse Accounting 58
Australian Environmental Labelling Association 211
Australian Environmental Labelling Standards 211
Australian Food and Grocery Council 118
Australian food value chain 118–9
Australian Forest Certification Scheme 193
Australian Forestry Standard 193, 231
Australian Green Procurement Network 211
Australian Institute of Landscape Architects 210
Australian Landscape Industry Association 230, 233
Australian Life Cycle Inventory 42

Australian National Land and Water Audit 132
Australian National Pollutant Inventory 110, 286
Australian National Weeds Strategy 208
Australian Native Landscapes 279
Australian Open Garden Scheme 141, 303
Australian Organic Farming and Gardening Society 25
Australian Pesticides and Veterinary Medicines Authority 279
Australian Urban Tree Evaluation Program 208
Australian Water Resources Assessment 2000 95
Austromyrtus dulcis 199
Austrostipa stipoides 299
avocado 198, 199

Backhousia citriodora 199
Baeckea 175
banana 198
Banksia 175
barley 114, 119
basal metabolic rate 69
batteries 68, 103, 108
Bauera 175
Beaufortia 175
benchmark 10–1, 36, 82, 211
berry saltbush *see Atriplex semibaccata*
Beth Gott 14
Betula pendula 249
Bill Mollison 26–7
biocapacity 38–41, 48–9, 292, 296
biochar 242
biodiversity 1, 16, 20, 22–3, 31, 36, 38, 51, 81, 94–5, 109, 113, 120, 127–32, 200–2, 253, 293, 305
    conservation 128
biodynamic gardening 24, 124
    preparations 24
bioenergy 53
bio-ethics 20
biogas 60, 62, 105, 123
biogeochemical cycles 73, 109, 293
biointensive calorie farming 23
    gardening 23, 69
biological carrying capacity 38
    fixation 110
    significance of forests 56
biologically productive area 38
    productive land 37
biomass 52, 57–8, 60, 62, 72, 113, 242

energy 58, 72
BioNova Natural Pools Australia 156
bio-retention systems 218, 219
biosphere viii, 2, 7–9, 37–8, 51–4, 73, 110, 128–31, 261, 291, 293, 306
biotechnology 18–21
Birrarung Marr 216
birth control 32, 34, 36, 54
blackberry *see Rubus fruticosus*
blackberry rust 283
black plague 16
blackwater 91
blueberry 198, 199
Blue Planet 73
blue water 76–7, 79, 91, 292
borage 282
Boronia 175
Bossiaea 175
Boston ivy *see Parthenocissus tricuspidata*
botanical gardens 28, 135, 168–70, 179, 199, 207–8, 211–3, 222–3, 260, 262, 268–71
Brachysema 175
Briggs and Stratton engines 256
Brockhoff, Fiona 298
Brundtland Commission 4
brush fencing 197
buffalo grass 209
building sector 64
built-in obsolescence 108
bull oak *see Allocasuarina luehmannii*
Bureau of Meteorology 90, 157, 159, 273
bush tomatoes *see Solanum* species
Business Council for Sustainable Development 111

Cactoblastis moth 282
calcium sulphate *see* gypsum
Callistemon 197
Callitris glaucophyllus 193
calorie farming 24, 28, 69
calorific value
    of food 121
    of wood 58
Cape Schanck home 305
canola 18–9, 22
car production 100
carbon
    calculator 58
    credits 64
    cycle 54–5
    emission sink 54
    farming 23
    labels 107
    land management 59
    neutral 66
    sequestration 57, 68, 294, 295, 306
    statistics 60
    trees 57–8
carbon dioxide *see* CO_2
carnation *see Dianthus*
Cartagena Protocol on Biosafety 19
catastrophic climate change 53
cation exchange capacity 242
cement and concrete 195–6, 236–7, 239
    aerated concrete 195
    CO_2 emissions 195
    Eco-Cement™ 195
    HySSIL 195
    pH 236
    recycled concrete 195, 239
Centre for Integrated Sustainability Analysis 41
Ceratopetalum gummiferum 209
CERES 121, 169
CH2 building *see* Council House 2
Chadwick, Alan 23
chemicals 108, 236, 252, 278–80, 284–7
    atrazine 278, 279
    communities 287
    environmental and health effects 284–5
    fertilisers 252
    fungicides 284
    fungicides, embodied energy 284
    greywater 252
    glyphosate 21, 242, 278–80
    herbicides 252, 278
    herbicides, amphibian development 279
    herbicides, embodied energy 278
    herbicides, surfactants 280
    herbicides, use of 285
    Nutrient Management Codes of Practice 286
    pesticides 252, 284
    pesticides, embodied energy 284
    pesticides in Australia 285–6
    pesticides, non-specific 284
    pesticides, use of 285
    pests and diseases 284–7
    simazine 278
    Sustainable Gardening Australia Chemical Rating System 286–7
    weeds 278–80
cherry 198
Chinese gooseberry 198
Chorizema 175
cider gum *see Eucalyptus gunnii*
Class A standard water 155, 175–8, 222–3, 226, 270, 275–6
climate change vii, 1, 6, 16, 51–6, 60, 78–9, 84, 134, 157–8, 162
    Australian regions 162
    gardens 183–5
    the water cycle 79
climate regulator 77
climate-neutral 66
Club of Rome 3
CO_2 sequestration 26, 55–6, 59, 68, 294
COAG 131
coast beard heath *see Leucopogon parviflorus*
coast teatree *see Leptospermum laevigatum*
Commission on Sustainable Development 129
community gardens 124–5, 215
companion planting 23, 28, 282
compost 196, 258–60, 273, 278, 279, 287, 301, 303
    Aerobin™ 258
    Australian Native Landscapes 279
    Australian Standard 278, 287
    humus 259
    importing 260
    nutrients in 259
    process 259
    systems 258–9
    Vertical Composting Unit 260
    weeds and diseases 241
composting toilets 146, 303
concrete *see* cement and concrete
Conference of the Parties 129
Connellan, Geoff 268
consumption ix, 1, 3, 9–12, 130, 293
    horticultural 293
Consumption Atlas of Australia 41
contrails 65
Cootamundra Wattle *see Acacia baileyana*
corn 19, 114, 290
Correa alba 299
corridors (green) 133, 135
Corymbia maculata 193
couch grass 209, 258, 266, 278
Council House 2 150, 226
Crocosmia × *crocosmiiflora* 209
crop rotation 126
CSIRO 193, 195, 239
cullet 103
cultivated landscapes 127
cultural services 127
Cyperus papyrus 222

David Holmgren 26–7
David Swann 298
Daviesia 175
decoupling resource use and economic growth 44
deep percolation 89, 92, 94, 217
deforestation 16, 55, 57, 59, 77
delivered energy 64
dematerialisation 99–101, 292
denitrification 110

Denmark and Greenland Geological Research Institute 279
Denver Water Department 179
Department of Environmental Protection, WA 280
desalination 71, 94
desertification 26, 130
design *see* garden design
Dianella revoluta 151
Dianthus 175
diet 18, 21, 41, 59, 69, 113–4, 117–8, 122–5
Dietes grandiflora 276
dipstick® 245
diquat 279
direct and indirect resource impacts 40
downcycling 195
drivers of environmental change 31–2
drooping sheoak *see Allocasuarina verticillata*
dynamisations 24

Earth Charter vii
Earth system governance 8
Echium plantagineum 207
eco-agriculture 22
eco-labelling 107, 211
ecological and systems thinking 293
ecological creditors 38
    debtors 38
    gardening 203, 305
    metabolism 100
Ecological Footprint 11, 37, 48, 291, 292, 294, 296, 301
    Australian 38
    global 38
    individual 39, 41–2
    household 39
    overshoot 38
Eco Logical Water Capturing System 170–1
economic externality 127
    growth 3, 31, 44, 292
Ecospecifier 231
ecosystem services 2, 8–9, 73, 127–8, 144, 292–4
Ecosystem Wellbeing Index 37
Ehrlich & Holdren 32
El Niño (ENSO) 60
embodied energy 52, 64, 67–72, 112, 125, 138, 147, 154–5, 179, 183–7, 237, 238, 247, 266, 278, 284, 304
    paving 184
    resources 40, 52, 123, 292
    teak garden chair 185–6
    wood decking compared with concrete 183
embodied water 52, 86, 79, 87, 96, 163, 191, 238

beef 42, 52, 84, 86
emissions 6, 36, 40, 42–3, 51–66
    Australia 60–2
    global 52–3, 63
    household 64–5
    individual 65
Enchylaena tomentosa 151
end-of-pipe management 10, 99
energy 51–72, 180–91, 234–5, 240, 251, 255, 256, 261, 278, 284, 291, 292, 306
    audit 65
    clothes drying 183
    content of foods 69
    effect of evapotranspiration 180–1
    efficient buildings 64, 181–2
    embodied 70, 155, 179, 183–7, 237, 238, 278
    evapotranspiration and energy use 180–1
    fluorescent light 188–9
    garden 66–72
    global 31, 52–3, 55, 62, 65
    halogen light 189
    household 64, 68
    individual use 65–6
    intensity 63, 65
    landscape design 138, 146, 180–91
    landscapes 180–91
    LED lighting 187–8
    lighting 187–90
    lighting and paving reflectance 189
    maintenance 187–91
    passive solar houses 181–2
    photovoltaic panels 181–2, 223, 224
    pumps 190
    reducing use 180–1, 183–91
    renewable and non-renewable 53, 146
    reprocessing materials 184
    re-using materials 184
    solar hot water 181–2, 223
    solar light 187
    spas 191
    sustainable lifestyle 8, 66, 145, 228, 306
    swimming pools 190–1
    tree shade and energy use 180
    trees 58–9
    units 70
    use in Australia 60–2
    use in buildings 64, 180
English broom 207
ENSO 60
Environment Protection and Biodiversity Act 1999 132

Environment Protection and Heritage Council 286
environmental accounting vii, 10, 292
environmental and economic weeds 21–3, 25, 29, 95–6, 116, 119, 126, 134, 144, 146, 151–4, 177, 179, 197–8, 202, 205–9, 241–3, 246, 248–9, 251–2, 256, 258, 260–1, 264–6, 276–84, 289, 301, 302
environmental flows 78, 87, 94, 158
Environmental Performance Index 44–5, 47
environmental services 144, 145 *see also* garden services
Environmental Sustainability Index 37, 44–5
environmentally friendly technology 32
environmentally sensitive urban design 222–4, 297
    bicycle parking 224
    bicycle routes 224, 225
    nature strips 225
    pedestrian permeability 223, 224
    porous pavement 228
    protecting existing trees 225
    purple pipes 222
    recycled water 222–3
    residential properties 225
    street trees 225–8
    structural soils 226–8
    urban forests 224–8
Environment Australia 286
Environment Protection Authority 236
EPA *see* Environment Protection Authority
equipment 234, 239–40, 252, 253–6, 258, 275, 303
    appropriate for task 255
    Briggs and Stratton 256
    chipper 258, 275
    electrical 255
    energy consumption 234, 240, 255
    energy in fuels 256
    energy source preference 255
    landscape maintenance 252, 253–6, 258, 275
    maintaining 240, 255–6
    maintaining blades 256
    maintaining engines 256
    non-powered 253–5
    oil disposal 256
    oil recycling 256
    petrol and petroleum fuels 255–6
    pollution 255
    powered 255, 258
    volatile organic compounds (VOCs) 255
Erica 207

erosion 235
erosion control 148–50
    and plants 148–50
    products for 149–50
    staggered block walls 151
    Water Sensitive Urban Design 150
ethical eating 125
ethnobotany 14
Eucalyptus camaldulensis 193, 249
Eucalyptus cladocalyx 209
Eucalyptus gunnii 199
Eucalyptus regnans 193
Eucalyptus species 193, 197, 222, 226
European Inventory of Existing Commercial Substances 109
Eutaxia 175
eutrophication 95, 130, 174, 247, 249
evaporation 89, 92–3, 305
evapotranspiration 74, 77–80, 89–93, 117, 134, 179–80, 217, 244, 250, 253, 269–70, 273
e-waste 101–2, 108, 111
extinction vii, 74, 94, 114, 130–1

family farms 17
family planning *see* birth control
farmers
    cooperatives 124
    markets 124
fauna 27, 28, 132, 133, 200, 201, 211, 266, 299, 306
Federal Office of the Gene Technology Regulator 19
Federation Square 216
feijoa 198, 199
feral animals and plants 16 *see also* environmental and economic weeds
Fertiliser Industry Federation of Australia Inc. 111, 286
fertilisers and fertilising 246–9, 251, 252, 260–1
    application 249
    controlled release 248
    effect on local ecology 249
    embodied energy 247, 261
    fish-based 260
    inorganic 260
    landscape maintenance 251, 252, 260–1
    nitrogen 109, 110, 247–8, 261
    nutrient content 260
    organic 260
    pH 260
    phosphorus 248, 261
    plant requirements 249
    potassium 248, 261
    organic versus inorganic 247
    soil nutrient levels 260
    soil nutrient testing 247, 260–1
    urine 248–9
    weeds 248
Ficinia nodosa 151
Ficus microcarpa var. *hillii* 226
field capacity 90, 92, 94, 269–72
Fifth Creek Studio 150–1, 154
fig 198
Fiona Brockhoff 298
fire-stick farming 14
fixed nitrogen 109
flash flooding 93, 95
flax lily *see Dianella revoluta*
flora 132, 154, 205, 206, 211, 212, 220, 306
food 113–26, 138, 197–200, 291, 292, 302, 303, 305, 306
    aid 116
    Australian food value chain 118–9
    Australian species 199
    climate change 117–8
    co-ops 125
    diet *see* diet
    energies 308
    energy 67, 69, 120, 198
    energy use in USA 198
    grocery products industry 118
    harvesting 287–8
    landscapes 197–200
    miles 120–1, 138, 145, 198
    plant maintenance 253, 257, 275, 287–8
    plant productivity 199
    plants 14–5, 68, 71, 114, 198–9, 209, 253, 275, 283, 317–8, 321
    plants and aesthetics 198
    production ix, 4, 18–9, 23, 41, 51, 77, 91, 115–9, 124, 138, 145, 197–8, 269, 303, 312
    production in home garden 126
    schools 215
    security 4, 20, 26, 73–4, 113–7
    self-sufficiency 124
    shortages 17, 114–5
    street trees 215
    sustainability 125
    trends in consumption 120
    waste 103, 105, 112, 123
Food Education Program 121
forests 41, 49, 55–60, 66, 72–3, 77, 81, 91, 127, 130–4, 148, 180, 186, 191–3, 196, 207, 210, 224–5
forest and woodland farming 27
Forest Stewardship Council Australia 193
fossil fuels 54, 66
fossil water 74, 76, 78
freshwater fisheries 78
Friends of the Earth 125, 128
fruit 275, 302

Fukuoka farming 24–5
Fukuoka, Masanobu 24–5
fusarium wilt 21
future developments for sustainable horticulture 304–5
Fytowall™ 150

garden
    activities 147
    consumption viii, ix, 138
    design *see* landscape design
    ecology 251, 297
    produce 67–8, 71, 197–200, 287–8, 302
    waste ix, 103, 105, 112, 211
garden services 144, 146, 180, 201
    air filtering 144
    climate moderation 144, 180
    food 144
    water purification 144
gardens for wildlife 127
garlic 198
gene technology 18–9, 21, 114
genetic
    engineering 18–9, 20–1
    erosion 18
    piracy 20
    resources 129
genetically modified food 18, 122
genetics 17
Genuine Progress Indicator 37
geochemical cycles *see* biogeochemical cycles
Geoff Connellan 268
Geoff Gurr 282
geomembrane 164
geosphere 130
geothermal energy 60
giant sensitive tree *see Mimosa pigra*
Gladiolus 207
glass 70, 101–8
global
    biodiversity 16, 38, 129
    community 4, 20
    food production 19, 77, 91
    Footprint 38
    hectare 37, 48–9
    industrial metabolism 99
    life expectancy 34
    net primary production 113
    sustainability 1, 48, 51, 127
    trade 35, 38, 79
    warming 55, 58, 130–1
    water cycle 73–7, 91, 156
    water use 74
Global Biodiversity Outlook 129
Global Environment Outlook 129
Global Strategy on Diet, Physical Activity and Health 123
globalisation 28, 101, 121, 123

glyphosate (Roundup™) 21, 242, 278–80
   in groundwater 279–80
GM crops 5, 13, 18–22, 24, 26, 74
   debate 13, 21
Good Environmental Choice Australia 211, 231
Good Wood Advisory Centre 192
Good Wood Guide 192, 193
gorse 207
Gott, Beth 14
grape 198, 282
grapefruit 198, 199
Green Building Council 211, 230
green design 102, 292
   economy 11, 291
   purchasing 29, 106–7, 231, 292
   waste viii, 28, 67–8, 71, 105, 112, 212–3, 258, 260, 275–9, 287, 301
   water 73, 76–80, 91, 166
greenhouse gas 6, 36, 52, 58, 61–7, 77, 102–3, 108, 112, 118–9, 130–1, 154, 163, 181, 185–6, 191, 195, 198, 255, 258, 266–7
Greenpeace 128
Green Revolution 17–8
   second 18
green roofs 152–4, 218, 288–9
   Charles Sturt University, Albury 152
   climate 152, 153
   elements 153
   extensive 152
   Fifth Creek Studio 154
   Germany and Switzerland 152
   green roof at Parliament House, Canberra 152
   guidelines 153
   habitat creation 154
   intensive 152
   Landscape Research, Development and Construction Society 153
   maintenance 288–9
   Marine Science Consortium 152
   plant selection 153–4
Green Screen® 150, 151
green walls 150–2, 289
   Fytowall™ 150
   maintenance 289
   VersiTank® 250 150
green waste 258, 275, 277–8, 279, 301 *see also* green wealth
green water 292
green wealth 258, 275, 276
greywater 73, 84, 86, 89–91, 96–7, 138, 146, 158, 172–8, 211, 252, 301, 303
   biological oxygen demand 174
   Biolytix® 176

chemicals in 173–5
do's and don'ts 172
eutrophication 174
irrigation 172, 173
hair conditioner 175
laundry detergents 173–5
nitrogen 174–5
peat (sodium ion adsorption) 175
pH 174
phosphorus 174–5
salinity 174
shampoo 175
soap 175
sodic soils 174
soil interactions 175
toxicity for plants 175
treated 175–7
treatment systems 146
ultrasonic washing machines 175
untreated 172–5
wetland processing 176
Gross National Product 37
Gross Pollutant Traps see litter traps
groundwater 18, 74, 76, 79, 81, 88, 91, 131, 173, 189, 230, 248–9, 261, 279
Gurr, Geoff 282
gypsum 239, 241, 257

hard landscaping 147–8
   around street trees 148
   for healthy plants 147–8
   over structural soil 148
   permeable paving 148
   soil-filled trenches 148
Hart, Robert 27
hawthorn 207
hazardous waste 103
heat of decomposition 58, 67, 71
heirloom seeds 25, 28, 125–6
herbicide resistance 21
high energy foods 28, 72
holistic horticulture 24, 26, 28, 293
Holmgren, David 26–7
Homebush Olympic site 226, 228
horticultural primary production 296
horticultural sustainability
   accounting 295–7
   sustainability audit 296
horticultural therapy 124
Horticulture for Tomorrow (H4T) 119
household
   consumption patterns 39, 41, 63, 84
   Ecological Footprint 39
   ecology 39
   energy usage 68
   income 63, 85, 104
   metabolism 39
human

commerce 127
consumption ix, 1, 3, 9–12, 74, 94, 130, 291
   economy 9, 82, 293–4, 297
   energy 68–9
   equivalent (H-e) 69
   food chain 69
   impacts on Australian environment 131
   impacts on global environment 130
   labour 67–70, 234, 239
   nitrogen production 109
   pathogens 16, 178
   population 3, 18, 28, 33–4, 36, 291
   population doubling times 34
   population explosion 3, 28, 33
   resource use 3
Human Development Index 37, 48
Human Wellbeing Index 37
hunter-gatherer vii, 9, 15
hydro 52–3, 60, 62
hydrological cycle 73
hydroponics 124
hydrosphere 73
Hypocalymma 175

Impatiens 175
Independent Science Panel 116
indigenous Australians 13–5, 253
indigenous biodiversity, land for nature 294–5
indigenous fauna 201, 299
Indigenous Flora and Fauna Association 246
indigenous plants 204, 299
indigenous vegetation 299
individual
   Ecological Footprint 39, 41–2, 49, 294
   emissions 42
   energy use 42
   energy use, direct and indirect 44, 66
industrial agriculture 9, 17, 20, 22, 116, 125, 292
   energy production 55
   metabolism 99
   nitrogen fixation 110
   waste 43, 103, 130
Industrial Revolution vii, 3, 16, 33
industrial sites for parks 215–6
infiltration 77–9, 87–93, 147–50, 167–70, 174, 179, 217–22, 226–8, 235, 240, 245, 257, 271–3, 288, 300
infiltration trenches 218
infrastructure 146–7
inorganic mulches 196–7
inorganic nutrients 116
Input-Output Analysis 37, 39, 43, 66

Integrated Pest Management 20, 213, 283, 285
Integrated Product Policy 100
Intergovernmental Agreement on the Environment and the Council of Australian Governments 131
Intergovernmental Panel on Climate Change 128
international trade 21, 79, 122, 205
Intsia retusa 193
invasive plants 16, 205–6 *see also* environmental and economic weeds
irrigation 9, 16–8, 31, 59, 74–7, 81–2, 87–94, 97, 119–20, 134, 138, 143, 151–2, 155–65, 170, 172, 177–9, 243–5, 252, 253, 262, 267–75, 301, 303, 305
   application efficiencies 268
   Class A pan 270, 275
   classification of plants species 268
   compaction 272–3
   conserving water 267
   crop coefficient 270
   crop factor 269, 270, 271
   delivery rate 271–2
   delivery rate calculation 272
   delivery rate measuring 271
   drip 178, 244–5, 268, 271, 301
   efficiency 178, 268
   embodied energy 179
   Enviroscan 271
   evaporation 272, 273, 275
   evapotranspiration 269, 270, 272, 273, 274
   field capacity 269, 272
   food plants 269, 275
   frequency 269
   guide 270
   how 268
   how much 269–70
   infiltration rate 271, 273
   indigenous plants 275
   irrigation efficiency 262
   landscape coefficents 271
   maintenance 252, 253, 262, 267–75, 301, 303
   microspray 178, 268
   monitoring conditions 178–9, 267
   ornamental plants 270–1
   percolation 269
   permanent wilting point 269
   plant grouping 177
   plant quality 269, 273–5
   plant selection 177
   plant water demand 269, 270–1, 273
   rainfall 269, 273, 274
   refill point 269
   root zone 269, 271
   run-off 269, 271
   scheduling 262, 270
   seasonal changes 253, 273
   slopes 271, 273
   soil layering 272–3
   soil management 273
   soil 'tank' 269
   soil type 271, 272
   soil water availability 269–70
   soil water movement 270
   sprinklers 268
   subsurface rock 272–3
   tensiometer 270
   transpiration 270, 272, 273
   water holding (storage) capacity 269, 272, 273
   water restrictions 178
   water storage capacity *see* water holding capacity
   water tension 270
   water use efficiency 268
   weather 269
   wetting pattern 178
   when 269
   zones 243–4, 267

Jacksonia 175
Journal of Green Building 211
jumbo waste 102

Kakadu plum *see Terminalia ferdinandiana*
kangaroo grass *see Themeda triandra*
Karkalla 297, 298–303
Karkarook Park 220
keystone species 133
kikuyu 209, 258, 266, 278
kiwi fruit *see* Chinese gooseberry
knobby club rush *see Ficinia nodosa*
Kunzea pomifera 199

Lactuca 175
Lake Erie 174
Lanfax Laboratories 173
La Niña 60
La Trobe power industry 84
Land
   clearance 6, 16–7, 205
   degradation vii, 78, 113, 130–1
   management regimes 14–17, 51, 54, 59–60, 213
   ownership 13
landfill 55, 100–8, 112, 185, 189, 194–6, 238–9, 252, 258, 267, 278, 289, 305
landscape construction 229–50, 297, 304
   Alternative Technology Association 231
   Australian Landscape Industry Association 230, 233
   build less 237
   cement and concrete 236–7, 239
   chemicals 236
   choosing contractors 233
   communication between stakeholders 232
   composts 240–1
   concreting guidelines 237
   ecolabelling 231–2
   Ecospecifier 231
   embodied energy 237
   energy 234–5
   equipment 234, 239–40
   erosion 235
   fertilisers and fertilising 246–9
   fertilising, soil nutrient testing for 247
   Good Environmental Choice Australia 231
   green purchasing 231
   irrigation system, drip 244–5
   irrigation system installation 243–5
   irrigation system zones 243–4
   LCA 233, 238
   locating services 239
   matching designs to sites 232–3
   minimising site damage 239–40
   minimising waste 238–9
   monitoring new landscapes 249–50
   mulching 249
   planting 243, 246–7
   planting and plant quality 246
   planting, sourcing plants 246
   planting trees 246–7
   protecting topsoils 235–6
   protecting vegetation 236
   protecting water resources 236
   protection zones 233, 235
   recycling 238–9
   recycling, grinding building materials 239
   recycling pots 239
   rotary hoes 240, 241
   Santa Fe Green Building Council 230
   selecting materials 237–8
   sequencing works 233–4
   site deliveries 240
   soil 135–6, 239–43
   sustainable construction guide 230
   using knowledge and technology 230–1
   using sustainable designs 232
   Water Efficiency Labelling and Standards (WELS) 231–2

landscape design 137–213, 297, 298–303, 304
   activity areas in garden 141
   balance 140–1
   biodiversity 138, 200–2
   botanical gardens 211–3
   budget 141
   cement and concrete 195–6
   CH2 building *see* Council House 2 building
   climate control with plants 150–4
   climate moderation *see* garden services
   concrete *see* cement and concrete
   construction drawings 143
   contractors 210
   Council House 2 building 150
   crib walls 151
   design development 142
   design implementation 209–10
   design process 139–44
   ecological considerations 144
   eco-roofs *see* green roofs
   ecosystem services *see* environmental services
   efficient use of space 146
   energy and landscapes 138, 146, 180–91
   embodied energy 183–7
   engineering with plants 148–54
   environmental design considerations 144–56
   environmental services 144, 145 *see also* garden services
   erosion control 148–50
   food in landscapes 197–200
   framing 140
   fundamentals of landscape design 138–9
   gabions 151
   garden activities areas 147
   garden services 144, 146, 180, 201
   garden uses 141
   general considerations in landscape design 139
   green roofs 152–4
   green walls 150–2
   hard landscaping 147–8
   infrastructure 146–7
   LED energy 187–8
   Life Cycle Assessment 163, 184–6
   lighting 187–90, 201–2
   maintenance 141
   major features 154–6
   materials 138, 146–7, 191–7
   mulches 196–7
   Park Güell, Barcelona 157
   passive solar houses and vegetation height 181, 182
   permeable paving 148, 167–72
   plant growth 145–6, 147–8
   planting plan 143
   plant selection 202–9, 299
   plastic 194–5
   porous paving *see* permeable paving under landscape design
   practical steps for landscape design 141–4
   preliminary questions to guide design 139–41
   principles 139, 140–1
   proportion 141
   protecting soil quality 145–6
   repetition, simplicity and rhythm 140
   resources and infrastructure 146–7
   roof gardens *see* green roofs, intensive
   self-sustaining gardens *see* ecological considerations
   sequence and focal points 140
   site analysis 142
   site choice 145
   site plan 142
   soil 145–6, 174
   specifications for contractors 209–10
   stabilising slopes 148–50
   street trees 148
   style 139–41
   sustainable water use 156–80
   sustainability factors to consider 144
   swimming pools 154–6, 190–1
   timber 192–4
   transition and contrast 140
   unity 140
   useful organisations 210–3
   vegetation and solar panels 181–2
   VersiTank® 250 green wall 150, 151
   water 138, 144, 146, 156–80, 191, 222–3
   wire mesh wall panels *see* Green Screen®
landscape maintenance 251–90, 297, 301, 304
   appropriate for needs 252
   biodiversity 253
   changing gardens 252–3
   chemicals 252, 278–80, 284–7
   compost 258–60, 278, 279, 287
   controlling growth 275–80
   diseases *see* pests and diseases
   energy 251, 255
   equipment 252, 253–6, 258, 275
   fertiliser and fertilising 251, 252, 260–1
   food 253, 257, 275, 287–8
   fruit 275
   garden ecology 251
   green roofs 288–9
   green walls 289
   green waste 258, 275, 277–8, 279
   green wealth 258, 275, 276
   greywater 252
   hydrophobic soils *see* non-wetting soils
   inappropriate plant placement 276
   irrigation 252, 253, 262, 267–75, 301, 303
   lawns 275, 289
   maintenance programs 251
   maintenance schedules 253
   pests and diseases 280–7
   materials 251, 289–90
   monitoring 252, 253, 283
   mulching 251, 264–7
   new gardens 252
   non-wetting soils 257
   old gardens 252–3
   pests and diseases 251, 252, 280–7
   potted plants 288
   pot recycling and re-using 289–90
   pruning 275
   rain gardens 288
   rainwater 252
   recycling 275, 276
   resources of site 252
   scheduling 276, 277
   seasons 253
   soil quality 251, 256–67, 271, 272, 273, 278, 279, 287
   solar buildings and appliances 276
   special garden features 288–9
   structures and fittings 290
   trees 252
   vegetables 257, 275
   water 251, 252
   weeds 252, 264, 276 *see also* environmental and economic weeds
   wetting agents 257
Landscape Research, Development and Construction Society, Germany 153
Lantana camara 209
laundry detergents 172–5
lawns 289
LCA *see* Life Cycle Assessment
leachate 102–3
Lechenaultia 175
lemon 198, 199
lemon myrtle *see Backhousia citriodora*
Leptospermum laevigatum 299
lettuce *see Lactuca*

Leucopogon parviflorus 299
levels of biological organisation 8, 129
    human organisation vii, 6, 8, 11, 32, 37–8, 51, 128–9, 291
    Life Cycle Assessment (LCA) 37, 41, 163, 184–6, 231, 233, 238, 260, 266, 304
        landscape materials 184, 185
        schematic representation 185
        teak garden chair 185, 186
Life Cycle Costing 37, 186–7
life expectancy 34, 36, 48, 225, 291
light brown apple moth 282
lighting 187–90, 201–2, 304
    effect on ecology 201–2
    and paving reflectance 189
    fluorescent 188–9
    halogen light 189
    LED lighting 187–8
    solar light 187
light pollution 201, 202, 304
lilac pipes *see* purple pipes
lilac water 77, 91
lime 198, 199
Limits to Growth 3
litter trap 219, 220
Living Planet Index 38
Living Planet Report 38
Lobularia maritima 282
local food ix, 51, 124–5, 138, 145
local sourcing 51, 114, 116, 118, 120, 121, 123, 124, 125, 126, 138, 144, 145, 149, 183, 186, 192, 196, 197, 198, 201, 209, 230, 231, 235, 237, 246, 251, 253, 267, 287, 299, 300, 301
locavore 125
Lomandra sp. 218, 222, 226
loquat 198
low carbon economy 107
Lynbrook 219, 220

Macadamia integrifolia 199
macadamia 198
Malthus, Thomas 36
mandarin 198, 199
mango 198
marigold *see Tagetes*
market gardens 22–3, 123, 198, 275
Masanobu Fukuoka 24–5
mass production 17, 36, 100, 124
Material Flow Accounting 101
material flows analysis 99, 101
materials 99–112, 191–7, 251, 289–90, 291, 304, 306
    and trade 100
    brush fencing 197
    cement and concrete 195–6
    choosing 191–7

compost 196, 258–60, 273, 278, 279, 287
dematerialisation 99–101, 292
glass pebbles 197
inorganic mulches 196–7
Integrated Product Policy 100
Intsia retusa 193
landscape design 138, 146–7, 191–7
landscape maintenance 251, 289–90
merbau *see Intsia retusa*
mulches 196–7
National Packaging Covenant 100
organic mulches and composts 196
plastic 194–5
pot recycling and re-using 289–90
product stewardship 100
radially sawn timber 193
reducing use 146–7
river pebbles 196–7
selecting 237–8
Smartwood™ 193
structures, fittings furnishings 290
timber 192–4
timber longevity 193
treated pine 193–4
waste 103 *see also* waste
zero waste 99, 101, 292
Mathis Wachenagel 37
MCPA 279
Mediterranean-style diet 123
medlar 198
mega-diverse ecosystems 132
Melaleuca 197
Melaleuca nesophila 209
Melaleuca lanceolata 299
Melaleuca quinquenervia 208
Melaleuca uncinata 197
Melbourne Water 179, 216–7
merbau *see Intsia retusa*
methane 55, 58, 102–3, 115, 118–9, 258
micro-organisms 18, 96, 110, 146, 155, 169–77, 200–1, 219, 235, 242, 256, 260, 264, 266, 284
midgen berry *see Austromyrtus dulcis*
Millennium Declaration 4
Millennium Development Goals 4
Millennium Ecosystem Assessment 2, 127, 129
Mimosa pigra 207
mineral reserves 2
mining 11, 55, 62, 81–2, 84, 116, 226, 237, 256
    waste 103
mixed farming 28

Mollison, Bill 26–7
Monash University 262
monocultures 18, 116, 125
montbretia *see Crocosmia* × *crocosmiiflora*
moonah *see Melaleuca lanceolata*
Morgan, Paul 305
Mornington Peninsula National Park 299
mountain ash *see Eucalyptus regnans*
mountain pepper *see Tasmannia lanceolata*
Mount Annan Botanic Garden 213
mulch and mulching 67, 93, 103, 112, 126, 149, 161, 170, 179, 194, 196–7, 222, 226, 231, 234–6, 239, 245, 249, 251–2, 257–8, 264–7, 273, 275, 278, 281, 288, 301
    benefits 264
    bushfires 266
    cultivation 266
    depth 266
    embodied energy 266
    evaporation from soil 265
    frost 266
    inorganic 196–7, 264
    nutrient competition 266
    organic 196, 264
    rainfall barrier 266
    soil compaction 266
    soil fauna 266
    soil structure 266
    soil temperatures 265
    sourcing 267
    weeds 265
mulberry 198
municipal solid waste 105
muntries *see Kunzea pomifera*
Murray-Darling 81–2, 118
    statistics 118

National Action Plan for Education for Sustainable Development ix
National Action Plan for Salinity and Water Quality 82
National Association of Sustainable Agriculture Australia 26
National Chemicals Taskforce 110, 286
National Government Waste Reduction and Purchasing Guidelines 108
National Greenhouse Strategy 131
National Heritage Trust 84, 131
National Land and Water Resources Audits 132, 286
National Oceans Policy 131
National Packaging Covenant 100
National Strategy for Ecologically Sustainable Development 131

National Strategy for the Conservation of Australia's Biological Diversity 130–1
National Water Reform 131
natural
　agriculture 25
　capital 2, 127
　economy 9, 293, 298
　farming 25
　gardening 25
　resources 293
nature strips 225, 226
naturalised alien plants 15, 153, 206
Neolithic Revolution 15–6
Netafim 305
Net Primary Product 198, 294
Newly Industrialised Countries 101
nitrates 95, 109, 110, 247–8
nitrification 110
nitrifying bacteria 110
nitrogen
　cycle 110–1, 261
　fertiliser 109, 110, 247–8, 261
　fixation 110
nitrous oxide 55, 109, 163, 242
non-government organisations 128
North Atlantic Oscillation 60
NSW Christmas Bush *see Ceratopetalum gummiferum*
nuclear power 53
nutrient leaching 116
Nutrient Management Code of Practice 111, 286

obesity 120, 122–3
OECD 36, 41, 61
Ogalalla Aquifer 78
olive 198
one planet consumption 48
Open Garden Scheme 141
organic agriculture 26, 28, 292
　foods 25, 125, 211
　gardening 25, 26
Organic Farming and Gardening Society 25
organic matter 257–60, 262
organic mulches and composts 196
Organisation for Economic Co-operation and Development *see* OECD
organisational levels (scales, contexts or frames of reference) 6
overfishing 74
oxygen 54, 79, 88, 96, 109–10, 146, 148, 155, 174–5, 226–7, 236, 240, 242, 265
ozone depletion 130, 185

packaging 23, 42, 100, 103, 108, 118, 120–1, 125, 186, 192, 238, 261, 267, 278, 284, 287

paints 100, 103, 108, 184, 236–7, 239
paraquat dichloride 279
Park Güell, Barcelona 157
parsley 198
Parthenocissus tricuspidata 153
passionfruit 198
passive solar houses 181, 182
Paterson's curse *see Echium plantagineum*
Paul Morgan 305
paw paw 198
peach 198
peak oil 114
pear 198
periwinkle 207
permaculture 25–7, 126, 292–3
permeable paving 148,167–72, 218, 227, 228, 238, 300
　bonded gravel 169–70
　gravel 168
　maintenance of 169
　mulches 170
　nutrients in run-off water 169
　pavers 168
　paving 168
　porous asphalt 169
　porous concrete 168–9
　wooden decking 168
　WSUD 218
persimmon 198
pesticides ix, 17–9, 21, 23, 25–6, 71, 95–6, 100, 103, 112, 144, 193, 201, 252, 279, 283–7
pests and diseases 18, 71, 116–7, 120, 122, 202, 208, 251–2, 280–7
　Australasian Biological Control Association 283
　beneficial organisms 281, 283
　biological control 282–3
　chemicals and community 287
　chemicals and environment 285–6
　chemicals, national management 286
　chemicals, SGA rating system 286–7
　chemicals, synthetic 283, 284–5
　common pests and diseases and their control 280
　companion planting 282, 283
　crop rotation 281
　cultural techniques 281–2, 283
　degree days 282, 283
　ecological engineering 282
　garden or glasshouse ecosystem 283
　guide to controlling pests and diseases 280–2
　hygiene 281, 283
　Integrated Pest Management 283
　IPM *see* Integrated Pest Management

　irrigation 281
　key to 280
　nutrition 281
　pest cycles 282
　physical barriers 283
　plants for location 281
　predators and parasites 282
　quarantine 281
　resistant cultivars 280–1, 283
pH 174, 236, 241, 243, 247
phosphates 95, 174, 247–8
phosphorus fertiliser 95, 102, 109, 116, 146, 169, 173–5, 241, 247–9, 261
photosynthesis 9, 53–4, 67, 109, 120
photosynthetic bacteria 54
photovoltaic panels 63, 66, 69, 146, 181–2, 223, 224, 255
Phytophthora cinnamomi 281
Pimelea serpyllifolia 299
Pilea 175
Pittosporum undulatum 209
plant growth
　hard landscaping 147–8
　soils 145–6
planting 243, 246–7
　plant quality 246
　sourcing plants 246
　trees 246–7
plants
　and climate control 150–4
　domestication 15
　green roofs 153–4
　green walls 150–1
　hard landscaping 147–8
　potted plants 288
　quality 246
　selection *see* plant selection
　sourcing 246
　stabilising slopes 148–50
　variety protection 20
plant selection 25, 93, 138, 144, 151, 153, 161, 177, 200, 202–9, 213, 243, 273, 275, 299, 305
　Australian National Weed Strategy 208
　Australian Urban Tree Evaluation Program 208
　breeding 305
　characteristics for sustainability 202–3
　disease resistance 305
　ecological gardening 203
　energy expenditure 208
　environmental and economic weeds 205–8
　garden thugs 208
　indigenous plants 204, 299
　native versus exotic 204–5
　naturalised plants 206
　plant adaption 203–4

strategies to combat weeds 207–8
summary 208–9
weeds 205–8 *see also*
environmental and economic weeds
weeds and gardens 207
what is a weed? 206
plant tolerances 305
plant-based celluloid 100
plastics 100–4, 106, 108, 119, 126, 139, 150, 163–4, 168, 173, 179, 184, 191, 194–5, 209–10, 235, 237, 239, 242, 244, 248, 259, 262, 265, 288–90
HDPE 194
PVC 194–5
recycled 195, 235
pleasure gardens 16–7
plum 198
Poaceae 207
pomegranate 198
population 11, 32
Australian 3, 36
control *see* birth control
global 33, 34, 35–6
global by country 35–6
global, developed and undeveloped world 35–6
global rate of change 35–6
global urban and rural 35–6
porous paving *see* permeable paving
post-industrial society 54
potable water 91
potassium fertiliser 247, 248, 261
potatoes 282
pots 139, 289–90
poverty vii, 2, 4, 17–8, 21, 32–3, 47, 114, 123, 294
power generation 58, 102
machinery 66, 70, 119
prickly pear 282
prickly spear grass 299
primary consumers 116
primary production 53, 294
Process Energy Analysis 37
producers 9, 72, 306
Product Energy Analysis 37
product stewardship 100, 104
toxicity 100
production and consumption 6, 8–11, 118, 120–1, 293–4
Protea 175
protection zones 233, 235
public gardens 87, 258
Pultenaea 175
pumps 190
purple pipes 222, 223
purple water 77, 91

quality of life 48, 106, 210, 230, 286, 291–2

quandong *see Santalum acuminatum*
quince 198

radiant energy 54, 181
rainbow chard 198
rainfall 157–8, 159, 162, 164, 170–1
and climate change 157–8, 162
collection for Australian cities 159
Eco Logical Water Capturing System 170–1
on gardens 159
optimising rainfall use 159–62
from paving 159
from roof 159, 303
from synthetic turf 164
Rainforest Alliance™ 193
rain gardens 218, 221–2, 288
Aurora residential estate 221, 222
design details 221
drinking fountain 222
Melbourne 221–2
Royal Botanic Gardens Melbourne 222, 223
Victoria Harbour 222
recycled water 172–9, 222–3
Aurora residential estate 223
recycling 24, 59, 71, 77, 83–6, 101–10, 118, 120, 144, 146, 163, 172, 185–6, 192–6, 211–2, 229, 233, 238–9, 251–2, 256–7, 275–8, 289–90, 301, 302, 303, 305
grinding building materials 239
pots 239, 289–90
water 172, 257
Reduce, Reuse and Recycle 99, 111, 292
Rees, William 37
refrigerants 100
refrigerated storage 121, 124
Regional Forest Agreements 131
Register of Critical Habitat 132
regulating services 127
renewable
energy 68–9
resources 293
Renew: technology for a sustainable future 211, 231
resource
accounting 292
consumption vii, 47, 51, 185, 291, 296
depletion vii, 3, 11, 18, 101
intensity 43, 111, 286, 292
productivity 292
wars/conflict 74
retarding basin 85, 218 *see also* bio-retention
rice 15, 18, 21, 63, 82–4, 114, 119, 122–3

river red gum *see Eucalyptus camaldulensis*
RMIT University 175
road transport 61, 239
Robert Hart 27
Rome Declaration of World Food Security 113
rotary hoes 240, 241
Royal Botanic Gardens Cranbourne 212–13
Royal Botanic Gardens Melbourne 179, 212–3, 222, 223, 262, 268, 269, 270, 271
Royal Botanic Gardens Sydney 213, 260
rubbish 68, 94, 99,103, 105, 258, 279, 305
Rubus fruticosus 207
ruby saltbush *see Enchylaena tomentosa*
Rudolph Steiner 24
ruminant emissions 118–9

salinisation 16
Salvation Jane *see Echium plantagineum*
Salvia 175
Sanctuary: sustainable living with style 211
sandy soils 88, 178, 239, 241–2, 250, 258, 262–3, 268, 270, 276, 301
Santa Fe Green Building Council 230
Santalum acuminatum 199
SA Water Corporation 213
Scion, New Zealand 239
sea box *see Alyxia buxifolia*
Sentec Pty Ltd 270, 271
septic systems 95
settled communities 15–6
SGA *see* Sustainable Gardening Australia
sewer mining 84
shifting agriculture 15
showy honey myrtle *see Melaleuca nesophila*
simazine 278
Slow Food 124–5
Smartwood™ 193
Smog 109, 130
social equity 22, 35
sodification 18, 130–1
soft engineering 85
soil 145–6, 174, 235, 236, 239, 240–1, 242, 243, 247, 251, 256–67, 268, 271, 272, 273, 278, 279, 287
additives *see* amelioration
aeration 257, 258, 272
aggregates 257, 259
agrichar *see* biochar
alleviating compaction 240
Amazonian dark earths 242
amelioration 239, 240–2
biochar 242

clay 240–1, 242, 257, 258, 262, 264, 271, 272, 273
compaction 235, 236, 240, 241, 243, 257, 266, 273
  alleviating 240
compost 258–60, 273, 278, 279, 287
cultivation 257, 264, 266
deep ripping *see* ripping
degradation 115
drainage 262–4, 272, 273
environmental services 145
erosion 257, 264
fertility 260–1
fertility and plant grouping 146
fertility and plant selection 146
gypsum 239, 241, 257, 264, 272, 273
gypsum test 264
humates 241–2, 259
hydrophobic (non-wettable) 262, 263
importing topsoil 243, 256
improvement 92–3, 303
landscape construction 135–6, 239–43
landscape design 145–6
landscape maintenance 251, 256–67, 271, 272, 273, 278, 279, 287
loam 272, 273
micro-organisms 146, 235, 256, 260, 262
mulch and mulching 258, 264–7, 273
non-wettable *see* hydrophobic
nutrient testing 247
organic composts 240–1
organic matter 257–60, 262
organic matter and non-wettable soils 262
organic matter recycling 258
pH 241, 243, 247
pH and plant growth 174
pores 257, 273
preparation 240–3
protecting quality 145–6, 235–6
quality 145–6
quality and garden zones 146
reservoir 91–2
ripping 240, 243, 264
sandy 241, 242, 257, 258, 262, 268, 271, 272, 273, 301
sodicity 257
sodium ions 257
soil-less mixes 243
solarisation 242, 266
structure 26, 88, 116, 146, 149, 172, 174–5, 196, 146, 235–6, 239–41, 247, 256–9, 266, 272, 281, 287
terra preta see Amazonian dark earths
using equipment on 239
vibra moles 264
water absorbing agents 273
water and soils 257, 258, 262–4, 269, 273
water holding capacity 241
water storage 262, 269
water storage crystals 241, 262
wetting agents 262, 273
zeolites 242
Soil Association 193
soil-plant-atmosphere-continuum 93
Solanum species 199
solar
  agriculture 72
  cities 64
  horticulture 72
  hot water 223
  power 146, 166, 305
  radiation 61, 68, 269
  rubbish bin 305
solvents 100, 179, 194
sorghum 115, 119
spas 191
sports fields 87, 164
spotted gum *see Corymbia maculata*
Standards Australia 193, 196
staple foods 114
start of pipe management 10, 31, 99, 130
State of the Environment Reporting 103
stationary energy 63, 66
Steiner, Rudolph 24
St John's Wort 207
stomata 89, 270
stormwater 84–5, 87, 91, 93–6, 134, 157–8, 160, 164–70, 212, 216–21, 236–7
  drains 87, 93, 158, 167, 218, 237
  pollutants 95–6
  retention basins 94
street trees 148, 221, 222, 225
  for food 215
  life expectancy in Montreal 225
  structural soils 226–8
structural soil 148, 226–8
Substance Flow Analysis 37
subsurface drains 88–9, 92–4, 145, 170, 172, 178, 231, 245, 262, 268
sugar gum *see Eucalyptus cladocalyx*
sullage 91
superfarms 17
super-weeds 21
supporting services 127
surface run-off 87, 93
sustainability
  audit 296
accounting vii, 6–7, 10–12, 31–49, 51, 87, 291, 293, 295
  definition 4
  governance 7
  management 4–10
  management hierarchy 6–7
  science viii–ix, 11, 54, 127, 130, 292
  transformation 3, 291, 293
sustainable
  agriculture 22–9, 116, 118, 124
  development (definition) 4
  food production 116
  gardening viii, 28–9, 213, 230, 286
  gardening defined 1
  gardening future developments 304–5
  horticulture 27, 297
  living 1, 306
  water use in gardens 95–6
sustainable garden design *see* landscape design
Sustainable Gardening Australia viii, 28–9, 191, 230–1, 260, 286–7
sustainable horticulture 293
future developments 304–5
Sustainable Landscapes 213
sustainable living 291–3
sustainable living beyond the home 228
Sustainable World Initiative 116
swales 218
Swann, David 298
swapping 72, 99
sweet alyssum *see Lobularia maritima*
sweet pittosporum *see Pittosporum undulatum*
swimming pools 154–6, 190–1
  BioNova Natural Pools Australia 156
  chemicals 155
  embodied energy 155
  energy for running 190–1
  filters 155
  natural 155–6
  water loss from 155
synthetic chemicals 1, 70, 100, 108, 112, 252, 278–80, 284–7, 303
systems 24
  open and closed 293, 296

2,4-D 279
Tagetes 175
Tamarind 15
Tamarindus indicus 15
tanks 159–66, 223, 303
  Action Tanks 164
  architectural function 163
  bladder systems 164, 165
  Cape Schanck home 305
  concrete 163

concrete house slabs 165
connection to garden and household 165
disease 166
fibreglass 163
first flush diverters 166
flooding 160
insect proofing 165–6
Life Cycle Assessment 163
mains 165
maintenance 166
modular 163, 164
optimising rainfall use 159–62
plastic 163
pumps for 166
Rain Reviva 165
Rainstore® 163–4
roof gutter storage 165
spouting 166
spouting, Smartflo system 166
steel 162–3, 303
Tankmasta® 164
types 162–5
underground 163–4
under house *see* bladder systems *or* concrete house slabs
water efficient appliances 161
Waterwall® 163
wet system 165
WSUD 218
tank size 159–62
tank size and soil type 159–60
Tasmannia lanceolata 199
Terminalia ferdinandiana 199
terra nullius 13
terra preta see Amazonian dark earths
Tetragonia tetragonioides 199
The Australian National Pollutant Inventory 110
The Convention on Biological Diversity 129
Themeda triandra 151
Thomas Malthus 36
threatened fauna 132–3
Thryptomene 175
thyme rice flower
timber 192–4
    Australian Forest Certification Scheme 193
    Australian Forestry Standard 193
    Forest Stewardship Council 193
    Forest Stewardship Council Australia 193
    Good Wood Advisory Centre 192
    Good Wood Guide 192, 193
    longevity 193
    merbau *see Intsia retusa*
    radially sawn 193
    Smartwood™ 193
    treated pine 193–4
    treated pine, ACQ 194

treated pine, LOSP 194
treated pine, Tanalith E 194
Towards Sustainable Horticulture 213
treated pine 193–4
    ACQ 194
    LOSP 194
    Tanalith E 194
trees 56–60, 252
    carbon content 57–8
    carbon sequestration 55–9
    energy content 58–9
    fresh and dry weight 57–9
Triple Bottom Line Accounting 43
trophic levels 116

United Nations Brundtland Commission Report on Sustainable Development 4
United Nations Conference on the Human Environment 129
United Nations Convention on Biodiversity 19
United Nations Environmental Program 128
United Nations Millennium Declaration 4
United Nations Millennium Development Goals 4
University of Melbourne 268, 270, 286
University of Sydney 282
urban
    agriculture 113, 124–5, 294
    development 130, 220
    habitat 200
    landscape 1, 7, 49, 59, 77, 124, 294, 305
    planning 215
urban forests 148, 180, 224–8
    protecting existing trees 225
    residential properties 225
    street trees 225–8
urban horticulture and landscapes 1, 7, 8, 49, 51, 59, 73, 77, 141, 200, 216, 305, 306
urbanisation 35–6
    and water 77

vegetable patch 113–4, 124
vegetables 257, 275, 303
vegetarian 125
Veg Out Community Gardens 214, 215, 216
Verbena 175
VersiTank® green wall 150, 151
VicUrban 219, 220, 221
VOCs *see* volatile organic compounds
volatile organic compounds (VOCs) 255

Wachenagel, Mathis 37

Warrigal greens *see Tetragonia tetragonioides*
Waste 101, 238–9
    Australia 101, 103–5, 108
    batteries 68, 103, 108
    chemicals 108
    construction 238–9
    electronic 102–3, 108, 111
    glass 103
    hazardous 103
    household 103–5
    industrial 43, 103, 130
    jumbo 102
    mining 103
    oils 103
    organics 103
    packaging 103
    plastic 103
    stream categories 101
    wood, paper, pulp, cardboard 103
    tyres 103, 105, 108, 147, 170, 186
    water
water 73–95, 251–2, 291–2, 306
    and trade 78
    black 90–1, 172, 175–6, 226
    blue 76–9, 91, 292
    calculating water resource 159
    Class A 175, 178, 222–3
    Class A *see also* Class A standard water
    colour coding 76–7, 91
    consumption 76, 80–5, 96, 191
    credits 127
    cycle 73–80, 85, 91, 156, 216, 293
    design for sustainable use 156–80
    and electricity generation 84
    embodied 191
    features 179–80
    flows in the garden 87–97
    footprint 80, 294
    garden services 144
    green 71, 76–80, 91, 166, 292
    greywater 73, 84, 86, 89–91, 96–7, 138, 146, 158, 172–8, 211, 252, 301, 303
    household water resources 158, 159
    irrigation 177–9
    intensity 84
    landscape design 138, 144, 146, 156–80, 191, 222–3
    lilac *see* purple water
    outflows in garden 92
    pollution 6, 18, 114, 129, 237 *see also* garden services
    potential 92
    productivity 77, 83, 87, 91, 268, 275
    protecting water resources 236

quality 74, 79, 81, 82, 87, 94–5, 110, 155, 176, 178, 213, 216, 218, 261
quality monitoring 95
rainfall 157–8, 159, 162, 164, 170–1
recycling 172–9, 222–3
site's natural resource 156–72
soil reservoir 166–7
soil reservoir from contouring 167
soil reservoir from downpipes 166, 167
soil reservoir from permeable paving 167
storage in soil *see* soil reservoir
stormwater from landscape 166–72
stormwater from roofs 158–62
storing stormwater 162–6
sustainable use 156–80
table 76, 88, 92, 261, 272
tanks 159–66, 218, 223
tank size 159–62
tank size and soil type 159–60
tanks, optimising rainfall use 159–62
tank types 162–5
trading 82
white 77, 91
water efficient appliances 161
WSUD 150, 168, 216–21, 228
Water Conservation Garden 179
water features 179–80
Water Efficiency Labelling and Standards (WELS) 231–2
water and soils 257, 258, 262–4, 269, 273
    drainage 262–4
    water absorbing agents 273
    water storage 262, 269
    water storage crystals 262
    wetting agents 262, 273
water use in Australia 80–7
    Melbourne 85

rainfall and availability 80
rural 82
state 85
urban 83, 95
water use
    environmental impacts 94
    garden 87–97
    global 74
    household 85–6
    individual 85–6
waterlogging 18, 88, 93–4, 134, 257, 272
Water Sensitive Urban Design 150, 168, 216–21, 228
    Aurora residential estate 221
    bio-retention systems 218, 219
    geotextile material 218, 219
    green roofs 218
    Gross Pollutant Traps *see* litter traps
    infiltration trenches 218
    integration with landscape 217
    litter traps 219
    natural amenity 218
    natural systems protection 217
    peak water flows 218
    permeable paving 218
    principles 217–8
    rain gardens 218, 221–2
    subsurface wetlands 221
    swales 218
    tanks 218
    urban vegetation 216
    water quality 216, 218, 220
    wetlands 219–21
watershed 94
Waterwatch Australia 95
Watsonia 207
wattleseed *see Acacia* species
weeds *see* environmental and economic weeds 205–8
well-being 1–2, 8, 16, 26, 48, 127–8, 291
Western Center for Urban Forest Research and Education 180

wetlands 74, 80, 85, 93–4, 124, 150, 157, 207, 219–21
    inlet zone 219
    macrophyte zone 220
    open water zone 220
wheat 15, 18, 79, 86, 114, 119, 242, 264
white correa *see Correa alba*
white cypress pine *see Callitris glaucophyllus*
white water 77, 91
whole farm planning 22
William Rees 37
wind 5, 26, 53, 60, 62, 93, 131, 134, 139, 142–4, 146, 158, 177–181, 190, 202, 233–7, 243, 246, 253, 255, 257, 267–70, 285
    Australian wind resource map 61
wind erosion 131, 257
Woodmark 193
World Business Council for Sustainable Development 286
World Conservation Monitoring Centre 129
World Economic Forum 44
world food shortage 17
World Health Organisation 123
World Meteorological Organisation 128
world trade 51, 73, 77, 100–1
World Wide Fund for Nature 38, 128
Worldwatch Institute 128
WSUD *see* Water Sensitive Urban Design
Wurundjeri Walk 219

Xeriscape 179

Yale Centre for Environmental Law 44
Yandilla Park 305

zeolites 242
zero waste 99, 111, 292